A TEXTBOOK OF IMMUNOLOGY AND IMMUNOTECHNOLOGY

A TEXTBOOK OF IMMUNOLOGY AND IMMUNOTECHNOLOGY

B. ANNADURAI
M.Sc., Ph.D., D.H.Ed.
Professor, Centre for Bioinformatics and Biotechnology
Head, Department of Plant Biology & Biotechnology
C. Abdul Hakeem College, Melvisharam
Vellore District, Tamil Nadu

S. CHAND
PUBLISHING
S Chand And Company Limited
(ISO 9001 Certified Company)

S Chand And Company Limited
(ISO 9001 Certified Company)
Head Office: Block B-1, House No. D-1, Ground Floor, Mohan Co-operative Industrial Estate, New Delhi – 110 044 | Phone: 011-66672000
Registered Office: A-27, 2nd Floor, Mohan Co-operative Industrial Estate, New Delhi – 110 044
Phone: 011-49731800
www.**schandpublishing.com**; e-mail: **info@schandpublishing.com**

Branches

Chennai	:	Ph: 23632120; chennai@schandpublishing.com
Guwahati	:	Ph: 2738811, 2735640; guwahati@schandpublishing.com
Hyderabad	:	Ph: 40186018; hyderabad@schandpublishing.com
Jalandhar	:	Ph: 4645630; jalandhar@schandpublishing.com
Kolkata	:	Ph: 23357458, 23353914; kolkata@schandpublishing.com
Lucknow	:	Ph: 4003633; lucknow@schandpublishing.com
Mumbai	:	Ph: 25000297; mumbai@schandpublishing.com
Patna	:	Ph: 2260011; patna@schandpublishing.com

First Edition 2008
Reprints 2009, 2011, 2013, 2014, 2016

Reprint 2021

ISBN: 978-81-219-2807-6 **Product Code:** H6IMN68BIOL10ENAA08O

PRINTED IN INDIA

By Vikas Publishing House Private Limited, Plot 20/4, Site-IV, Industrial Area Sahibabad, Ghaziabad – 201 010 and Published by S Chand And Company Limited, A-27, 2nd Floor, Mohan Co-operative Industrial Estate, New Delhi – 110 044

PREFACE

Immunology remained for many decades as a discipline almost completely isolated from the rest of biology. Till recently immunology had been treated the discipline concerned chiefly with production of vaccines. The scope of immunology has enlarged tremendously during the last 25 years or so. The efforts are made to exploit the potential of immune systems in the development of industry and medicine. Thus immunology is vital to economy of any country. The importance of immunology has been advanced considerably in the Indian scenereo of higher education for the past 10 years. There have been many changes in the higher education with immunology being an important component of college and university curricula in Biology, Biochemistry, Biotechnology, Microbiology, Molecular Biology etc.

Immunology is advancing every day in the 21st century. I have made an attempt to arrange the process of immune system in a sequential manner. I did not try to give advance knowledge to higher learner, but to some extent it will satisfy the researchers. However this book is also useful to the students of Biology, Zoology, Veterinary Science and Medicine.

There are about 28 chapters, which will be gain an overview of the current status of many diverse and interesting aspects of advances in immunology.

I would like to thank the management, the Principal and correspondent of C. Abdul Hakeem College, Melvisharam for their constant help and kind cooperation.

I am indebted to my elders Prof. V.M. Kalimullah, M.Sc., M.Phil., former Principal, Dr. R. Puvanakrishnan, Deputy Director and Head of the Department of Biotechnology, Central Leather Research Institute, Dr. S.C. Dhar former Head of the Department of Biochemistry, CLRI, Prof. D.B. Motlag, former Head of the Department of Biochemistry, University of Madras, Prof. S. Mahalingam, Registrar (Retd) Thiruvalluvar University, Vellore, Mr. K.M. Govindarajan, Chairman, KMG College, Gudiyattam for their help and continuous encouragement given to me during the preparation of this text book. I thank all my students of this district in various colleges and neighbouring districts who took notes and gave the copy of my lectures.

I am grateful to my colleagues Profs. M/S C.Rajalakshmi, P.Amudha and R. Saraswathi, Prof. Thomas John, for discussions and Mr.J.Yuvaraj. for drawing the neat computer diagrams. Above all I am thankful to Mr. Navin Joshi, General Manager, Mr. A.M. Arunachalam, Regional Manager and Mr. R.M. Nath, Manager, Kolkata Branch of S. Chand Company Ltd. for their encouragement and neat execution of the book in the present form.

I wish to thank all my friends who are teaching the subject and students who are my master to comment constructively in improving the text.

B.ANNADURAI

CONTENTS

CHAPTER 1

HISTORY OF IMMUNOLOGY

Immunology is a branch of microbiology and is defined as the study of the mechanism of defence of the body against harmful invading agents.

Smallpox was the deadly human disease which killed thousands of people. The survivors were scarred with disfiguring pock marks which remained throughout their life. As early as the eleventh century, the Chinese discovered, that persons who survived an attack of small pox did not get the disease a second time. Therefore they started the practice of infecting young children with small pox. Scabs from the pocks on the skin of an infected person were either put into the nose of infants or rubbed on to scratch on the skin. Children who survived after the resulting disease became resistant to small pox. But many died as a result of this crude practice. The Chinese later discovered that when the small pox scabs were selected from mildly affected individuals the effect of the disease was much milder. Thus, they eventually succeeded in reducing the hazards of the procedure to fairly low levels.

The news of protecting children against smallpox gradually spread to Central Asia and finally to Turkey. In 1718, Lady Mary Montagu, the wife of the English ambassador in Constantinople amidst strong opposition, used this technique for her children and succeeded.

The technique is now known as variolation (variola means small pox in Latin), spread rapidly in England. The death of Queen Anne's son and heir due to smallpox created an awareness of the technique among the ruling classes who readily adopted the same.

The adoption of this procedure spread very rapidly in America, where George Washington ordered all his troops to be inoculated in 1776. Nevertheless, variolation was a dangerous procedure, and many deaths have occurred during its practice.

In 1774, an English farmer named Benjamin Jestly, used dried scab material from cattle infected with cowpox to variolate his children with a hope to protect them. Cowpox is a skin disease of cattle that is caused by a virus closely related to smallpox virus. It also mildly affects the humans. In 1798, Edward Jenner, a physician, learned about the use of cowpox scabs from humans and decided to investigate the matter further. Jenner conducted many experiments and confirmed that exposure to dried cowpox scabs from humans could indeed safely protect humans against smallpox. He, unlike Jestly, published his results and received credit for the discovery. Jenner's technique of vaccination (vacca means cow in Latin) rapidly replaced variolation as the preferred method of protecting humans against smallpox.

In 1958, the World Health Organization (WHO) decided to eradicate smallpox by means of mass vaccination. This was effective that by 1980, smallpox had become the first infectious disease to be eradicated from the earth.

In spite of the discovery of vaccination, against smallpox in the late eighteenth century, it was not suspected that this type of approach could be used to protect against other infectious diseases. Although, a few unsuccessful attempts were made to use cowpox to cure unrelated diseases. Almost 100 years later that another significant advance occurred.

In 1879, in France, Louis Pasteur was studying the bacterium that causes a disease of chickens called fowl cholera. The bacterium is now called *Pasteurella multicida*. Pasteur possessed a culture of the bacterium that, when injected into chickens, consistently caused infection that killed them. One afternoon he told his assistant, Charles Chamber land, to infect some birds with the culture. Since it was late in the day and he was about to go on vacation, Chamberland decided to postpone the experiment until he returned. As a result, the chickens eventually received an injection of the bacterial culture that had remained on the bench for several weeks. The inoculated chickens remained healthy. Pasteur then decided to inject these chickens with second dose of bacteria from fresh culture. To Pasteur's surprise, the birds survived this second dose without becoming ill. Pasteur with remarkable insight, recognized that the phenomenon was identical in principle to vaccination. By injecting his chickens with aged culture of bacteria (a vaccine) he had protected them against the disease caused by a fresh culture of the same organism.

Once he established the general principles of vaccination, Pasteur tried to apply it to other infectious diseases. He first produced a vaccine against anthrax, a disease caused by a bacterium called *Bacillus anthracis*. Pasteur found that he could not make this organism safe by ageing it on the laboratory bench, but he could do so by growing it at an unusually high temperature. Pasteur then conducted a public experiment that convincingly showed that administration of his heated anthrax culture would protect sheep, cattle and goats against a subsequent lethal dose of anthrax bacteria. A few years later he developed an effective vaccine against rabies. Thus he showed that the general principles of vaccination. Thus he can be applied to diseases even other than smallpox and this approach could be used to protect animals and human beings against the infections. Louis Pasteur can, therefore, be considered as founder of the science of immunology.

Although Pasteur was remarkably successful in developing effective vaccines, he had little knowledge of the concept of the mechanisms involved. He suggested that the organisms in the vaccines removed essential nutrients from the body and thus prevented the subsequent growth of the disease causing agents. It was in Berlin about ten years later, in 1890, that Emil von Behring and Shibasaburo Kitasato demonstrated that the protection induced by vaccination was not due to the removal of the nutrients but was associated with the appearance of protective factors in the blood. They called these factors as antibodies. Within a few years, Paul Ehrlich had proved that antibodies could protect animals against foreign toxins other than those found in bacteria (for example, the toxin ricin from the castor bean), and another German, Richard Pfeiffer, had shown that antibodies could clump and then destroy *Vibrio cholerae* bacteria, the causative agent of cholera in human beings.

The antibodies could make a clump of bacteria was demonstrated and immediately it was applied by Isidore Widal to the diagnosis of enteric fever (typhoid and paratyphoid fever). Thus serum from an infected individual would make the bacterial clump, whereas serum from an unaffected individual would not.

In 1894, Emile Roux showed that the patients suffering from diptheria could be cured by an injection of serum from the blood of a horse immunized against diptheria (The horse serum thus contained antibodies to diptheria toxin). This method was tried for other diseases also but none had succeeded in their attempts except against tetanus. This technique is known as passive immunization, is still and it used to protect persons against tetanus after they have received a deep wound. The horse serum contains antibodies against tetanus toxin.

Further studies on the properties of antibodies led in 1893 to the observation that heating an antibody containing serum to 56°C caused denaturation and loss of its ability to kill bacteria. Since antibodies are not destroyed by heating, another component must be responsible for this activity. The heat labile component that acts with antibodies to kill bacteria is now called as complement.

Eli Metchinikoff in 1882 demonstrated first in invertebrates and later in mammals that certain cells could eat foreign material. He proposed that these cells which he called Phagocytes (eating cells) were the body's main defence against invading microorganisms. Further more, he suggested that antibodies were little significance. This proposal upset lot of people and provoked violent controversy. The conflict was resolved only in 1904 when Almorth Wright in England and Joseph Denys in Belgium showed that antibodies could coat bacteria and promote their destruction by phagocytes (this phenomenon is now called as 'opsonisation')

The history of immunology, since the beginning of twentieth century, is complex but can be outlined by summarizing the Nobel prize awarded to immunologists. The first Noble prize in medicine was awarded to Emil von Behring in 1901 for his work on the production of antibodies against toxins (antitoxins). In 1905 Robert Koch was awarded the prize for his studies on tuberculosis, one of his major contributions to our knowledge of this disease was his discovery of the tuberculin reaction. This is an immunologically mediated inflammatory skin reaction that is a very useful diagnostic test.

The Noble prize of 1908 was shared by Ehrlich and Metchinikoff. Paul Ehrlich's contributions to immunology were diverse and he developed many concepts, including the idea that the immune system would not normally mount a response against a normal body components. In 1913,Charles Richet won the prize for his discovery of anaphylaxis, while trying to immunize dogs against a toxin from sea anemones. Richet and his colleague Paul Porter observed that very small non toxic doses of this toxin made some dogs collapse and die. We now know that this reaction is caused by certain antibodies. Jules Bordet, the discoverer of complement, won the prize in 1919.

Karl Landsteiner was awarded a Noble prize in 1930 for his demonstration of blood groups, the complex carbohydrates found on the surface of red cells. These molecules are important in immunology because they can stimulate a life threatening immune response. If blood is transfused into the wrong person. Landsteiner's research eventually resulted in the development of successful blood transfusion procedures. For the next 30 years, the progress of immunology appeared to be slow. It was not until 1960 that Peter Medawar and Macfarlane Burnet jointly won the prize for their studies of the ways in which the immune system recognizes foreign material while avoiding reacting against itself. Burnet developed the first rational explanation of this ability of the immune system to identify foreign material-clonal selection theory, which Medawar later confirmed experimentally. The concept of clonal selection lies at the core of modern immunology.

In 1972, two protein chemists, Rodney Porter and Gerald Edelman, received the prize for demonstrating the chemical structure of antibody molecules and incidentally, showing how they bound to foreign material. One third of the 1977 prize was awarded to Rosalyn Yallow, for her studies on the technique of radio immunoassay. By combining the use of radioisotopes with antibodies that bind specifically to hormone molecules, Yallow showed that it was possible to develop remarkably sensitive tests to detect low levels of these hormones in body fluids.

Table 1.1. Milestones in immunology

	Scientist	*Discovery*
1.	1774- Benjamin Jesly	Used dried scrab material
2.	1798- Edward Jenner	learned about use of cowpox (Antigen)
3.	Louis Pasteur	coined the term vaccine
4.	1879.-Louis Pasteur (Founder of immunology)	Studied bacterium that causes a disease called fowl cholera, founder of immunology.
5.	1890,Emil von Behring and Shibasaburo Kitasato	Used the term antibody
6.	Richard Pfeffer	Antibody could clump
7.	Isidore Widal	He applied clumping in antibody.
8.	1894 Emil Roux	Patient suffering from diphtheria could be cured.
9.	1882-Eli Metchnikoff	Demonstrated first in invertebrates and later in mammals- phagocytosis.
10.	Almorth Wright and Joseph Denys	Opsonisation
11.	1901-Emil von behring (1st Noble prize in medicine)	Production of antibodies
12.	1905-Robert Koch (Prize)	Studied about tuberculosis. Discovery of tuberculin reaction
13.	1908-Ehrlish and Metchnikoff	Immune system could not normally cause response against normal body component.
14.	1913-Charles Richet (Prize)	Anaphylaxis
15.	Jules Bordet-1919 (Prize)	Complement
16.	Karl Landstuner-1930 (Nobel prize).	Blood groups.
17.	1960-Peter Medwar Mayfarlane Burnet	Immune system recognize foreign material while avoiding reacting itself.
18.	Burnet	Clonal selection theory
19.	1972.Rodney Porter Gerald Edelman (Prize)	Chemical structure of antibody. Radio Immunoassay.

20.	Rosalyn Yallow	Existence of genes, identify the molecules found on the surface of human cells.
21.	1980(Nobel Prize) George Snell Jean Dauiset	Inheritance of animal
22.	1984.Kohler Milestein (Nobel Prize Jerne (Nobel Prize).	Monoclonal antibodies, theories on specificity, development, control of Immune system.
23.	Susumu Tonegawa-1987 (Nobel Prize).	Molecular studies on Ab.
24.	Murray Nobel Prize.	1st surgeon: Organ transplant
25.	Donnal Thomas	Bone marrow transplant

In 1980 Nobel prize in medicine was shared by three immunologists who had made significant contribution in immuno genetics. George Snell developed genetically defined mouse strains and determined the existence of genes that control the process of graft rejection. Jean Dausset developed techniques to identify the molecules found on the surface of human cells. This enabled transplantation surgeons to match organ donors with recipient so that the immune response leading to graft rejection is minimized. Baruj Banacerrai studied the inheritance of an animal ability to make antibodies to defined molecules. In this way he and his co-workers discovered that genes regulate an individuals' ability to respond to a specific foreign molecule and so determine resistance and susceptibility to disease.

The 1984 Nobel prize in medicine was awarded to Niels Jerne, Georges Kohler, and Cesar Milstein. The prize was awarded to Kohler and Linstein for discovering the principles of the production of monoclonal antibodies and to Jerne for his provocative theories on the specificity, development and control of the immune system.

The 1987 Nobel prize in Medicine was awarded to Susumu Tonegawa for his molecular studies that showed how antibodies were generated so that they could combine with a vast array of foreign molecules.

The 1990 Nobel prize was awarded to Donnal Thomas and Joseph Murray for their pioneering work on organ transplantation when he grafted a kidney between identical twins. Thomas was the first to perform a successful bone marrow transplant in a patient who had first been irradiated to remove leukamia cells. Only three scientists won Nobel prizes between 1919 and 1972 but nine between 1980 and 1990, for their contribution in immunology.

QUESTIONS

1. Define Immunology.
2. Write short notes on

   (*a*) Variolation (*b*) Immunization (*c*) Prophylaxis (*d*) Inoculation (*e*) Attenuation (*f*) Phagocytes (*g*) Humoral immunity (*h*) Opsonization (*i*) Anaphylaxis (*j*) Complement fixation (*k*) Clonal selection theory
3. Describe in chronological order the history and development of immunology.
4. Mention the contributions of the immunologists

   (*a*) Edward Jenner (*b*) Louis Pasteur (*c*) Metchnikoff (*d*) Almroth Wright (*e*) Emile Roux (*f*) Paul Ehrlich (*g*) Robert Koch (*h*) Alexander Fleming (*I*) Karl Landsteiner.
5. How does the 20th century immunologist pave way for the future of immunology ?

CHAPTER 2

ARCHITECTURE OF THE IMMUNE SYSTEM

2.1. INTRODUCTION

The immune system consists of a network of lymphoid organs, tissues and cells (e.g., cytokine growth factors of regulatory agents, cytokinetic substances and antibodies.) The lymphatic system includes blind ended lymph capillaries which accumulate lymph as they receive fluid and protein molecules from the spaces between cells. The factors responsible for the formation of lymph are hydrostatic pressure, osmotic pressure, diffusion and also the secretory ability of the lymphatic capillary epithelium. From the lymphatic capillaries, the lymph passes into the lymphatic vessels which transport it to the large veins at the root of the neck. The larger lymphatic vessels converge to form large lymphatic trunks, thoracic duct and right lymphatic duct.

Lymphatic tissues contain cells that has phagocytic microorganisms. The lymph, if they come across more pathogens then they can destroy. The lymphatic tissues can become sites of infection. Thus swollen lymph nodes and tonsilites are common signs of many infectious diseases. All lymphatic organs contain large numbers of lymphocytes which originate in the bone marrow and are released into blood and lymph. Their life span may range from a few weeks to years. They are either dispersed to various lymphatic organs or remaining in the blood and lymph.

2.2. PRIMARY LYMPHOID ORGAN

The organs in which the progenitor cells become converted into lymphocytes are primary lymphoid organs. They contain identifiable but predominantly non functional lymphocytes. The two organs which are considered as the chief primary lymphoid organs, are (*i*) Thymus (*ii*) Bone marrow. The secondary lymphoid organs are (*i*) Lymph nodes and (*ii*) Spleen.

In addition, four other organs are also considered as important lymphoid organs. They are

(*i*) Payer's patches
(*ii*) Appendix
(*iii*) Tonsils and
(*iv*) Adenoids.

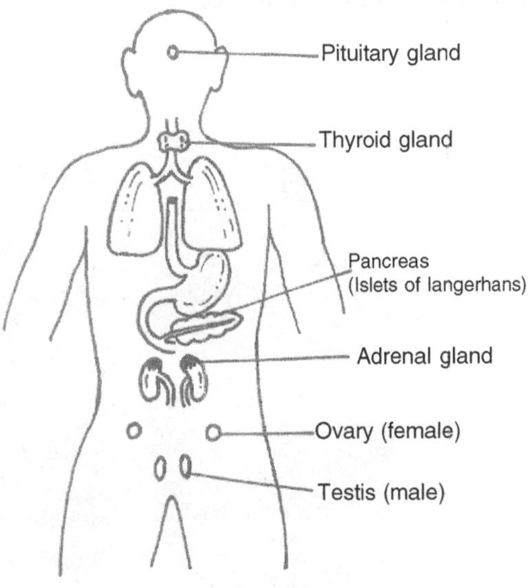

Fig. 2.1

During Matuation T cells change their phenotype. Differentiation markers of functional significance is either acquired or lost during the progression from stem cells to T cells.

2.3. THYMUS

The thymus is considered as the most critical organ of the immune system. It is responsible for the antigen independent maturation and development of T lymphocytes that effect cell mediated immunity and regulate most humoral and cell mediated responses.

2.3.1. Section of Thymus enlarged

The thymus is a bilobed, greyish, lymphoepithelial organ, located just above the heart and beneath the breast bone. It consists of a reticular network filled with a mass of lymphocytes and a small number of epithelial cells. In adults the thymus consists of many lobules, each containing an outer region or cortex and a medulla or central region. The medulla contains mature virgin T cells that have emigrated from the cortex and layers of comparatively large epithelial cells. Polypeptide hormones produced by the epithelial cells

(A)

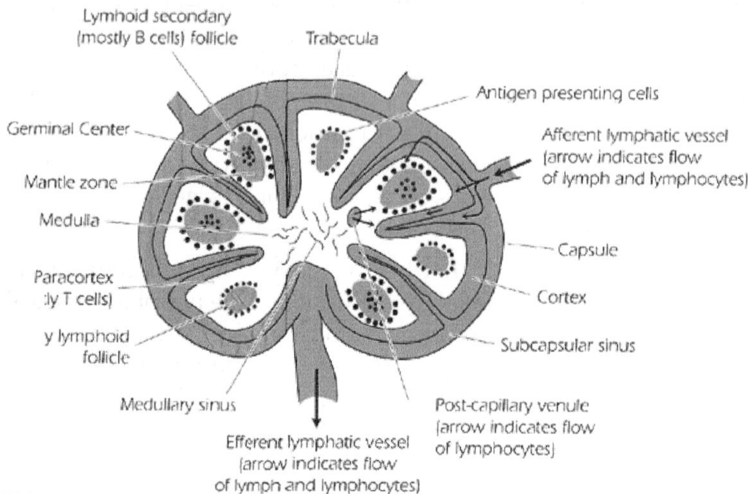

(B) A PORTION ENLARGED

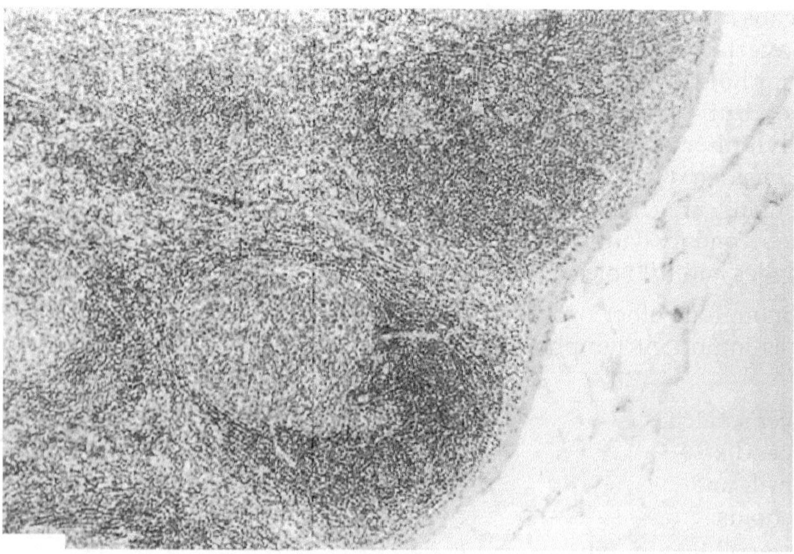

Fig. 2.1

appear to provide the microenvironment that promotes antigen independent differentiation of primitive thymocytes.

Other Lymphoid Organs. They are considered as single entity, referred as Gut associated lymphoid tissue (GALT). Now it is termed as Mucosal associated lymphoid tissue (MALT), should include lymphoid tissue found in the respiratory tract, gut and urinary tract. MALT forms a separate enter connected secretary system within which cells committed to IgA or IgE synthesis circulate.

Sequential changes occur as last during T cell matunation in the membrance antigens. At First stage, cell expresses CD_4^- and Cd_8^- phenotype. Here the cells are capable of giving rise to other lineages. During the second stage, the cells expresses CD_4^+ and CD8+ phenotype. In the third stage, they become either CD_4 T cells or CD_8 T cells. Immature cell as indicated markers are absent as they become nature T cells. At this stage of maturation in Thymus, T cells receptor diversity and positive and negative selection occurs.

Progenitor lymphocytes migrates through the blood stream from the bone marrow to the thymus for the maturation process. Maturation involves many divisional steps and

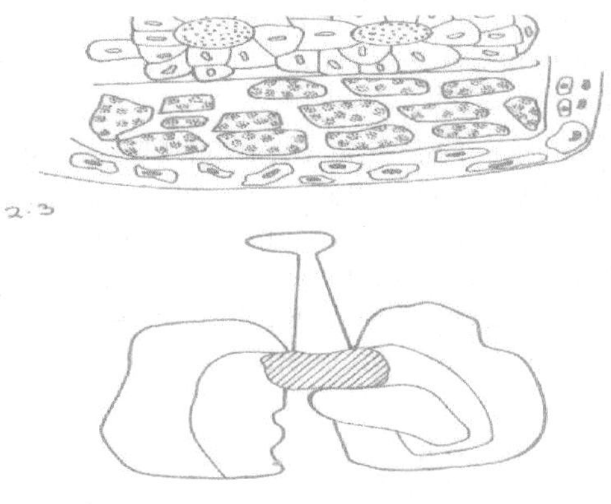

Fig. 2.3. Section of thymus enlarged

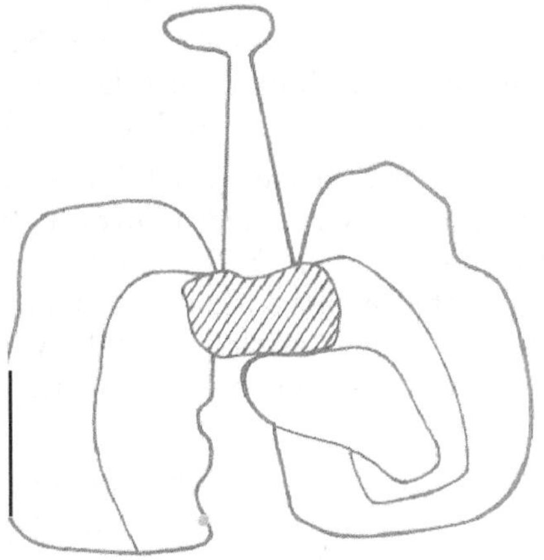

Fig. 2.4. Thymus

only the emergent progeny are capable of carrying out T cell related functions. The thymus is one of the most active sites of lymphocyte proliferation. However, most cells die in the process. Only 5% of the cells produced actually leave the thymus as viable cells.

The thymus reaches peak activity in childhood and attains its largest size at puberty. Thereafter, it gradually becomes smaller and is extremely small in old age. Early in life the thymus is crucial for T cell maturation and if the thymus is removed soon after birth (neonatal thymectomy) cell mediated immune responses fail to develop. Blood lymphocyte levels are drastically reduced and the response to most antigens is either poor or absent. Impaired thymic development may be associated with immune deficiency disorders, cancer and autoimmune diseases.

2.3.1.1. Ontogeny of T-Cells Lymphocytes

The T-cell progenitors i.e. the stem cells formed in the bone marrow during the course of hematopoiesis begin to migrate to the thymus at the 11th day of gestation in mice and 8 or 9th week of gestation in humans. The progenitor cells are attracter to the thymus by a chemotactic factor secreted by the thymic epithelial cells. Upon entry into thymus the progenitor T-cells undergo differentiation and also rearrangements of the germ line (nucleic acid present in embryo) present in T-cells receptor and expression of various surface markers.

The progenitor T-cells begin to express the membrane receptors CD3, CD6, CD1, CD2, and CD7. These "early thymocytes" also contain the enzyme terminal deoxy ribonucleotide transferase (TDT) adds the nucleotide sequences at the N-terminal of germinal DNA. Under the influence of interleukin secreted by macrophages and through the interaction of the CP2 receptor Leukocyte Addition Factor (LAF) is secreted by the stomal cells. A massive expansion of the cortical thymocytes occurs.

The earlier thymocytes differentiate into two subsets of T-cells namely T-cell receptor 1 and T-cells receptor -2 based on the types of *TCR*. These earlier fatal thymocytes (formed on 14th day) lack detectable CD4 and CD8 receptors and they are inferred to as double negative cells. The double negative thymocytes differentiate along one of the two developmental pathways. The immature thymocyte which expresses the T-cell receptor and CD3 complex as well as CD4 and CD8 receptor are referred to as double positive thymocytes, where as the other groups which only expresses the TCR CD3 complex and not the CD4 and CD8 receptor remain as double negative thymocytes. These thymocytes then undergo a selection of TCR specificity which again involves two processes namely (1) Positive selection (2) Negative selection.

The thymic stromal cells which include the epithelial cells, macrophages and dendritic cells play very important role in positive and negative selection. These thymic stromal cells express the high levels of class II and Class 1 MHC molecule.

The thymocytes which have receptors capable of binding with self MHC molecule are positively selected. However the thymocytes which either have very high affinity receptors for self MHC molecules do not interact with self MHC molecules and are eliminated as antiself cells. These antiself cells which do not interact with MHC molecules migrate to the thymic medulla where they are susceptible to the influence of various hormones like corticosteroides and leads to programmed cell death.

These thymocytes which enter the medulla may produce rearrangements of y and d chain genes. (These genes are responsible for the production of a polypeptide chain on the surface of the thymocytes) develop into double negative CD3+, d, y cells which amounts for 0.5 to I% of thymocytes.

This thymocyte sub population can be detected by the 14th day of gestation and reaches the maximum between 17th and 18th day and decline until birth. At the 16th day of gestation, the double positive cells can be detected expressing both CD4, and CD8 receptors. But not expressing the L1 B and T-cells receptors.

The L1 B and T-cell receptors appear on the 17th day and (reclines) maximum at birth. These double positive thymocytes that express the CD3 and B receptor complex that survive the thymic selection undergo differentiation into two population of cells, namely CD4+ helper cells and CD8+ cytotoxic suppressor T-cells.Cd4+ cells represent 10% of total T cell population. CD8+ T cells represent 5% of the thymus population. However the CD4+, CD8+ constitutes 8% of population. These matured (a, b, CD3+) T-cells move towards the periphery of medulla and then pass through the post venulary capsules to the peripheral lymphoid organs. Thymic epithelial cells are called nurse cells. These cells produce hormones required for differentiation, proliferation and maturation.

2.3.2. Thymosin

It is protein with lymphocytopoietic capacity .The major effect of it is to enhance maturation of pro thymocytes to mature T-cells.

2.3.3. Thymopoietin

This was originally termed as thymine, and thymopoietin is produced by epithelial cell.

2.3.4. Thymic factor

It is a low molecular weight peptide. It increases cyclic AMP level within cells.

2.3.5. Thymic humoral factor

It is not antigen specific and has a molecular weight of 45,000 to 60,000 kd.

2.4. BONE MARROW

In adults, red bone marrow serves as the major source of all blood cells including lymphocytes. It is found in the cavities of most bones in the body including the skull, ribs, sternum, femur and spine. However, the distribution of red marrow diminishes with age. Red bone marrow is compartmentalized into hemopoietic (blood forming) and vascular areas. The principle architecture of hemopoietic (blood forming) marrow consists of sinusoids arranged peripherally around a central vein. Within these sinusoids are found the precursors (blunt cells) of the different blood cell lines and their more mature descendants. For e.g, a particular sinusoids might contain several myloblasts (granulocyte progenitor cells) surrounded by larger numbers of more mature granulocytes. This pocket reflects proliferation and maturation of granulocytes from the precursors pool. Bone marrow, thus contain various intermediate and mature forms of erythrocytes, monocytes, granulocytes, lymphocytes and megakaryocytes.

The generalized progenitor cells are capable of yielding any of the blood cell lines. This ultimate progenitor is called a stem cells. This remarkably renewable pluripotent stem cell is believed to have a lymphoid morphology and its existence is inferred from its ability to repopulate all of the blood cells in a lethally irradiated animal.

Bone marrow is not only the source of all blood cell classes but it also provides the microenvironment for the antigen independent differentiation of B cell. Moreover, bone marrow serves as a secondary lymphoid organ where mature, virgin, antigen reactive lymphocytes (T and B cell) may respond to antigen trapped by antigen presenting cells, such as macrophages. Thus, bone marrow may provide an antigen processing environment, as does the spleen.

2.4.1. Bursa of Fabricus

The bursa of fabricus is responsible for the antigen independent maturation of B cells involved in humoral immunity. The bursa is a lympho epithelial organ seen as an intestinal pouch in birds but not in mammals. It is lined with epithelial cells which cover outer cortical and inner medullary by areas that are packed with B cells and plasma cells. If the bursa is surgically removed by a procedure known as bursectomy, birds may have normal cell mediated immune response but they express little. If any antibody producing capability like the thymus, the bursa starts to shrink or atrophy at puberty. The organ serves as the haemopoietic inducing microenvironment for progenitor B cells in birds.

The mammalian equivalent of the bursa of fabricus appear to be the red bone marrow. Thus, the two functional compartments of immune maturation in humans and all mammals are the thymus and the bursa equivalent which process T and B lymphocytes respectively.

2.5. T CELL MATURATION

The lymph nodes, spleen adenoids, tonsils, lymphoid patches of the gut and the appendix are all where mature T and B lymphocytes may have an opportunity to bind antigen and undergo further antigen dependent differentiation. It is here that the active immune responses begins. All of these lymphoid organs are capable of trapping antigen which normally involves large tissue phagocytes knows as macrophages. These peripheral lymphatic organs concentrate antigens from all parts of the body since the system contains recirculating lymphocytes, which pass from the blood through the lymphoid tissues and back to the blood. It virtually ensures that circulating antigen reactive lymphocyte with the proper receptors will encounter these antigens. Further more, the size of the lymphocytes are available in various sizes to "fit" any type of antigen. Thus, the traffic of lymphocytes between the blood, lymphoid tissues and lymph enables antigen reactive cells bind to the antigen and in turn, be recruited to sites where the action is occurring.

During the course of normal lymphocyte circulation, lymphocytes leave the blood and accumulate in lymphatic tissues, to set up B and T cell areas. Here they reside temporarily before returning to the circulation. The normal steady input, and output of cells among lymphoid organs is disrupted during a primary immune response and dramatically so following a secondary immune response. This is seen from a sharp fall in cell output within 24 hours of the time an antigen becomes localized in the lymph nodes (or spleen).There is a depletion of antigen reactive cells from the circulating pool of lymphocytes. This temporary cessation of lymphocyte recirculation is called lymph nodes shut down. Approximately three days later activated cells are released into circulation. In this regard, the system further serves to transport products and cells of the immune response to the tissues through the blood stream.

The actual sequence of T-Cell maturation within the thymus is not completely understood. After progenitor T-cell formed during haematopoiesis enter the thymus , they are thought to multiply rapidly within the condense; this rapid proliferation of thrombocytes is coupled to an enormous rate of cell death. A small subset of some mature thymocytes are then thought to migrate from the cortex to the medulla where they continue to mature and finally leave the thymus via post capillary venules.

A small sub-population of cortical thymocytes can mature and leave the thymus without entering the medulla.

Both cortex and medulla of the thymus are cross and by a three dimensional stroma cell network composed of epithelial cells inter digitating dendritic cells and macrophages, which make up the frame work of the organ and contribute to thymocyte maturation. Many of the stroma cells physically interact with the developing thymocytes. Some thymus epithelial cells in the outer cortex, called nurse cells have long membrane processes that surround as may as 50 thymocytes, forming large multicellular complexes.

Other cortical epithelial cells have long inter connecting cytoplasmic processes that form a net work and have been shown to interact with numerous thymocytes, as they traverse the cortex.

Interdigitating dendritic cells, which are coated at the junction of the cortex and medulla, also have long processes that interact with developing thymocytes.

2.5.1. Maturation and selection of T-cells

Thymic epithelial cells secrete several hormonal factors necessary for the differentiation and maturation of thymocytes into the various types of mature -T-cells.

Four such hormonal factors that have been characterized all a1 thymosis, b4-thymosis, thymopoietin and thymulin. Thymic stromal cells also secrete interleukin-7 (IC- 7) which stimulates growth of thymocytes.

In the course of thymocyte maturation within the thymus, the antigenic diversity of the T-cells development is generated by a series of random gene rearrangement. After developing thymocytes begin to express antigen binding receptors they are subjected to a two step selection process, so that only T-cells recognizing antigenic peptides in the context of self-MHC molecules are released from the thymus thymic stomal cells which express high levels of Class – I and II MHC molecules, play a role in this solution process.

During this selection process, any developing thymocytes that are unable to-reorganize self-MHC molecules (or) that have the high affinity for self antigen and self

MHC are eliminated by programmed cell death. Thus only those cells whose receptor re organizes a self MCH molecule and foreign antigen are allowed to mature.

2.6. BONE MARROW: B-CELL MATURATION

In birds, lymphoid organ called the begin of fabricus is the primary site of B-cell maturation, There is no bursa in mammals and no single counterpart in humans of Primary lymphoid organ. Regions of bone marrow and possibly of other lymphoid tissues serve as the bursa equivalence where B-cell maturation occurs.

Immature β-cells, proliferate and differentiate within the micro environment of the bone marrow. Stromal cells within the bone marrow interact directly with the B-cell and secrete various cytokines that are acquired for B-cell developmental process.

Similar to thymic selection during T-cells maturation, a selection process within the bone marrow eliminate B-cells with self reactive antibody receptors.

2.7. SECONDARY LYMPHOID ORGAN

They are the organ and tissues in which mature, immunocompetent lymphocytes encounter trapped antigens and all activated into effector cells. In mammals the lymph nodes and spleen and mucosal associated lymphoid tissues (MOR) constitute the secondary lymphoid organs.

2.8. SPLEEN

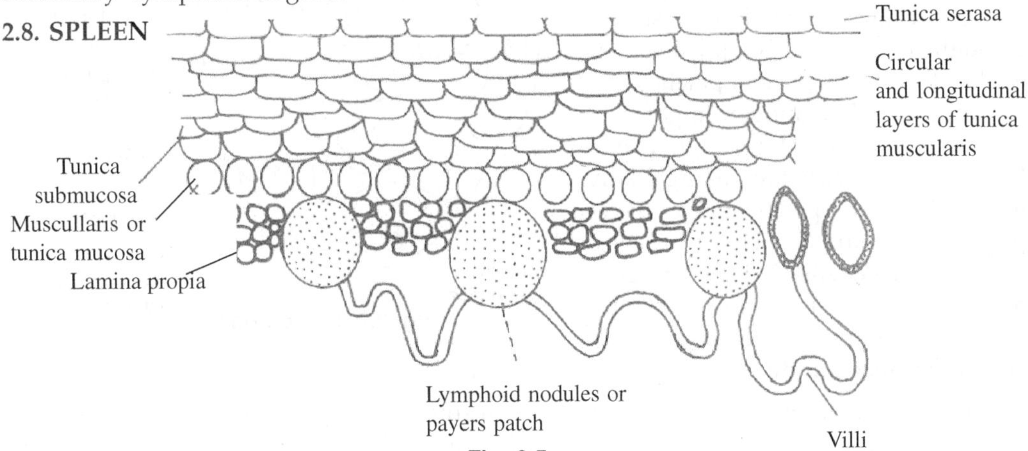

Fig. 2.5

Spleen is a large, encapsulated, lymphoid organ. By means of the evident macroph-ages, it filters out antigens that enter the blood stream. The cortex of the spleen contains packed lymphocytes with germinal centers and a medulla containing a variety of cell types. Basically, there are two types of tissues in the spleen

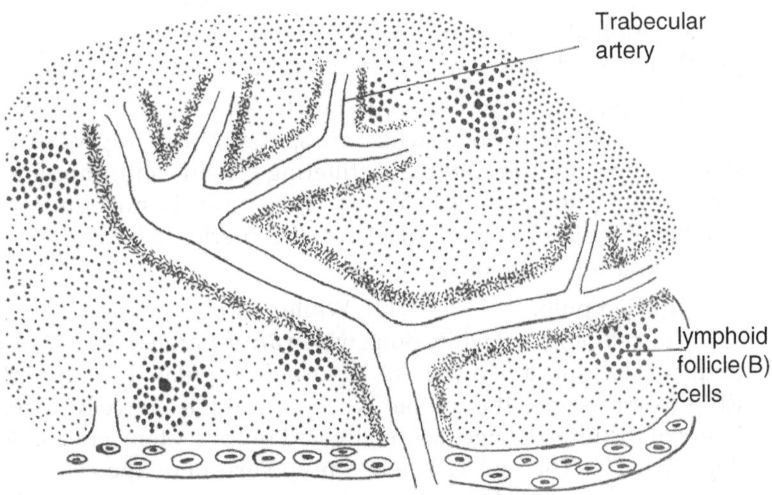

Fig. 2.6. The

referred to as the lymphoid or white pulp of the cortex and the erythroid or redpulp of the medulla. T Cells are diffusely packed in the white pulp, with the follicles.

2.8.1. PRIMARILY COMPOSED OF B CELLS

Most lympho-cytes enter and leave the spleen by way of the blood stream. The white pulp forms a sheath around entering blood vessels. There by permitting circul-ating B and T cells to enter and leave the spleen by traversion the walls of small incoming arteriole and outgoing veins. The red pulp is involved in removing old red blood cells and is reserve site for haemopoiesis during pathological states.

2.9. LYMPH NODES

Lymph nodes are small, bean shaped organs that act as filters. They are located at major junctions of the network of lymph flow through lymphatic channels (lymphatic vessels).

They consist of an outer cortex and an inner medulla. If antigens enter the body through the respiratory organs of gastro intestinal tracts, they must pass through the regional lymph nodes as well as the tonsils peyer's patches and the appendix. The antigen might pass via lymphatic channels through several lymph nodes with increasing likely hood of being trapped by macrophages particulate matter such as bacteria might be phagocytised. These antigens have the opportunity to react with immigrate B and T cell entering are already localized in different area of the lymph nodes.

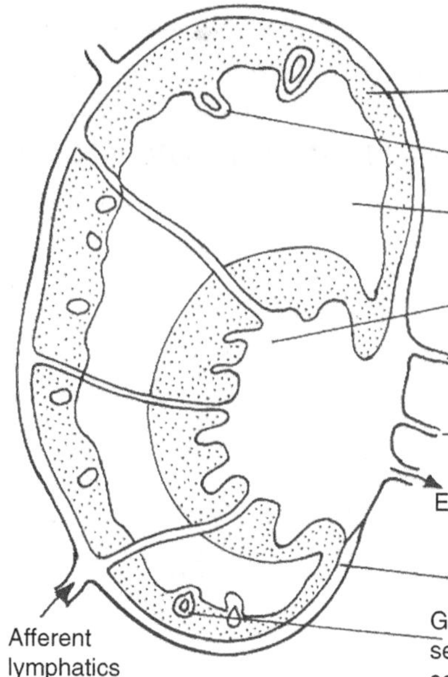

Fig. 2.7. Symph nodes

B cells and plasma cells are accumulated in subcapsular follicles while T cells occupy spaces between them in the diffuse cortex. If an animal has been stimulated by antigen, rapidly dividing B lymphocytes can be seen in the germinal center of each follicle. The size of the germinal center is proportional to the intensity of antigenic stimulation and can be greatly enlarged during a secondary immune response.

The spleen is large void, secondary lymphoid organ situated high in the left abdominal cavity. Unlike lymph nodes which are specialized to trap localized Ag from regional tissue spares. The spleen is adopted to filtering blood and trapping blood born Ag and thus term respond to systemic infection. The spleen is surrounded by a capsule that such a number of projection (trabeculae) in to the interior to form a compartmentalized structure. The compartment are of two types.

(1) Red pulp. (2) White pulp.

The red pulp and white pulp which are separated by a diffuse marginal zone. Spleenic pulp consists of a network of sinusoid populated with macrophages and numerous RBCs (erythrocytes).It is the site where old and defective RBC are destroyed and removed. Many of the macrophages within the red pulp contain energy RBC or iron pigment from degraded Hb. The spleenic white pulp surrounds the arteries, forming a periarteriolar lymphoid sheath.(PALS) populated mainly by T-lymphocyte. The marginal zone located peripheral the PALS, is rich in B-cells organized into primary lymphoid follicle. The initial activation of B and T cells takes place in T cells rich PALS

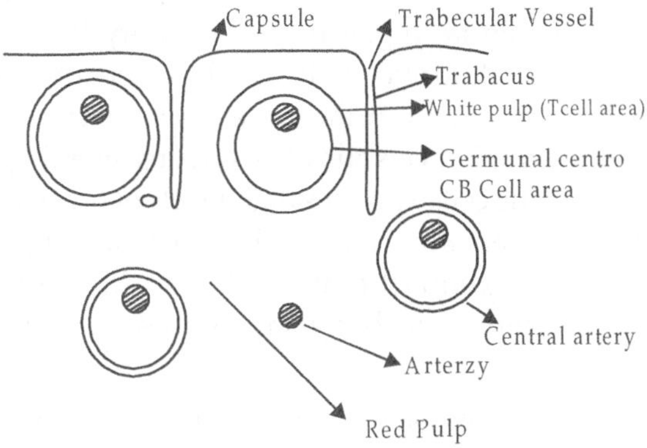

Here inter digitating dendritic cells capture Ag and present it with class II MHC molecule to thymus cell. Once activated, these thymus cells can activate B-cell, the activated B-cells together with some TH-cell, then migrate to primary follicle in the marginal zone. Upon antigenic challenge, these primary follicle develop into characteristic secondary follicle containing germinal centre (like those in lymph made)where rapidly dividing B-cells (centroblast) and plasma cells are surrounded by dense cluster of concentrically arranged lymphocytes.

The effects of spleenectomy of the immune response depends on the age at which spleen is removed. In children spleenectomy often leads to an incident of bacterial species caused primarily by Streptococcus pnemoniae, *Neisseria meningitis and Haemophillus influenza*. Spleenectomy in adults has less adverse effect, although leads to some blood born bacterial infection.

Unlike the lymph node, the spleen is not supplied by afferent Lymphatic drainage, the tissue space. Instead, blood born Ag are carried into the spleen to the spleenic artery, which empties into the marginal zone. As Ag enters the marginal zone it is trapped by interdigitating dendritic cells, which carry the Ag. to periarteriolar Lymphoid sheath. Lymphocyte in the blood also enters the sinus in the marginal zone and migrate to the periarteriolar lymphoid sheath.

2.9.1. Lymph nodes

Lymph nodes are encapsulated bean shaped structures containing a reticulate network packed with lymph nodes, macrophages and dendritic cells clustered at junction of the lymphatic vessels, lymph nodes are the first organized lymphoid structure to encounter antigens that enters the tissue spaces. As lymph particulate through nodes, any particulate antigen that is brought in with the lymph will be trapped by the cellular network of phagocytic cells and dendritic cells (follicular and interdigitating) the overall architecture of lymph nodes provides an ideal micro environment for lymphocytes to effectively encounter and respond to trapped antigen.

A lymph node can be divided into three roughly concentric regions, (*i*) cortex (*ii*) para cortex (*iii*) medulla . Each of which provides a distinct micro environment. The outer most layer, the cortex contains lymphocytes (mostly B-cells), macrophages and follicular dendritic cells arranged in primary follicles(In the absence of Ag activations, a lymphoid follicle called a primary follicle) following antigenic challenge. The primary follicles enlarge into secondary follicles, each containing a germinal centre. These B-cells activation and differentitation in to plasma and memory B-cells occur in the germinal centers of lymph nodes.(So children with B-cell deficiency the cortex lacks primary follicles and germinal centre). Beneath the cortex is the paracortex, which is populated interdiging largely with T-lymphocytes and also contains inter degenerating dendritic cells. There interdigitating dendritic cell expresses high levels of Class-II MHC molecules, which are necessary for antigen presentation in to T cells. The paracortex some time is referred to as thymus dependent area, in contrast to the cortex, which is thymus independent area. The inner most layer of lymphnode, but many of these are plasma cells article sensing molecules.

As antigen is carried into a regional node by a lymph it is trap, processed and presented together with class-II MHC molecules by interdegitating dendritic cells in the places counter resulting in thymus cell activation. The initial activation of B-cells is also thought to take place with in the T-cells rich paracortex. Once activated, thymus B-cells forms small loci consisting largely of proliferating B-cells at the edges of the paracentres. Some B-cells within the loci differentiate into plasma cell secreting IgM and IgG. There fore such maximum size within 3-5 lays of Ag challenge.

With in 4-7 days of Ag challenge, a few B-cells, together with few thymus cells migrate to primary follicle, at the centre. It is not known that what cause these cells to migrate to the primary follicles. Within a primary follicle, cellular interaction between follicular dendritic cells, B-cells, and thymus cells takes places leading to a development of a secondary follicle with a germinal centre, trap Ag complexed with Ab and retain the Ag-Ab complex on the membrane for a longer period of time. Ag trapped on the membrane of these cell is thought to be particularly effective in activating B-cells. Follicular dendritic cell also produce growth factors that are essential for B-cells activated. The activated B-cells in the germinal centre divides rapidly and either differentiate into plasma and memory B-cells or die by programmed cell death. Some of the cell death occuring within the germinal centre is due to the process of selection to remove self reactive B-cells. Plasma cell leaves the germinal center and migrate to the medulla where they secrete large quantities of Ab.

Lymphatic vessels pierce the capsules of a lymph node at numerous site and empty lymph into subcapsular sinus. Lymph coming from the tissue percolates slowly inwards through the center, paracortex and medulla, allowing phagocytic cells and dendritic cells to trap any bacteria or particulate material (Eg Ag-Ab complex) carried by the lymph.

Following infection or introduction of other antigens into the body, the lymph leaving a node through its single efferent lymphatic vessel is enriched with Ab newly secreted by medullary plasma cells and also have the 50 fold higher concentration of lymphocytes than the efferent lymph.

The lymphocyte in lymph leaving a node is due in part to lymphocyte proliferation with in the node in response to Ag. Most of the these, however represent the blood born lymphocyte that migrate into the node by passing between specialized endothelial cell linking post capillary venules of the node factors released in lymph nodes during Ag structure are thought to facilitate this thymus lymphocyte migration.

QUESTIONS

1. Define immune system.
2. Explain the primary lymphoid organs.
3. What is a thymus ?
4. Draw the structure of the section of thymus and explain.
5. Explain the significance of bone marrow.
6. What is Bursa of fabricus ?
7. Explain the T cell maturation.
8. Explain the B cell maturation.
9. What are secondary lymphoid organs ?
10. Explain the Lymph nodes.
11. Explain the spleen and its importance.
12. Describe the location of spleen in Human system.
13. Describe the components of spleen.
14. Write short notes on.
    (*i*) Red pulp (*ii*) White pulp (*iii*) PALS/S

CHAPTER 3

IMMUNE RESPONSE

3.1. INTRODUCTION

Specific immune responses are divided into humoral and cellular responses, Cellular immunity can only be transferred by cells. The best examples of cellular immune responses are the rejection of an organ graft by lymphoid cells and graft versus host tissue (where transferred cells attack an immunologically compromised recipient).

The vital role played by lymphocytes in the types of immune response was demonstrated by cowans and co workers. They cannulated and drained rat thoracic ducts to obtain a cell population comprising more than 95% lymphocytes. They showed that these cells could transfer the capacity both to make antibody and to reject skin grafts. This transfer of immune responses firmly established the lymphocyte has the key cell involved in both types of specific immune responses.

Since cellular immune responses are dependent on an intact thymus, the lymphocytes responsible are known as thymus dependent (T) cells. Similarly, antibody producing lymphocytes, which are dependent on the bursa equivalent (the bone marrow), are known as B Cells. In response to antigen stimulation, B cells will mature to antibody secreting plasma cells.

3.1.1. Origin of lymphocytes

Because lymphocytes are the pivotal cells of the specific immunologic response, on understanding of their origin movement, processing and general function is necessary before their biology can be presented in more detail, undifferentiated lymphocytes are initially derived from bone marrow stem cells, they are produced in the bone marrow stalk cells. They are produced in the bone marrow at a very high rate (109 cells / day). Some lymphocytes migrate through the circulation into the secondary lymphoid tissue (thymus, spleen and lymph nodes) where they produce lymphocyte colonies.

The undifferentiated lymphocytes that migrate to the thymus undergo special processing and become T cell or T lymphocytes. (the letter T represents the thymus). Some T cells are transported away from the thymus and enter the blood where they comprise 70 80% of the circulating lymphocytes. Other T cells tend to reside in various organs of the lymphatic system such as the lymph nodes and spleen. This thymus dependent differentiation of T cells (or thymocytes) occurs during early childhood and by adolescence the secondary lymphoid organs of the body generally contains the full complement of T cells.

B cells or B lymphocytes differentiate in the fetal liver and adult bone marrow, (the letter B was originally derived from the Bursa of fabrics), a specialized appendage of the cloaca of chickens where these lymphocytes differentiate). B cells are distributed in the blood and constitute 20 to 30% of the circulating lymphocytes. They are also seen in various lymphoid organs along with the T cells.

There is a population of lymphoid cells that do not have characteristics of either T or B cells. These are called Null cells or Third population cells because they lack the specific surface markers of B or T cells and can be distinguished from them by the presence of cytoplasmic granules. It is currently believed that this population of cells contains the majority of natural killer and antibody dependent cytotoxic T cells. Although these cells are probably of bone marrow origin, their exact lineage is uncertain.

3.1.2. Function of lymphocytes

As an overview, plasma cells respond to antigens by secreting antibodies into the blood and lymph. Antibodies are glycoproteins produced by plasma cells after the B cells in their lineage have been exposed to antigens. Antibodies are specifically directed against the antigen that caused their formation. Because antibodies are soluble in blood and lymph fluids, they provide humoral (latin humora liquid) Immunity or Antibody mediated immunity, the humoral immune response defends mostly against bacteria, bacterial toxins and viruses that enter the body's various fluid systems.

T cells do not secrete antibodies. Instead, they attack (host cells that have been parasitized by microorganisms) tissue cells that have been transplanted from one host to another and (*iii*) cancer cells, they also produce lymphokines, chemical mediators that play specific augmenting and regulatory roles in the immune system. Since T cells must come into close physical contact with foreign cells or infected cells in order to destroy them, they are said to provide Cell mediated immunity.

Natural killer (NK) cells destroy tumor cells, virus and other parasites infected cells, and play in regulating the immune response.

Among circulating lymphocytes approximately, 20 to 30% are antigen reactive B cells, 65 to 75% are T cells and less than 5% are null cells (Killer cells which kills the antigen). i.e. not identifiable as either B or T cells on the basis of surface markers). The number of lymphocytes in the body is enormous, Perhaps, over a trillion cells constituting approximately 1% of the body's total weight.

Specific binding of antigens by membrane bound receptors can stimulate the transformation of small lymphocytes into larger forms with more abundant cytoplasm, and larger numbers of mitochondria and the ribosomes. Some large B lymphocytes can undergo antigen dependent differentiation into plasma cells which are filled with endoplasmic reticulum indicative of active antibody synthesis. Other large lymphocytes may revert back to small cells following antigen stimulation and function as memory lymphocytes. T cells can undergo an analogous activation by exposure to antigen. Identification of specific population and sub populations of cells that participate in the immune response has become critical in immunology.

T-lymphocytes are critically involved in both humoral and cell mediated immune response. They are, therefore intimately related to our well being. Indeed, the very quality and quantity of life may be dependent on proper numbers of healthy T cells. If T cell formation or T cell populations once formed are adversely affect the results genetic defects, disease, drugs, radiation or the ageing process, individual may become especially susceptible to certain virus, intracellular parasitic malignant and autoimmune diseases.

T cells may be detected, isolated and accessed largely due to the surface markers expressed which permit identification, pluripotent stem cells in the bone marrow give rise to progenitor T cells, which migrate to the thymus. Once they enter the cortex or outer portion of the thymus they are called T-lymphocytes. As the T cells mature, the nature of distribution of surface markers (including CDs and TCR's) change in particular. The characteristics of TCR's on individual T cells change during antigen independent differentiation and ultimately contribute to the antigen specificity displayed by mature T lymphocytes. These surface markers and receptors provide a powerful fool for studying the differentiation process, since through them T cells can be identified at various stages of development. All members of the class of cells known as T cells whether mature or immature including sub populations in or out of the thymus share antigen that are not present on B lymphocytes. In mice, this antigen is known as theta or Thy 1. The human counter part markers are CD5, CD3 and CD 2. Within T cell populations, there are subpopulations, for eg. Ly 2,3 in murine (mouse) and CD4 or CD 8 in human. Furthermore, T cells express histocompatability antigen that are distinct for each individual and that can vary on T cell sub sets.

Major histocompatability antigens are products of a cluster of linked genes, known as the Major histocompatibility complex (MHC). Genes if the I region (immune region of the murine MHC they exert dramatic control over immune responses, Immune response (IR) gene products (Ia antigen may be displayed on lymphocytes as well as macrophages and influence or restrict T and B cell activities).

Several subclasses of T cells have been identified that can be discriminated on the basis of their roles in immune responses, for eg. some T cell regulate humoral and cell mediated responses by providing help, whereas other T cells may supplies a response if necessary. T helper cells (TH) and T suppressor cells (TS) possess distinct antigens which identify helpers nor suppressors. But serve as cytotoxic effector cells (TC), while still other T cells (TD) mediated delayed type hypersensitivity reaction by recruiting and regulating a variety of non specific blood cells and macrophages. Both the TC and TD cells may be directly involved in the rejection of tissue implants that are incompatible to the host. Thus, T cells share a common class antigen Thymus of 1 mouse or CD 5, CD 3, CD 2 (Human) and express the TCR's that distinguish them from B cells. In addition they display MHC antigen representing cells. Moreover, they express different antigen during the maturation process (differentiation antigens) Further more, the different types of T cells exhibit subset antigens reflecting the specific functions.

3.2. EFFECTOR RESPONSES IN THE HUMORAL BRANCHES

The humoral immune response, which is one equally adapted to the elimination of extra cellular pathogens, is characterized by the production of large number of Ab molecules the specific for Ag and Antigenic determinants on a foreign pathogen. The potential diversity generated by this branch of the immune system can provide different Ag binding specificities. This enormous Ag binding diversity is coupled with conservation of constant region sequence, which conservation of confers biological effector function on Ab molecule.

3.2.1. Antibodies perform this function by following mechanism

- Activating the complement system, resulting in lysis of microorganism.
- Acting as opsonin enhancing phagocytosis of microorganism.
- Binding to and neutralizing bacterial freezing.
- Binding this and inhibiting their ability to infect host cells.

Binding to potential pathogens at mucus membrane surface three by preventing coconization.

Binding to Fc receptor on NK cells or macrophages in ADCC, conferring specificity for Ag on these other wise non- specific cells.

3.2.2. Primary and secondary responses

The first contact of an individual with an exogenes antigen generated the humoral response characterized by production of Ab presenting plasma cell and memory B cells. The kinetics of primary response are measured by serum Ab level, vary depending on the nature of Ag, the route of Ag administration, the presence or absence of adjuvants and species or strain being immunized.

In all cases, however, a primary response is characterized by a lag phase, during with native B cells undergo clonal selection in response to the Ag and differentiated in to plasma cells and memory cells. The log phase as followed by logarithmic increase in serum Ab level, which reaches a peak, plateus for a variable time, and then decline. In case of an Ag such as sheep red blood cells, lag phase lasts 3-4 days.

$$\text{B cell} \frac{\text{lag phase}}{3\text{-}4\,\text{days}} \text{plasma cell} + \text{M cels}$$

During the humoral response, IgM is secreted initially, after followed by IgG. Depending on the persistence of the Ag, immune response can last for varying period, sometimes only a few days and sometime several weeks.

The memory B cells formed during a primary response stop dividing and enter the lag phase of the cold cycle. These cell have variable, life span and persist does the life of the individual. The existence of long lived enquiry cells accounts for a phenomenon is called as original antigenic which was first observed when the Ab response to influenza vaccines was monitored. Monitoring revealed the immunization with an influenza vaccine of one strain elicited an Ab response to that strain, but paradoxically also elicited an Ab response of greater magnitude to another influenza strain that the individual had been exposed to during childhood, it was as of the memory of the first Ag exposure had left lifelong imprint on the immune system. This phenomenon can be explained by the presence of memory cell population, eluted by influenza strain encountered in childhood, that inactivated by cross reading epitopes on the vaccine strains. These presence generates a secondary response characterized by Ab with higher affinity for the earlier strain.

The capacity to develop on two humoral response depends on the existence of population of memory B cells and memory T cells. Ag activation of these memory cell results in secondary Ab response that can be distinguished from response in several ways. The two response has a shorter lag period.

3.2.3. T cell accessory membrane molecules

Recognition of Ag MHC complexes is mediated solely by the TCR CD3 complex. A variety other membrane molecular play an important accessory role in Ag recognition and T cell activation.

Some of these membrane accessories strengthen the interaction between T cells and Ag presenting cells on target cells.

Some act in signal transduction and come exhibit both function.

3.2.4. CD4 and CD8 receptors

T cells can be sub-divided based on their expression of CD4 or CD8 membrane molecules. They are T_H, T_C and T_S.

$CD4^+$ T cells recognize Ag in association with Class II. MHC molecules are largely function as helper cells.

Where $CD8^+$T cells recognize Ag in association with class I MHC molecules and largely function as cytotoxic cells.

3.2.4.1. CD4

CD4 is a 55 Kd monomeric membrane glycoprotein that contains four extracellular Ig like domain (D1D4), a hydrophobic transtrem, regain and a long cytoplasmic tail containing a three serine residue which can be phosphorylated.

The monomeric CD4 molecule contains four Ig fold domains, where each chain in the CD8 molecules contains one Ig fold domain.

3.2.4.2. CD8

CD8 generally is present as disulphide lined Ab heterodimer and less frequently as can a homodimer.

Both the α and β chains of CD8 are small glycoproteins, of approximately 30 to 38Kda.

Each chain consists of a single extracellular Ig like domain, a hydrophobic transmembrane region and a cytoplasmic tail containing 25 to 27 residues, several of which can be phosphates.

3.2.5. Properties

Both CD4 and CD8 possess two key properties that allow their membrane molecules to be classified as co receptors.

(1) Recognition of peptide MHC complexes.

(2) Signal transduction.

The Extra cellular domains of CD4 and CD8 bind to the membrane bound proximal domains of MHC molecules on APC's or target cells.

CD4 binds to the F2 domain of class II MHC molecules where as CD8 binds to the three domain of MHC class I.

This interaction apparently function to the acid affinity of the interaction β T cells receptor and a peptide MHC complex. The binding of T cells receptors to peptide MHC complexes has been shown to be augmented about 100 fold by the presence of CD4 as CD8 on the membrane.

3.2.5.1. Signal transduction property

The signal transduction property of CD4 and CD8 is mediated through their cytoplasmic domains.

Both CD4 and CD8 are non-covalently associated with the protein tyrosine kinase attack. This interaction which is essential for effective signal transduction is similar to interaction of the B cells coreceptor with the kinases.

Recent evidence suggests that during Ag recognition by a T cell TCR molecules first binds to a dimer of the class II MHC molecule.

The membrane distal domain of CD4 then binds to the B2 domain of the class II MHC molecule, forming a ternary complex of TCR, CD8 and MHC. Once bound to the MHC molecules the CD4 molecules is thought to conformation change enabling its membrane proximal domain to interact with the membrane proximal domains of an adjacent CD4 molecule that has complexed in a similar manner with a TCR and MHC molecule.

Eventually a tetrameric structure even larger oligomeric structure will be determined on the membrane. This cross linkage of the TCR's on a T cells, forming a lattice type structure, may be necessary for transmembrane signaling events. It is not known whether CD8 binding to class I molecules forms of similar oligomeric structure or not.

3.2.6. Other accessory membrane molecules

In addition to CD4 and CD8 T cells possess several other accessory molecules including CD2, LFA 1, CD28 and CD45. Unlike the CD48 and CD8 co –receptors, the other accessory molecules do not interact with peptide MHC Complexes rather these additional accessory molecules bind to other ligands present on APC's on target cells.

As T cells are activated the strength of adhesion, but some of these accessory molecules and their respective ligands has been shown to then there by prolonging the association between the interacting cells and providing time for directed secretion of various cytokines or lytic enzymes. Like CD4 and CD8 some of these of other accessory molecules also function as signal transducing molecules.

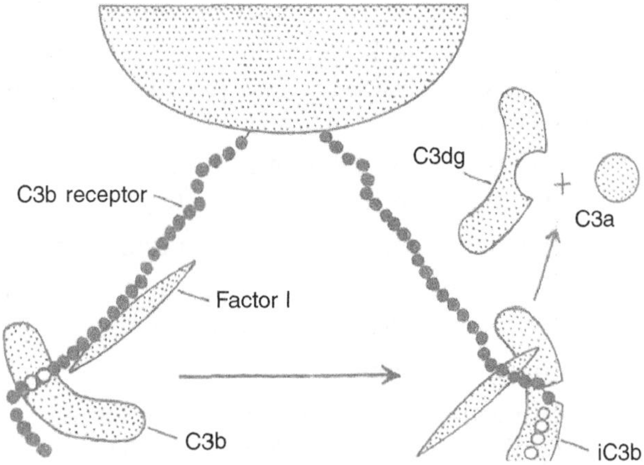

Fig. 3.1.

The role of these accessory molecules treated by the ability of monoclonal Ab's block T cell activation.

The interactions between various thymus and Tc cells accessory membrane molecules and their ligands of Ag presenting cells and target cells, respectively.

3.2.7. T cell activation

The central event in generation of both humoral and cell mediated immune response is the activation and clonal expansion of T cells (Activation of Tc cells, which is generally similar to T cell Activation). T cell activation is initiated by interaction of the TCR CDS complex with a processed antigenic peptide bound to a class II MHC molecule on the surface of an APC. This interaction and the resulting activating signals also involve a variety although the interaction between the T cells receptor antigenic peptide and an MHC molecules is central to development of both humoral and cell mediated responses.

The TCR recognized the complexes consisting of a peptide MHC molecules exhibiting both the peptide and MHC molecules.

3.2.8. Evidence that TCR above recognized peptides complex

Discovery of CD3 and CD4 and CD8 raises the possibility that these molecules associated with the TCR might been necessary for Ag recognition by T cells. Experiment of Kappler Marock out line that the notion, that a signal receptor on the T cells recognizes peptide complex to a self MHC molecules however their experiment does not the possibility that the other membrane molecules associated with the T cells receptor or MHC molecules.

A definitive experiment proving that the α, β, T cells receptor alone recognizes both peptide and MHC. These experiment is a similar approach to that of the early Kapplar Marrack experiment but it involve gene transduction inserted marker experiment but it involve gene transduction instead of cell fusion. Functional TCR a and b chain from a Tc cell a clone must specific for one hapten on H 2 target cells, were transfected into other cytotoxic cells clone specific for a different hapten on Ag H-2K.

The transfection cells express both the own T cells receptor and the new T cell receptor.

Cytolysis assays showed that the transfected cells recognize both hapten MHC complex for which the original; Tc clones. The specific however the transfected cells did not recognize on target cells of a different MHC haplotype, in other words transfection of TCR genes.

Alone transfected reactively to a particular Ag and a particular MHC molecule of accessory molecule on the Thymus cell and APC Interaction of thymus cell which Ag initiates a cascade of biochemical event may induce the resting TH cell to enter the cell cycle (Go to G1 transition) and culminates in expression of the high affinity receptor for Ic 2 and secretion of IL 2.

In response to IL 2, the activated TH cell progress through the cell cycle, proliferating differentiating into memory cells with effector cells.

Following interaction of Thymus cells with Ag numerous genes are activated.

The gene products that appear can be divided into 3 categories depending on how early they can be detected following genes.

3.2.9. Immediate genes

Expressed in 1/2 hr. of Ag recognition en code a number of transcription factors including FOS, Myc, C jun, NF AT and KB.

Early genes: Expressed within 1–2 hr of Ag recognition encode Ic 2, IL 2R, IL 3, IL 6, IFN and numerous other proteins.

3.2. 10. Late genes

Expressed more than two days after Ag recognition encode various adhesions molecules.

3.2.11. TCR coupled signaling pathways

Although the precise mechanisms that link Ag recognition by the T cell receptor to gene activation are not fully understood a number of signaling events common to may cells have been shown to occur in the Thymus cell.

Although the interaction between the T cell receptor and antigenic peptide and an MHC molecules is control development of both humoral and cell mediated responses.

The TCR recognized the complexes consisting of a peptide MHC molecules exhibiting both the peptide and MHC molecules.

Evidence that TCR alone recognises peptide MHC molecule

Discovery of CD3, CD4, and CD8 rise the possibility that these molecules associated with the TCR might be necessary the notion that a single receptor on the T cell recognises peptide complex to a self MHC molecule, however their experiments does not rule out the the possibility that the other membrane molecule associated with the T cell receptor might contribute to the recognition of either the peptide or MHC molecule.

A definitive experiment proving that the d, P, T cell receptor alone recognises both peptide and MHC. These experiment is a similar approach to that of the earlier capplar merrack experiment but it involve gene transduction, instead of cell fusion. Functional TCR α and β chain from a Tc cell alone specific for one hapten on target cells were transfected into other cytotoxic T cells clone specific for a different hapten on Ag $H\text{–}2^K$. The transfected cell express both the own T cell receptor and the new T cell receptor.

Cytolysis assays showed that the transpected cells recognize both happen MHC complex for which the original TC clones the specific, however the transfected cells did not recognise on target cells of a different MHC haptotype. In other words transfection of TCR genes alone transfected to a particular Ag and a particular Ag and a particular MHC molecule.

These events included activation of Interior rapid specific phospholipase C (PLCQr1), hydrolysis of plasma membraneinositol phospholipase in intracellular Ca2+ levels and action of various protein kinases that subsequently, phosphorelate a number of prokins involved in mediating the activation signal. TCR itself has share cytoplasmic domains which are unable to mediate signal transduction. Instead signal transduction is mediated by the cytoplasmic domains of the CD3 CD5, by the coreceptor CD4 or CD8 and by various accessary molecule like CD2 and CD45.It appears that signal transduction is accomplished by a series of protein Plation event catalysed by protein kinases and dephosphorylation events and catalysed by protein phosphatases.

The cytoplasmic domains each CD3 chain contain a sequance motif called ITAM that is through interaction with proteins, tyrosine, kinase signal transduction. Two protein trikinase ZAP 70 have been shown to be associated with the cytoplasmic domains of CD4 and CD8 with the protein tyrosine kinase. It is Hypo that TCR aggregation complexed with the coaggregation of the coreceptors, CD4 or CD8 and the kinases in one place facilitating their activation sensitisation of the protein tyrosine kinases appears to require the activity of CD45. This membrane molecules has a cytoplasmic tail with two protein tyrosine phosphatase domains, which are thought to catalyze dephosphorylation step is to activate these two trgrosine kinases so that they though begin to prosphory late the and beta chains of CD3 phospholipase with and other cellular substrate, phosphorylation of PLCr allows it to hydrolyse phosphatidyl ionositol 4, 5 bis P [P_1 P_2] a membrane phospholipid into two important products IP3 and DAG.

Once formed, Ip3 and DAG initiate 2 necessary signaling pathways. In one pathway Ip3 triggers an T in intracellular Ca2+ and the subsequent activation of calmodulin dependent phophatase called calcineurin, with dephosphorylation the inactive cytosolic form of the T cell specific nucleare factor NFAT. In the other pathway DAG activates protein kinase C, which is than phophorylates various cellulase substrate and mediates realise of numerals factor NDCD. Both buclea factor enter the nucleus where they particles in activation of various genes including the IL2 gene third signaling pathway is generated by the CD28 B7 interaction This pathway works together with the TCR mediated signaling pathways to activate a protein kinase designates, that is involved in the phosphorylation of the nuclear factor Cjun of the DNA binding ptns involved Tn cell activation only ND AT is T cell specific the others are formed in many types of cells. Following Ag recognition by TH cells, NF ATC then translocates to the nucleus where it is dimirized with Ap1 the resulting complex then binds to the IL2 enhance of dephosphorylation of NF ATC PO4 is inhibited at it is by cycloperin A and FK 506, than Thymus cells activation is blocked and the immune response potent reduced because of their pert immune suppresive capability cytoperin 'A' FK 506m have proven to be extremely effective in prolonging graft survival in transplant reciepients.

The population of memory B cells specific receptor a given Ag is considerably larger than the population of corresponding native B cells converting in part for overcoming of the differences between the primary and secondary response, furthermore memory B cells are, more early activated than native B cells. The processes of affinity maturation and class sutehing are responsible for the higher affinity and different types exhibited in a secondary response. The higher levels is Ab coupled with the overall high affinity provide an effective host defense again reinfection. The changes in isotypes that occurs in the secondary response produces Antibodies whose biological effecter functions are particularly suited to eliminate a given pathways 397 to 398.

3.2.12. Comparison of primary and secondary immune responses

Property	Primary response	Secondary response
Responding B cell	Native (virgin) B cell	Memory B cell
Lag period following Ag administration	Generally 4-7 days	Generally 1-3 days.
Time of peak response	7-10 days	3-5 days.
Magnitude of peak Ab response	various depending on Ag	Generally 100 – 1000 times higher than primary response.
Iso type produced	Ig M predominates early	Ig G predominates in the response.
Ags	Thymus dependants thymus	thymus dependent
	lower	higher.

3.2.13. Viral neutralization by humoral Antibody

Antibodies specific for viral surface Ags are often crucial in containing the speed of producing acute infection and in protecting against reinfection. Most of virus express surface receptor molecules that enable them to initiate infection by binding specifically to host cell membrane molecules. For e.g. influence virus binds to sialic acid residues in cell membrane glycoproteins and glycolipids retrovirus binds to in the cellular adhesing molecules (ICAMS) and epsin mar cirus binds to type 2 complement receptor on B cells. If Ab in produced to the viral receptor it can block in faction although by presenting binding of viral particles to host cells. Secretory Ig A in mucous Secretions plays an important role in host defense against virus by blocking viral attachment to mucosal epithelial cells. The advantage of the attenuated polio vaccine, is that it induces production of secretory Ig A, which is effectively blocks attachment of peliouys along gastrointestinal tract.

Viral neutralization by Ab sometimes involves mechanism that operate following viral attachment to host cells. In some cases Antibodies may block viral penetration by binding to Epitopes that are necessary to mediate fusion of the viral envelop with the plasma membrane. If the induced Ab is of complement activation more lysis of enveloped viral can endure. Ab complement can also agglutinate virals particles and function as a opsonizing agents to facilitate Fc or C3b receptor mediated phagocytosis of the viral particles.

3.2.14. Polio of Thymus cells in humoral response

Activation of B cells by soluble protein Ags requires involvement of Thymus cells. In this case binding of Ag to the mg doses not induce once effect competence signal as additional interaction with membrane molecules on the thymus cell needed to induce B cell activation. In addition a cytokine mediated progression in required for B cell proliferation.

3.2.15. Formation of T B conjugate

Following binding of Ag by MIG on B cells the Ag is internationed by receptor mediated endocytosis i.e., processed with in the Endocytic pathway into peptides Ab binding also provides complete signal I and appears to induce the B cells to induce class MHC molecules and co stimulate ligand bromines. Expression of both of these membranes enhance the ability of the B cell to function as an Ag presenting cell in thymus cell activation. The Antigenic peptides produced with in the endocytic processing pathways associated at class II MHC molecules and are present on the B cell membrane to the thymus cell inducing its activation.

3.2.16. Cell mediated Immune response

It is host defenses that are mediated by Ag specific T cell and vaccines non specific cells of the immune system. It protect agents intra cellular gatuate viruses and cancer and is response for graft rejection. Transfer of prime T cells confers this type of immunity of the recipient.

Generation of the cell mediated immune response:

3.2.17. Contact dependent Help mediated by CD40 CD40 L interaction

Natural B cells express the membrane glycoprotein CD40 which balance to a family of cell surface proteins that regulates cell proliferation and programmed cell death by apoptosis. On activation thymus cells express a hygain for CD known an CD40L. Interaction of CD40L with CD40 on the B cell delivers couptance signals 2 to the B cell the signal 1 generation by Ig M cross linkage constitutes effective complex signals that derives the B cell to G1 The complete signal transducted Via intracellular signals pathways leading to changes in gene expression.

Although B cells stimulated with membrane from activated thymus cells are able to proliferate. They fail to differentiate unless lymphokines also are present. This finding suggested that both a membrane content signals and cytokine signals are necessary induce B cells publicity and differentation generated, are B cells beings to express membrane receptor for various cytokines. These vuretores the bind the cytokines release in a diverted fashion toward the B cells from the interesting T4 cell, thereby generation f progression signals. These thymus cells derived cytokines (1L 2,4 and 5) provide progression signals leading to B cells protection. As activated B cells proliferated then differentiation even occur formation of plasma and memory B cells surfling and affinity maturation.

3.2.18.T cell receptor complex

Membrane bound Ig on B cells associated with another membrane protein the Ig A, Ig B heterodimer, to form the B cell Ag receptor.

Similarly the T cell membrane receptor associated with CD3, forming the TCR CD3 membrane complex

∗ In both cases the accessory molecules is involved in signal transduction after interaction of T cell with Ag.

∗ The first evidence suggesting that the T cells receptor is associated with another membrane molecules came from experiments in which fluarevent Ab to the receptor was shown to "co cap" another membrane protein designed CD3.

CD3 is a complex of five invant polypeptide chains that associated to from dimer in a heterodimer of gamma and epilosilon chains heterodimer of delta and epilosilon chains and a homodimer of zeta chains (ss) on a heterodimer zeta and eta chains (sn)

∗ The sn chains which are endorsed by the same gene, differ in their can beta terminal ends due to difference in RNA splicing of the primary transcript.

∗ The a b or c TCR heterodimer determines the ligand binding specific where as the CD3 dimer and required for expression for expression of the T cell receptor and for signal transduction.

The r,s and chains of CD3 and membrane the Tg super family each containing an Ig like extracellular extra cellular domain followed by a trans membrane regions a cytoplasmic domain of more than 40 amino acids.

* The transmembrane region of all the CD3 polypeptide chains contains a negatively charges aspartic acid residence.

* These negative charged groups may enable the CD3 complex to interact with one on 2 positively charged aminoacid that are present in the transmembrane region of each TCR chain.

3.2.19. ITAM (Immune receptor type specific based activation molecule)

The cytoplasmic tailed of the CD3 chains a might be called the immune receptor type specific based activation molecule (ITAM)

This modifies found in a number of other to us including the Ig A, Ig B heterodimer of the most B cell receptor complex F receptor from Ig E and Ig G.

3.2.20. Role

The ITAM sites have been shown to interest with tyrosine kinase and to play an red in signals transmission.

(*i*) In CD3, the s chain each contain a signals copy of ITAM

(*ii*) The s and n chain contain a copies and these occurs more rapidly reaches a magnitude and lasts for a longer duration. In addition, the two response is characterized by secretion of Ab c a higher affinity for Ag and is types other than the Ig M predominate.

Time after immunization can complement and isotype of seem Ab following primary and Secondary immunization with Ag.

The population of memory B cells specific for a given Ag is considerably larger than the population of corresponding B cells, accounting in part for some of the differences between the primary and secondary response Further more memory B cells are more easily activated them to native B cells. The processes of affinity maturation and class switching are responsible for the higher affinity and different types exhibited in a secondary response. The higher levels of Ab coupled with two overall higher affinity provide an effective host defence against reinfection. The change in isotype that occurs in the secondary response provides Antibodies whose biological effecter functions are particularly switched to eliminate a given paterigen. 397 to 398

Comparison of primary and secondary Ab response property response secondary one responding B cell Naive cvirgin B cell memory B cell lag period following generally 4 – 7 days generally 1 – 3 days. Time of peak response 7 – 10 days 3 – 5 day magnitude of peak varies depending generally 100 1000Ab response. On Ag times higher than primary response.

Isotype produced Ig M predominates early Ig G predominates in the response.

Ags	Thymus dependents Thymus independent	Thymus dependent
Ab affinity	Lower	Higher

QUESTIONS

1. Explain the immune responses in Human beings.
2. Outline the T lymphocytes in immunity.
3. Explain the effector responses in the Humoral branches.
4. What are the primary and secondary response in immunity ?
5. Write an account on the origin of B lymphocytes.
6. Outline the biology of B lymphocytes.
7. Explain the various molecules present on the T cell surface.
8. Explain the genetic organization of the T cell receptor.
9. Write a note on CDB complex.
10. Explain the role of CD4 and CD8 molecules in T cell activation.
11. What is the part placed by CD2 molecule in T cell activation ?
12. Outline briefly the immunoglobulin gene family.
13. Write an account on suppressor T cell.
14. Write a note on Macrophages.
15. Explain oxygen dependent killing mechanisms by Macrophages.
16. Explain oxygen independent killing mechanisms by Macrophages.
17. Write short notes on.
    (*a*) Dentritic cells.
    (*b*) Natural killer cells.
    (*c*) Neutrophils.
    (*d*) Easinophils.
    (*e*) Basophils.

CHAPTER 4

IMMUNITY

4.1. INTRODUCTION

The term immunity refers to the resistance exhibited by the host towards injury caused by microorganisms and their products. Production against infectious diseases in only one of the consequences of immune response, which in its entirely as concerned with the regulation of the body against any foreign antigen.

Immunity against infections disease is of different types.

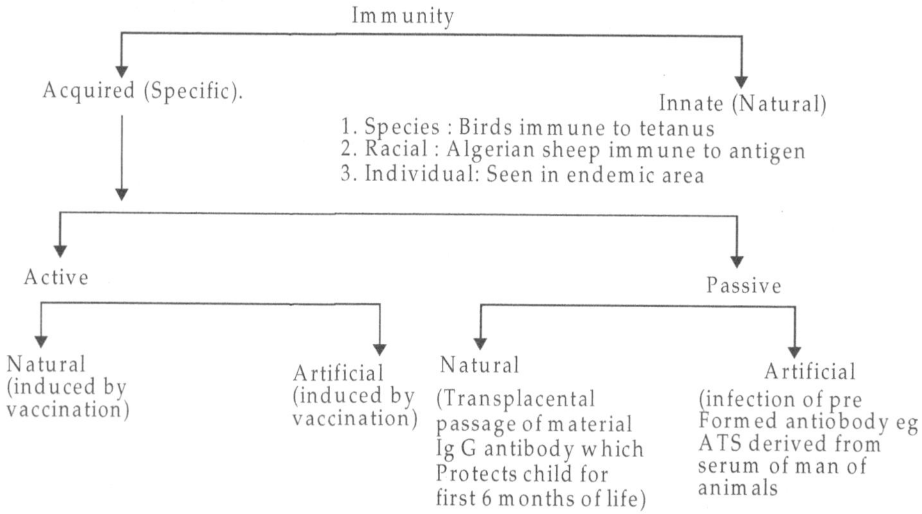

Immunity can be classified as Acquired immunity and Innate immunity.

Acquired immunity is again classified in to Active and passive immunity. Again Active immunity is classified in to Natural and Artificial. The Passive immunity is also classified in to Artificial and Natural.

Innate immunity is classified in to species, Racial, and individual immunity.

4.1.1. Acquired immunity

It supplements the protection provided by Innate immunity. It is present only in Vertebrates. The resistance acquired during life of an individual is known as acquired immunity.

Naturally all organisms are provided with natural immune response to a foreign invader. But immunity is acquired during contact of the invader and is specific to the invader is known as Acquired immunity. This is discovered when Edward Jenner found out the vaccines. The persons once affected by life threatening disease such as bubonic plague or small pox did not die when they are affected again. The compound to which acquire immune response induced is termed as antigen.

Acquired immunity is of three types Active, passive and adaptive immunization.

1.**Active**: It is the resistance induced by an individual after effective contact with an antigen. It is by humoral or antibody mediated immunity (AMI) and cell mediated immunity). It is of two types, namely **Natural** and **Artificial**.

Natural: It is by natural infection by pathogens.

Artificial: Resistance developed by vaccination prepared by live, attenuated or killed microorganisms.

2. **Passive immunization**: It is by the transfer of specific antibody from an immunized individual. eg. Immunization passed from mother to child.

3. **Adaptive transfer (immunization)**: It refers to the transfer of immunity by the transfer of immune cells.

4.2. INNATE OR NATURAL IMMUNITY

Innate immunity is the resistance to infectious which an individual possess by virtue of his genetic and constitutional make up.

(i) It is not affected by prior contact with microorganism or immunization.

(ii) It may be nonspecific when it initiates a degree of resistance to infections in general or specific when resistance to a particular pathogen is concerned.

Innate immunity may be classified as

(i) Species immunity,

(ii) Racial immunity or

(iii) Individual immunity.

4.3. SPECIES IMMUNITY

It refers to the total or relative resistance to pathogens shown by all member of a species.

For instance all human beings are totally insusceptible to plant pathogens and to many pathogens of animal such as rinderpest or distemper. This immunity is something a person obtains as a birthright, for the reason that he belongs to the human species.

The mechanism of species immunity is not clearly understood, but may be due to physiological and biochemical differences between the tissues of the different host species, which determine whether or not a pathogen, can multiply in them. An early insight into the basis of species immunity was gained by Pasteur's experiments on astute in frog's, which are naturally resistant to the disease but become susceptible when their body temperature is raised from 25°C to 35°C.

4.4. RACIAL IMMUNITY

With in a species, different races may show difference in susceptibility to infection. This is known as **racial immunity**.

The classical example of which in the high resistance of Algerian sheep to anthrax, such racial differences are known to be genetic in origin and by selection and in breeding, it is possible to develop at will races that possess high degree of resistance or susceptibility to various pathogen.

It is difficult to demonstrate marked differences in immunity in human races, as controlled breeding is not possible in the human species. It has been reported that the Negroes in the USA are more susceptible than the Whites to tuberculosis. But such comparisons are initiated by external influences such as differences in Socio economic levels. An interesting instance of genetic resistance to ***Plasmodium falciparum*** malaria is seen some parts of Africa and the Mediterranean coast. A hereditary abnormality of cull silk ling, prevalent in the area confers immunity to infection by the malarial parasite and may have evolved from the survival advantage conferred by a malarial environment.

4.5. INDIVIDUAL IMMUNITY

The difference in innate immunity exhibited by different individuals in a race is known as **individual immunity**.

The genetic basis of individual immunity is evident from studies on the incidence of infectious diseases in twins. It is well documented that homozygous twins are susceptible to cepromatious leprosy and tuberculosis. Such correlation is not seen in heterozygous twins.

4.6. INFLUENCE OF FACTORS ON INNATE IMMUNITY

Several factors influence the level of innate immunity in an individual

4.6.1. Age

The two extremes of life early higher susceptibility to infectious diseases as compared to adults. The foetus in uterus is normally protected from maternal infection by the placental barrier.

New born animals are more susceptible to experimental infections than older ones.

Coxsaukie viruses cause fatal infection in suckling mice but not in the adult.

Old persons are highly susceptible to infection due to the gradual waning of their immune responses.

4.6.2. Hormonal influences

Endocrine disorders such as Diabetes mellitus, Hypothyroidism and Adrenal dysfunction are associated with an enhanced susceptibility to infections. This high incidence may be related to the increased level of carbohydrates in tissues.

Corticosteroids exert an important influence on the response to infection. They depress the host resistance by their anti-inflammatory and anti-phagocytic effects and by the suppression of antibody formation and hypersensitivity. They also have a beneficial effect in that they neutralize the harmful effects, of bacterial products such as endotoxins. The elevated steroid level during pregnancy may have a relation to the increased susceptibility of pregnant woman to many infections, the reported effect of stress in increasing susceptibility to infections may, in some, be due to the release of steroids.

4.6.3. Nutrition

The interrelation between malnutrition and immunity is complex but in general both humoral and cell mediated immune process are reduced in malnutrition.

Cell mediated immune responses such as the Mantoux test become negative in severe protein deficiency as in kwashiorkor. Because of its prevalence, malnutrition may well be the commonest form of immunodeficiency.

Paradoxically, there is some evidence that certain infections may not become clinically apparent in the severely ill nourished. Malarial infection in the famine sticken may not induce fever but once their nutrition is improved clinical malaria develops. It has also been reported that some viruses may not multiply in the tissues of severely malnutrition individuals.

4.7. IMMUNITY

Implied resistance to disease become infectious disease cause death in humans.

4.8. IMMUNE RESPONSE

When foreign substance likes bacteria, virus, fungi, or parasites are introduced in to the vertebrate host. The host either (*i*) Non-specifically clears the infectious agent using preformed components.

(*ii*) Produces specific cells or molecules directly against foreign invaders, the later responds is traditionally called immune response.

4.9. INNATE IMMUNITY

Operates non-specifically during the early phases of an immune response.

Specific recognition of pathogen initially activates non-specific cells and molecules. Specific immunoglobulins developed from human beings and animals.

Innate immunity serves the first line of defence and includes both external and

Internal non-specific response components of innate immunity

(*i*) Preformed {they were present before challenge}

(*ii*) Standardized {their response magnitude is consistent).

(*iii*) Without memory (the host does not realize it has been reexposed to the some invaders)

(*iv*) Non specific (It does not distinguish between}

4.9.1. Innate (Non-specific) Immunity

Innate immunity can be envisioned as comprising four types of defensive barriers **anatomy, physiological endocytic** and **phagocytic** and **inflammatory**.

Tissue damage and infection induce leakage of vascular fluid containing serum proteins with antibacterial activity and influx of phagocyte cells into the affected area.

4.9.2. Anatomic Barriers

Physical and anatomic barriers that tend of prevent the entry of pathogens and an organism's first line of defense against infection, the skin and the surface of mucous membranes, are included in this category because they provide effective barriers to the entry of most microorganisms.

The skin consists of two distinct layers–a relatively thin outer layer the epidermis and a thicker layer the dermis. The epidermis contains several layer of tightly packed epidermal cells. The outer epidermal layer consists of dead cells are filled with a water proofing protein called keratin. Old epidermal cells are removed from the surface and are replaced by new cells next to the dermis, as a result, the epidermis is completely renewed every 15-30 days. The epidermis does not contain blood vessels and epidermis causes instead of bathed in nutrients that differ from the underlying dermis. The dermis, which is composed of connective tissue, contains blood vessels, hair follicles, subcutaneous glands and sweat glands.

The subcutaneous glands are associated with the hair follicles and produce an oily secretion called sebum. Sebum consists of lecithin and fatty acids maintaining the pH of the skin between 3 and 5, which is inhibitory to the growth of most microorganisms. A few bacteria that metabolize sebum lived as common salts of the skin and are responsible for a severe form of acne. One acne drug isotretions (Acutance) is of vitamin A derivative that prevents sebum formation.

In fact skin is not only prevents the penetration of most pathogens but also inhibits most bacterial growth to its low pH. Breaks in the skin even small ones, resulting from wounds or abrasion are obvious route of infection. The skin also prevents from the biting insects (Eg: Mosquitoes, mites, ticks, fleas and sand flies). If these harbor pathogenic organisms, they can introduce the pathogen into the body as they feed. The protozoan that causes malaria, for eg is carried by mosquitoes who deposit it in humans when they take a blood meal. Similarly bubonic plague is spread by the bite of flies and Lyme disease is spread by the bite of ticks.

The conjunctive and the alimentary, respiratory and genital tracts and lined by mucous membranes, not by the dry, protective skin covering the exterior of the body. These membranes consist of an outer epithelial layer and an underlying connective tissue layer.

Although most pathogens enter the body by binding to are penetrating mucous membranes a number of non-specific defence mechanism serve to prevent this entry. For example saliva, tears and mucous secretions act to wash away potential invaders and also contain antibacterial or antiviral substances. The viscous fluid called mucus, which is secreted by epithelial cells of mucous membranes, entraps foreign microorganisms. In the lower respiratory tract and the gastrointestinal tract, the mucous membrane is covered by cilia, hair like processes projecting from the epithelial cells.

The synchronous movement of cilia propels mucous entrapped microorganisms from these tracts. In addition, non-pathogenic organisms find to colonize the epithelial cells of mucosal surfaces. These normal flora generally out compete pathogen for attachment sites on the epithelial cells surface and for necessary nutrition.

Some organisms have evolved ways to escape this defence mechanism these are likely to invade the body through mucous membranes. For eg. Influenza virus (the agent that causes flu) has a surface molecule that enables it to attach firmly to cell in mucous membranes, preventing the virus from being swept out by the ciliated epithelial cells. Similarly the organism causing gonorrhea has surface projections that allow it to bind to mucous membrane epithelial cells in the congenital tract. Adherence of bacteria to mucous membranes involves interaction between hair like protrusions on a bacterium called finebrine or pili, and certain glycoproteins or glycolipids that are only expressed by some mucous membrane epithelial cells. For this reason some tissues are susceptible to bacterial invasion, where as others are not.

The importance of anatomic barriers to host defence is illustrated by a group of mice described in a report in nature. These mice appeared to be immune to the parasite helminthes (worm) that causes schistosomiasis, a cleome and debilitating disease affecting more than 800 million people worldwide. After initial infection with this helminthes the mice developed portal hypertension. Similarly Schistosomiasis is observed in human being. However when mice were reinfected with the helminthes a second time, a very low yield of the helminthes was recovered and the mice appeared to be resistant to the infection. Because the mice land apparently developed immunity to the helminthes they were considered to be a potential animal needed for the study of Schistosomiasis in humans. After considerable findings were pouched into research on the mouse model the ability to clear the helminthes was found to have nothing to do with a specific immune response, instead of resulted from a complex anatomic reorganization in blood vessel architecture that occurred at the time of the second infection. The Nature article warned "Not to postulate immunological mechanisms where simple anatomy or physiological explanations might suffice".

Even when a pathogen includes the anatomy defence provided by the skin & mucous membranes, it still fewer other types of innate defenses including various physiologic, phagocyte and inflammatory services. Only by successfully evading these barriers can pathogens become established in a host.

4.9.3. Physiological barriers

The physiological barriers that contribute to innate immunity include temperature pH, O_2 tension and various soluble factors. Many species are not susceptible to certain diseases simply because their body temperature inhibits pathogens growth. Chickens for eg display innate immunity to anthrax because their high body temperature inhibits the growth of this pathogen, gastric acidity also provides an innate physiologic barriers to infection because very few ingested microorganisms can survive at the low pH of the stomach. One reason, new borne are susceptible to some disease that do not affect adult is that their stomach contents are less acidic than that of adult.

A variety of soluble factors also contribute to non-specific immunity. Among these soluble Proteins are lysozyme, interferon and complement. Lysozyme, a hydrolytic enzyme found in mucous secretions is able to cleave the peptidoglycan layer of the bacterial cell wall.

Interferon's comprises a group of proteins produced by virus-infected cells. Among the many function of the interferon's is the ability to bind to near by cells and induce a generalized antiviral state.

Complement is a group of serum protein that circulate in an inactive proenzyme state. These proteins can be activated by a variety of specific and non-specific immunological mechanisms that convert the inactive proenzymes into active enzymes. The activated complement components participate in a controlled enzymatic cascade that result in damage to these membranes of pathogenic organisms either by destroying the pathogens or depilatory their clearance.

4.9.4. Endocytic and Phagocytic barriers

Another improvement innate defence mechanism is the ingestion of extracellular macromolecules via endocytosis and of particulate material via phagocytosis.

These two internalization processes not only bring different types of extracellular material into the cell they also differ in several other ways.

In endocytosis, macromolecules within the extracellular tissue fluids are internalized by cells via the invagination (inward folding) and pinching off of the small regions of the plasma membrane (pH). The resultant endocytic vessels are small approximately 0.1 mm in diameter. Endocytosis occurs though one of two processes. Pinocytosis or mediated endocytosis.

Pinocytosis, non-specific membrane invagination internalizes macromolecules in proportion to their extra cellular concentrations. In receptor mediated endocytosis, macromolecules are selectively internalized after binding to specific are membranes receptors.

The endocytic vesicles formed by either forces fuse with coli other fare delivered to endosomes, which are intracellular acidic compartment that serve a sorting function The Audi interior of endosomes facilitates dissociation of macromolecular ligands from their receptors. The cattle are then recycled back to the cell surface. Three macromolecules contained within endosomes fuse with Primary lysosomes to form structures known as secondary lysosomes. Primary lysosomes are decided from the golgi complex and contain large of degradative enzymes, including proteases, nucleases lipases and other hydrolytic enzymes with in secondary lysosomes, the ingested macromolecules are then digested into small break down Products (eg. Peptides, nucleotides, and Sugars) which eventually are eliminated from the cell.

Phagocytosis involves the ingestion of particulate material including whole pathogenic microorganisms. In phagocytosis the plasma membrane empanels around the particulate material to from large vesicle called phagosomes. These vesicles are roughly 10-20 times larger then endocytic vesicle

The expansion of the membrane in phagocytes requires participation of microfilaments, which do not take part in endocytosis. Another difference between the two processes is that only specialized cells are capable of phagocytes, whenever virtually phagocytic cells include blood monocytes neutrophils and tissue macrophages.

One particulate material is ingested into phagosomes. The phagosomes fuse with lysosomes and the ingested material in than digested in the endocytic processing by Processers similar to that seen in endocytosis.

4.10. CONTAINING CLASS I AND II HLA MOLECULES

MHC polypeptides of both classes are initially expressed in the rough endoplasmic reticulum.

Class I puts sequentially bind endogenous peptide and

B- micro globulin in rough endoplasmic reticulum is then transported to the cell surface.

Class II proteases associated with (invariant chain) in the rough endoplasmic reticulum and so prevented from binding of endogenous peptides. They are translocated instead to an endosomal compartment where invariant chain dissociate & is replaced by exogenous peptides.

4.11. ANTIGEN PRESENTATION BY CLASS I AND II MAJOR HISTOCOMPATABILITY COMPLEX PROTEINS

Class I and II Major histocompatability complex proteins brave considerable structurally homologous and simple peptide binding site but they function differently in Ag presentation.

Class I proteins all presented in CD s 7 cell.

They present in immunogenic different 7 cells type. Class I presents CD8 7 cell and class II presents CD4 7 cells.

Another difference is that the two classes of Major histocompatability complex proteins associate with peptides that have been produced thought different protection pathway in this the cell, for Eg. Class I molecules bind peptide derived from endogenous synthesized proteins (i.e. viral proteins in virus infected cells), Class II molecules binds peptides derived from endogenous proteins (those in the external medium).

4.12. PEPTIDE BINDING BY CLASS I AND CLASS II MAJOR HISTOCOMPATABILITY COMPLEX

	Class I	Class II
1. Peptide binding domain	L1 / B1	L1 / B1
2. Nature of peptide binding	closed at both ends	open at both ends Cleft
3. General size of bound peptides	8-10 amino acids	13-18 aminoainds
4. Peptide involved in binding to MHC molecule.	Annular residues at both ends of peptides.	Annular residues distributed along the left of the peptides.
5. Nature of bound peptide Major histocompatability	Extended selective in which both ends interact complex cleft	extended structure that is held at constant elevation Above the floor of Major Histocompatability complex Cleft.

VACCINATION

There is an old saying 'Prevention is better than cure'. It holds good for vaccination. Edward Jenner before 200 years discovered the small pox vaccination. The basic principle of vaccination remains the same from Jenner's day. After 100 years Louis Pasteur (1870-80) discovered the deadly germ could also offer protection against the disease it caused, if the germ is altered or killed and then injected. This is the basis for developing 20 good vaccines. The Recent developments in Genetic engineering and molecular biology opens new era in vaccinology.

Immunization is the process by which individual resists and overcomesinfection by acquiring resistance. The resistance can be acquired either naturally or artificially. Resistance can be artificially acquired by the use of immunizing agents. There are two types of immunization.

1. Active Immunization
2. Passive Immunization

Active Immunization

Active immunization depends on the participations of immune system to produce resistance. This is al by giving

1. Live vaccine
2. Live or inactivated vaccine
3. Toxoids
4. Recombinant DNA vaccine
5. Synthetic Peptides
6. Internal image antibody
7. DNA vaccine

Passive Immunization

It is achieved by administration of preformed antibody obtained from humans or animals that have been actively immunized.

Vaccine

They are live or killed micro organism or their products, which are used for immunization. Different kinds of vaccines are known. They are killed/inactivated vaccine, live attenuated vaccine, sub unit vaccine, Recombinant DNA vaccine, synthetic peptide vaccine, internal image antibody vaccine or DNA vaccine.

Toxoids

When toxins is mixed with formalin, it is converted into toxoid. It retains its antigenicity but looses its toxicity Eg. Tetanus Toxoid and Diptheria toxoid.

Cold chain

Vaccines are thermolabile products, hence they require special storage and transportation arrangements to retain their efficiency. Since all vaccines return this efficiency at 2-8°C, they have to be maintained in the refrigerator or the box containing ice cubes.

QUESTIONS

1. Define immunity.
2. Explain different types of immunity.
3. Describe briefly on innate or natural immunity.
4. Explain about Racial immunity.
5. Write a note on individual immunity.
6. Explain hormonal influence on immunity.
7. Write the summary of non specific immunity.
8. Explain the anatomic barriers.
9. Explain briefly physiological barriers for immunity.
10. Write briefly the endocytic and phagocytic barriers.
11. Write briefly the class I and Class II HLA molecules.
12. Explain antigen presentation by Class I and Class II major Histocompatibility complex protein.
13. Write peptide binding class I and Class II major histocompatibility complex.
14. Write a note on acquired immunity.
15. With suitable example explain the mechanism of active immunity.
16. Explain the significance of local immunity.
17. What is Herd Immunity ?

CHAPTER 5

EPITOPES

5.1. INTRODUCTION

Understanding of epitopes, as antigenic determinants, is due to early studies of Karl landstainer. Antibodies formed in response to immunization with a native globular protein which do not tend to react with denatured preparations and this is consistent with the view that the majority recognize topographic (surface) structures (i.e., epitopes) which depend upon the confirmation of the native molecule.

For this reason, antibodies to native protein do not usually react as strongly with peptides having the same primary sequence, but brought together by the folding of the peptide chain in the native protein.

Antibodies to the whole molecule and to the isolated loop peptide do not react with the peptide after reduction of its disulfide bonds showing that the linear reduced peptide has lost the antigenic configuration.

Immune cells do not interact with or recognize, an entire immunogen molecule instead lymphocytes recognize discrete sites on the macromolecule called epitopes or antigenic determinants.

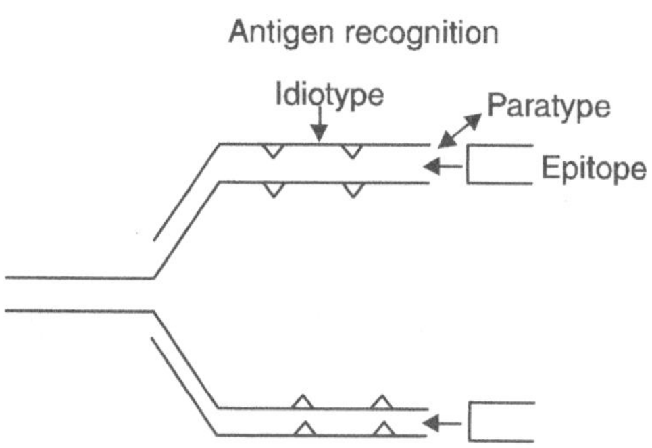

Fig. 5.1

Epitopes are the immunologically active regions of an immunogen that bind to antigen- specific membrane receptor on lymphocytes or a secreted antibodies. Studies with small antigens have revealed that B and T cells recognize different epitopes on the same antigenic molecule. For e.g., when mice are immunized with glucagons, a small human hormone of 29 amino acids, antibody is elicited to epitopes in the amino luminal portion, where as the T cells respond only to epitopes in the carboxy terminal portion.

The recognitions of antigens by T cells and B cells is fundamentally different. B cells recognize soluble antigens when binds to their membrane-bound antibody, because B cells bind antigen that is free in solution, the epitopes they recognize tend to be highly accessible sites on the exposed surface of the immunogen.

Most T cells recognizes only peptides combined with MHC molecules on the surface of antigen-presenting cells and altered self cells. Because most T cells recognizes

Antigen only when it is combined with an MHC molecules.

5.2. EPITOPE CHARACTERISTICS

1. An antibody that is specific for an antigen binds non-covalently to a region of the molecules surface known as epitope.

2. Naturally occurring epitopes are relatively small for example 4 to 6 aminoacid or sugar residues.
3. The convex epitope area might be only 2 λ 3nm across. At these dimensions, an epitope can fit into the antibody-binding site.
4. Thus, large proteins and oligosaccharides can express many different and/or identical repeating epitopes.
5. The antibody combining site or paratope is a cave pocket shaped to match the convexity of the epitope.
6. Linus Pauling showed that an epitope and its binding site must become closely related.
7. The forces responsible for binding include hydrophobic and van der Waals forces, which are spherical, symmetrical and hydrogen bridges, which are directional and require matching of the reactants. Electrostatic forces might also contribute, but they act at a distance.
8. Accordingly, formation of stable immune complexes normally occur only when the epitope and paratope fit "Jigsow Fashion"
9. The more hydrophobic regions of a protein surface are most likely to be antigenic. However the entire exposed surface of a protein can be antigenic, although only a few of these epitopes may actually evoke an immune response.
10. The epitope of coarse, must be readily accessible to an antigen receptor. Consequently the convex, protruding parts of the protein surface are probably the most prominent normally functioning epitopes.
11. The highest peaks on protein surfaces usually coincided with experimentally determined antigenic epitopes. These protruding parts of the molecular surface are generally most accessible to antibody molecules.
12. In addition, critical epitopes can be isolated inside large molecules, exposed only when the molecule unfolds during partial denaturation or processing. This occurs in catabolism of the molecule, catabolism of antigens occurs normally in Antigen Presenting Cells (APC'S) such as macrophages. However, if antigen is completely catabolised, epitopes are destroyed.
13. Not only in the position of an epitope with in a large molecule is important in determining its ability to induce an immune response, but the position of each sub-unit with in the epitope may also be important. For e.g.: each of the aminoacid residues comprising a given accessible epitope might unequally contribute to binding with an antibody paratope. Thus some components of an epitope are more immuno dominant than others.
14. Any aminoacid can contribute to a protein epitope. Since epitopes bind to antibodies through multiple non-covalent bonds which include interactions between charged groups and hydrophobic bonds residue whose side chain carry a charge (e.g., lysine and glutamic acid) or which enter into hydrophobic interactions (e.g., alanine, leucine, tyrosine valine) are common in epitopes.
15. Residues that are not part of the epitope might not bind to antibody, but might influence antigen conformation and affect epitope binding. Substitution of even a single aminoacid in an epitope can affect binding of antibody.

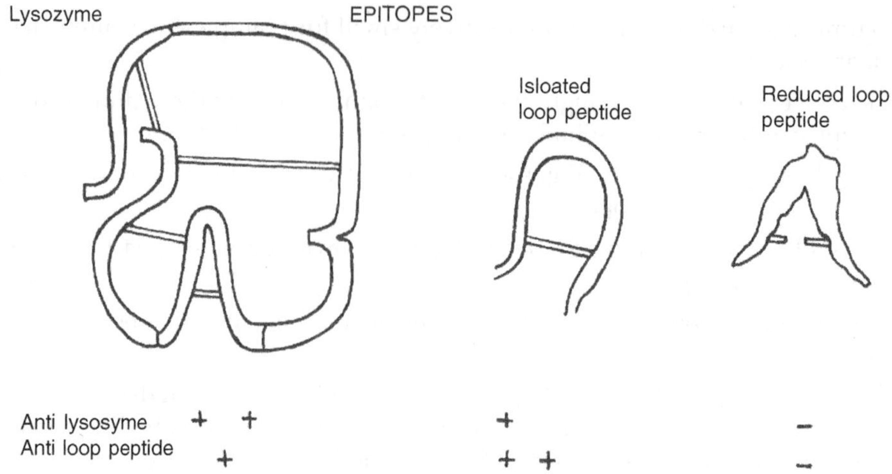

Specifficity g three dimensional contigration in a gloubular Pth lysozyme

Fig. 5.2

5.3. PROPERTIES OF B CELL EPITOPE

The conformation of the epitope recognized by B cells has been established by generalization emerged from studies. The ability to function as a B cell epitope is determined by the nature of the antigen-binding site on the antibody molecules displayed by B cells. Antibody binds to an epitope by weak non-covalent interactions, which operate only over short distances.

For a strong bond, the antibody's binding site and the epitope must have complementary shapes that place the interacting groups near each other. The size of the epitope that can be no longer than the size of the antibody binding site.

For any given antigen antibody reactions, the shape of the epitope that can be recognized by the antibody is determined by the shape assumed by the sequences of the aminoacid in the binding site and the chemical environment that they produce. Smaller ligands such as carbohydrates, small oligonucleotides, peptides and haptens often bind with in a deep pocket of an antibody. E.g., Angiotensin II, a small octapeptide hormone, binds within a deep and narrow groove of a monoclonal antibody specific for the hormones. Within the grooves, the bound peptide hormone folds into a compact structure with two turns, which brings its amino(N-terminal) and carboxy (C-terminal) termini close together. All 8 amino acids residues of the octapeptide are in van der Waals contact with 14 residue of the antibody groove.

An epitope structure was unidentified from X-ray crystallographic analysis of monoclonal antibody bound to globular protein antigens, such as hen egg white

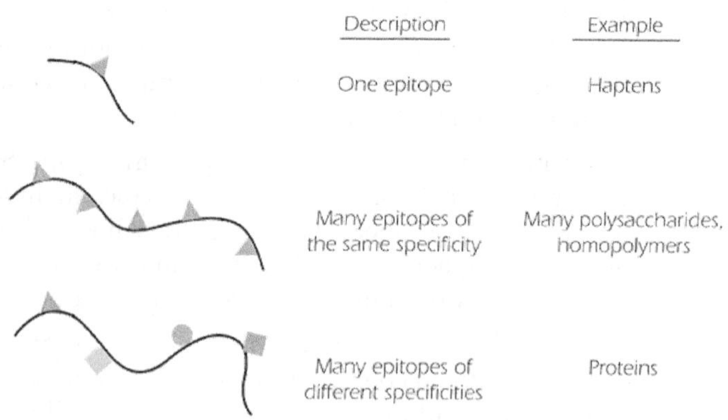

Fig. 5.3 (a)

lysozyme and new aminidase (an envelop glycoprotein of influenza virus). These antibody's make contact with the antigen across a large flat face. The interacting face between antibody and epitope is a flat or undulating surface in which protrusions on the epitope or antibody are matched by corresponding depressions on the antibody or epitope.

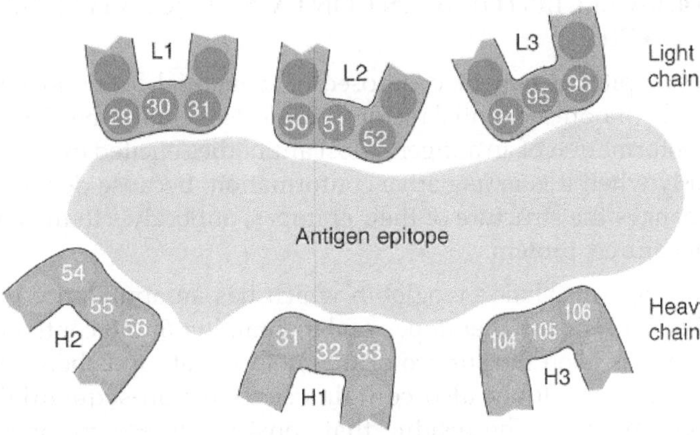

Fig. 5.3 (*b*)

These studies have revealed that 15-22 amino acids on the surface of the antigen make contact with a similar numbers of residues in the antibody's binding site; the surface area of this large complementary interface is between 650A and 900A. For these globular protein antigens, then the shape of the epitope is entirely determined by the tertiary conformation of the native protein. Thus globular protein antigens and small peptide antigens interact with antibody in different ways. Typically large areas of protein antigens are engaged by the antibody binding site.

1. The B-cell epitopes on native proteins are generally composed of hydrophobic amino acids on the protein surface, that are topographically accessible to membrane-bound or free antibody.
2. A B-cell epitope must be accessible in order to be able to bind to an antibody.
3. Aminoacid sequences that are hidden within the interior of a protein cannot function as B-cell epitopes unless the protein is first denatured.
4. Michael Sela demonstrated the importance of this topographical accessibility in experiments with synthetic branched co-polymers in which the accessible amino acids attached to the backbone polypeptide chain were varied.

One copolymer designated (T,G)A,C consisted of a polylysine backbone with poly D-alanine side chain whose N-termini were capped with variable amounts of L-glutamic acid and/or L-tyrosine. Antibody to (T,G)A,L reacted largely with the accessible lysosome and glutamic acid residues at the ends of the side chains.

Furthermore, the related synthetic copolymer A(T,G)L in which poly D-alanine residues were in the accessible terminal positions and the glutamic acid and tyrosine residues were less accessible did not react with the antibody to (T,G)A,L.

5. In general, regions that protrude on the surface of the protein, are recognized as epitopes. Because the residues are accessible they are often hydrophilic.
6. In the crystallized antigen-antibody complexes analyzed to date, the interface between antibody and antigen shows numerous complementary protrusions and depressions. Between 15-22 amino acids on the antigen contact the antibody, by 75-120 hydrogen bonds as well as ionic and hydrophobic interactions.

5.4. B-CELL EPITOPE CAN CONTAIN SEQUENTIAL OR NON-SEQUENTIAL AMINO ACIDS

Epitopes may be composed of sequential contiguous residues along the polypeptide chain or non-sequential residues from segments of the chain brought together by the folded conformation of an antigen. Most antibodies elicited by globular protein bind to the protein only when it is in its native conformation. Because denaturation of such antigens usually changes the structure of their epitopes, antibody's to the native protein do not bind to the denatured protein.

Sperm whale myoglobin which has an abundance of a-helical regions contain fine distinct sequential epitopes each containing 6 to 8 contiguous amino acids. Each of these epitopes is on the surface of the molecule at binds between the α-helical regions. Sperm whale myoglobin also contains several non-sequential epitopes or conformational determinants. The residue that constitute these epitopes are far apart in the primary aminoacid sequence but close together in the tertiary structure of the molecule. Such epitopes depend on the native protein conformation for their topographical structure. One well characterized non-sequential epitope is hen-egg white lysosome.

Sequential and Non-sequential epitopes generally behave differently when a protein is denatured, fragmented or reduced. For e.g., appropriate fragmentation of sperm whale myoglobin can yield fine fragment, each retaining one sequential epitope and the antibody can bind to each fragment.

On the other hand, fragmentation of a protein or reduction of its disulfide bonds often destroys non-sequential epitopes. For e.g. HEL has 4 intra side chain disulfide bonds, which determine the final protein conformation.

Many antibodies to HEL recognize 8 different epitopes, most of which are conformational determinants dependent on overall structure of the protein.

If the intra chain disulfide bonds of HEL are reduced with mercaptoethanol, the non-sequential epitopes are lost, for this reason, antibody native to HEL does not bind to reduce HEL. The inhibition experiment, demonstrate this point.

An antibody to a conformation determinant in this example, a peptide loop present in native HEL, was able to bind the epitope, only if the disulfide bond that maintains the structure of the loop was intact.

Information about the structural requirement of the antibody binding site was obtained by examining the ability of structural relatives of the natural antigen.

In this inhibition assay the ability of the closed loop to inhibit binding showed that the closed loop was sufficiently similar to HEL to be recognized by antibody to native HEL. Even though the open loop has the same sequence of aminoacid as the closed loop, it lacked the epitopes recognized by the antibody and therefore was unable to block binding of HEL.

5.5. B-CELL EPITOPES TEND TO BE LOCATED IN FLEXIBLE REGIONS OF AN IMMUNOGEN AND DISPLAY SITE MOBILITY

John A.Tainer and his colleagues analysed the epitopes on a number of protein antigens (Myohemeresthelin,insulin,cytochrome c,myoglobin and hemoglobin). By comparing the position of the known B-cell epitope will the mobility of the same residues

This analysis revealed that the major antigenic determinants in these proteins generally were located in the most mobile regions.

This investigation proposed, that site mobility of epitopes maximizes complementarity with the antibody's binding site, permitting an antibody to react with an epitope that it might not otherwise react with if it were rigid.

However because of the loss of entropy due to binding to a flexible site the binding of antibody to a flexible epitope is generally of lower affinity than the binding of antibody to rigid epitope.

5.6. COMPLEX PROTEINS CONTAIN MULTIPLE OVERLAPPING B-CELL EPITOPES, SOME OF WHICH ARE IMMUNODOMINANT

For many years it was a dogma in immunology that each globular protein had a small number of epitopes, each conformed to a highly accessible region and determined by the overall conformation of the protein is potentially antigenic. This has been demonstrated by comparing the antigen binding profiles of different monoclonal antibodies to various globular proteins.

For e.g., when 64 different monoclonal antibodies to bind to a panel of 10 different mammalian albumins,25 different overlapping antigen-binding profiles emerged. Suggesting that these 64 different antibodies recognized minimum of 25 different epitopes on BSA. Similar finding have emerged for other globulin proteins, such as myoglobin and HEL.

The surface of a protein, their must present a large number of potential antigenic sites. The subsets of antigenic sites on a given protein that is recognized by the immune system of an animal is much smaller than the potential antigenic protein, and it varies from species to species and even among individual members of a given epitopes.

Within an animal certain epitopes are recognized as immunogenic, but others are not. Furthermore, some epitope are called immunodominant, induce a more pronounced immune response than other epitopes.

The intrinsic topographical properties of the epitope as well as the animals. Regulatory mechanism influence the immunodominance of epitope.

QUESTIONS

1. Define Epitope.
2. Explain the characteristics of epitope.
3. What are the properties of B cell epitope ?
4. Explain the statement that B cell epitope can contain sequential or non sequential amino acids.
5. Explain that the B cell epitopes tend to be located in flexible regions of an immunogen and display site mobility.
6. Explain whether complex proteins contain multiple overlapping B cell epitopes, some of which are immunodominant.

CHAPTER 6

STRUCTURE AND FUNCTION OF SPECIFIC IMMUNOGLOBULINS

6.1. INTRODUCTION

All vertebrate animals have immunoglobulin molecules. They are synthesized by plasma cells. This was discovered by Paul Ehrlich, before a century.

During 1959-64 three scientists namely Porter, Edelman and Nisonoff by using papain digestion used the technique of fragmentation of immunoglobulin molecule led to the discovery of 5 main classes of immunoglobulin. They are largely confined to the broad and heterogenous band of gamma globulins observed on electrophoresis.

The antibody molecule is made up of two identical heavy and two identical light polypeptide chains held together by disulfide (S-S) bonds. The longer chains are known as heavy chains (H) and the shorter chains are known as light (L) chains. The H chains are different in structure and antigenicity. Depending on the structure of H chains gamma (y), mu (μ), alpha(α), delta (δ) and epsilon (ε). Immunoglobulins are classified into Ig G, Ig M, Ig A, Ig D and Ig E respectively.

IMMUNOGLOBULING (Ig G)

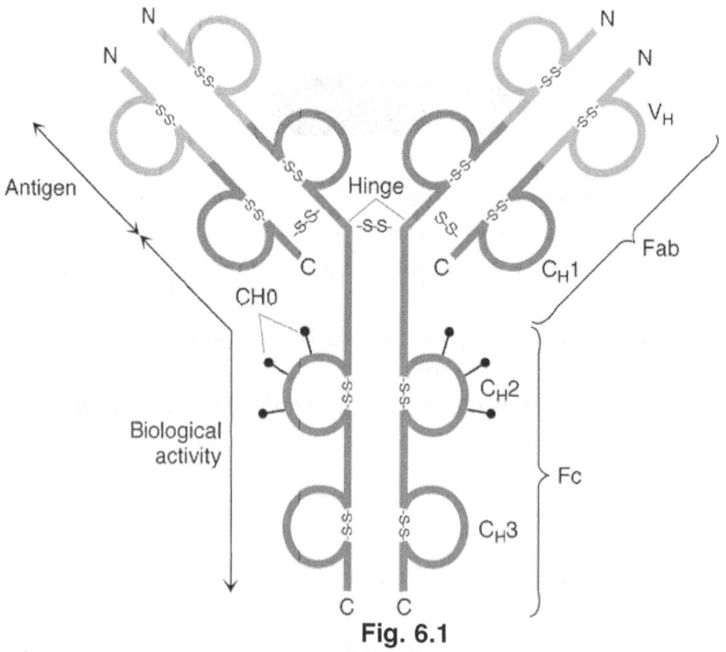

Fig. 6.1

The L chain are named after Korngold and Lappari as Kappa(κ) and Lambda(λ) who described them. K and L chains are in 2:1 ratio.

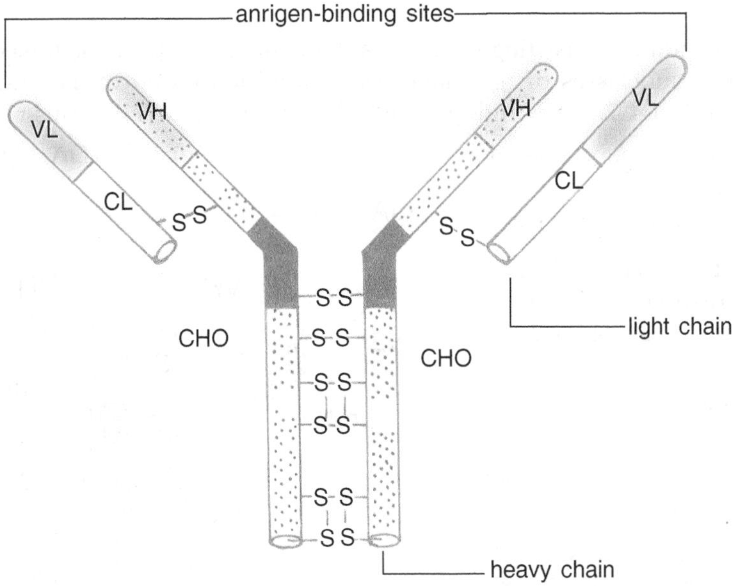

Fig. 6.2

6.2. IgG

Ig G is the major immunoglobulin in human serum, accounting for approximately 75% of the total normal serum immunoglobulin pool at a concentration of approximately 1200 mg/dl. It is the major antibody produced in the secondary immune response.

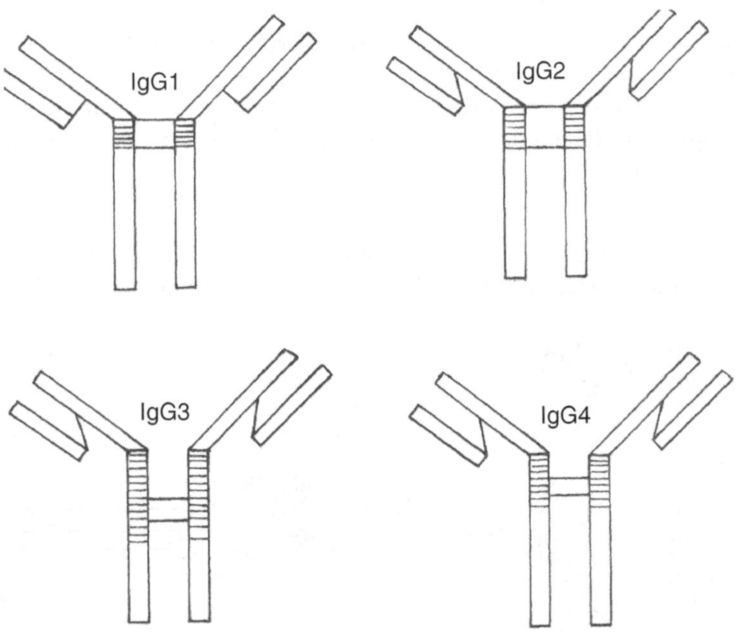

Fig. 6.3

6.2.1. Structure

Ig is a monomer, consisting of identical pairs of H and L chains linked by disulfide bridges. Four subclasses of IgG have been identified based on H chain differences: Subclasses IgG1, IgG2, Ig G3 and Ig G4 which correspond to 'H' chains r1, r2 r3 and r4.

1. Most IgG subclasses have a molecular weight of ISO KDA, IgG3 is slightly larger.
2. Most serum IgG is IgG1.
3. Most IgG has a half-life of approximately 21 days, Ig G3 has a half-life of only 7 days.

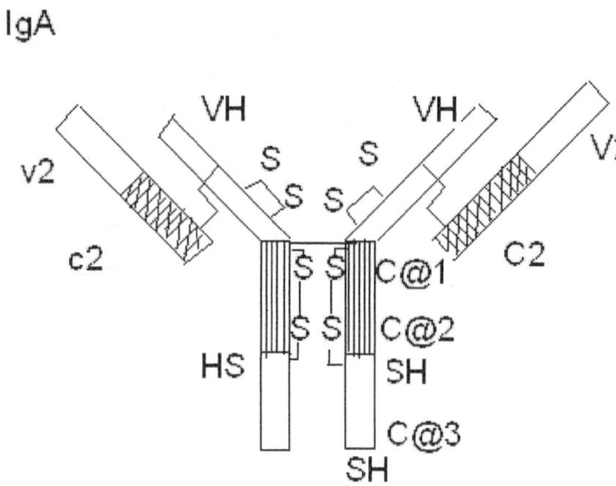

Fig. 6.4

6.2.2. Function

IgG is the only immunoglobulin that crosses the placenta, therefore, maternal.

IgG provides most of the protection of the newborn during the first months of life

IgG , Except for IgG4 molecules, binds complement by the classical pathway. The binding site for complement component C1q is in the CH2 domain.

Effective antitoxic immunity is exclusive IgG.

Ig G is the major opsonizing immunoglobulin.

6.3. IgA

IgA is present in two forms: one in the serum (serum IgA) and the other in various body secretions (secretory IgA). Ig A does not bind complement via the classical pathway. It has serum half life of 6 days . Ig A can be inactivated by an IgA protease produced by ***Gonococci, Meningococci, Pneumococci*** and ***Heamophilus influenzae***. Therefore IgA2 is more important in mucosal immunity against those pathogens.

6.3.1. Serum IgA

Serum Ig A, at a concentration of approximately 200 mg/dl, accounts for about 15% to 20% of the total normal serum immunoglobulin pool. In human, over 80% of serum Ig A exists in a monomeric form, the predominant subclass is Ig A. The other 20% of serum Ig A exists as dimers, trimers or tetramers. In these polymeric forms of Ig A, the monomeric units are linked by disulphide bonds and 'J' chains.

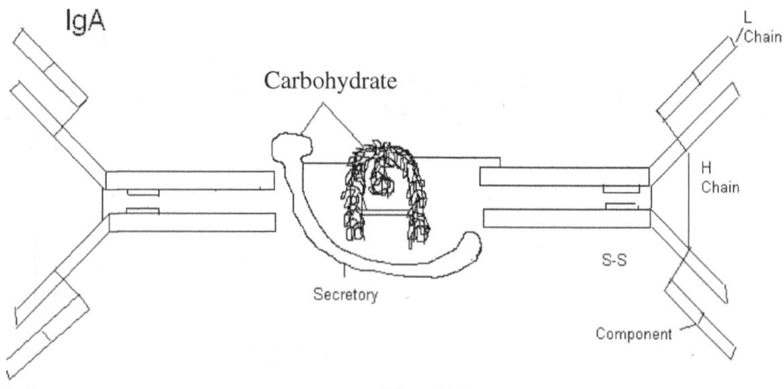

Fig. 6.5

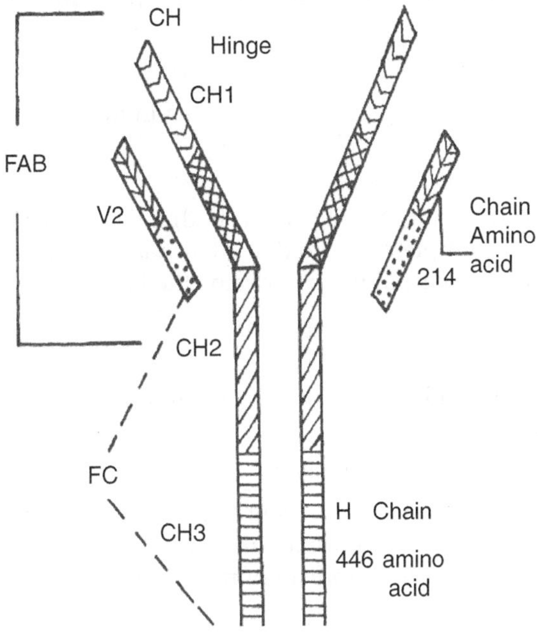

Fig. 6.6

6.3.2. Secretory IgA (S Ig A)

Secretory IgA is the predominant immunoglobulin in various secretions (eg: saliva, tears, colostrums and bronchaist, genitourinary and intestinal secretions).The dominant subclass in secretions is S IgA2.

6.3.3. Structure

1. These IgA molecule consists of 2 monomeric units with a J chain and secretory component.

2. Secretory component is a polypeptide chain with a molecular weight of approximately 70KDA. It is synthesized by exocrine epithelial cells and enables dimeric IgA to pass through the mucosal tissues into secretions.

3. As the complex is extracted, proteolytic cleavage of the receptor leaves a fragment. The secretory component attached by a disulfide bond to the Ig A dimer.

6.4. STRUCTURE AND BIOLOGICAL PROPERTIES OF IMMUNOGLOBULIN CLASSES.

6.4.1. Immunoglobulin IgG

It is a relatively stable four-polypeptide chain molecule found of two tight chains and two heavy chains of 'f' type. The light chain consist of a single constant Domain. Under single variable domain:

The heavy chain consists of four domains namely, a single variable domain and three constant domain. This is the predominant class of the total immunoglobulin and is upon four approximately 70% of the total immunoglobulin in human serum.

Normal serum concentration of Ig G is about 8-16 mg/ml. Ig G is of greatest amount in integral body fluids and is produced particularly during the secondary immune response. The molecular weight of Ig G is 150,000 and its sedimentation coefficient is 7s. IgG has a half-life of 25 days. The CHO content of Ig G is above 3%. There are four ontogenetically distinct Subclasses of IgG namely Ig G1, IgG2, IgG3 & IgG4.

Ig G is the major one form in h 67% of the total Ig G followed by IgG2—22%, Ig G3-7% and IgG4 - 4%.

The difference between the subclasses in the position and the number of inter chain Disulphide bridges.

The number is two for IgG (disulphide bridges) and IgG4 and four for IgG2 and 10-13 for IgG3.The large number of disulphide bond in IgG3 sinusoids with the long single region consisting of 62 amino acids and their results in the higher mol wt of IgG3. Under the Election microscope, the IgG loops like a 'Y' shape.

(*a*). IgG in the only immunoglobulin that crosses the human placenta and thus offers a passive protection of the new born for above 6-9 months.

(*b*). Ig G activates the classical complement pathway during Ag - Ab reaction.

(*c*). IgG neutralizes toxins and viruses. It enhances phagocytes and elimination of bacteria is called opsonization.

6.4.2. Immunoglobulin IgA

IgA is found only to a small extent in serum but predominantly seen in the external zero mucous secretion of the respiratory gastro intestinal tract. IgA is the second in abundance allowing the immunoglobulin about 10-15%.

Two sub classes of Ig A exist namely Ig A, and Ig A2. Ig A, has both inter chain and intra chain disulfide linkages, where as Ig A2 is completely devoid of inter chain disulphide bonding between the heavy and light chain. The partial the two drains the held together by non-covalent linkages.

6.4.2.1. Serum IgA

IgA is a monomer having a structure very similar to Ig G molecule? It has two heavy and two light chains. The light chain is in entive loop A.

The α chain of Ig A is differengt from chain of Ig G in having a greater

CHO content 10% of the molecule % in the aminoacid sequence. It has the molecular weight of 160,000.the sedimentation coefficient is 7S. The normal serum level of IgA is 0.6 to 4.2 mg/ml. It has the half of 6-8 days serum Ig A has more amount and of IgA than Ig A2.

6.4.2.2. Secretory IgA

It is nearly and equal mixture Ig A1 and Ig A2 in contrast to the serum IgA.

Human secretory Ig A is a much larger molecule than IgA molecule. Molecular weight is about 500,000. It occurs as a dermis formed of 2 moles of Ig A like by a secretory component or secretory piece and a joining chain. Both the secretory piece and joining chains are protein.

The secretory device is synthesized by the local epithelial which stabilizes the Ig A against proteolysis. It is bound by the both disulphide bond and non-covalent bonds to the Fc region of the IgA dimer. The secretory piece has a single protein chain mol Wt 70,000 the joining chain is also a poly peptide chains and molecular weight 15,000. This secreted by the plasma cell.

6.4.2.3. Biological properties of Ig A

Ig A does not fix complement, but is found to activate the alternative complement pathway, there by neutralizes local toxins and activate bacteriolytic activity. But this occurs only in the presence of lysozyme. IgA is also found to produce immunity gents tapeworm. IgA present in the coliseum protect the baby from intestinal pathogens, so breast feeding reducing CyI disease by reducing the entry of pathogens of the environment, which other wise occur, bottle feeding follow.

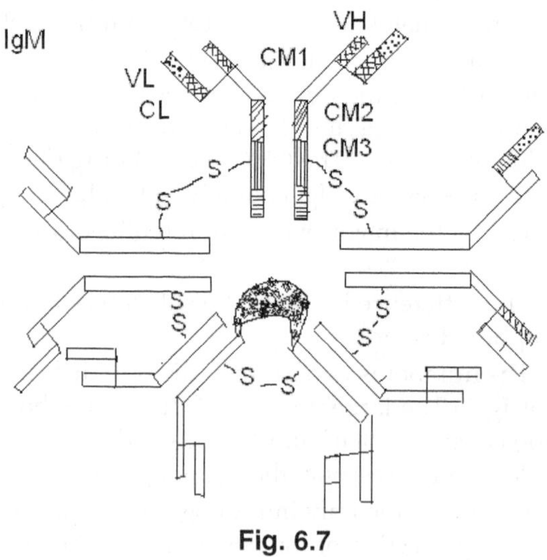

Fig. 6.7

6.4.3. IgM

Is the largest of the immunoglobulin. It is often refers as the immunoglobulin because of this higher molecular weight 950,000.They are polymers of usually 5 molecules (pentamers). Each with four polypeptide chains. The polymerization of this molecules occur in the presence of joining chain as in the case of secretary IgA. The joining chain is holding the five sub-units together at the Fc region and stabilizes the Fc sulphadryl groups. The serum level of IgM in very low above 5-2mg/ml This low serum level is due to its short half life is 5 days under low synthetic rate (5-8 mg/kg bodyweight/day).The CHO content is above 10-12 %.The sedimentation coefficient is 19S.There is no sub-classes for IgM. It is the earlier immunoglobulin to be synthesized by the fetus starting by about 5th month of the fetal life.

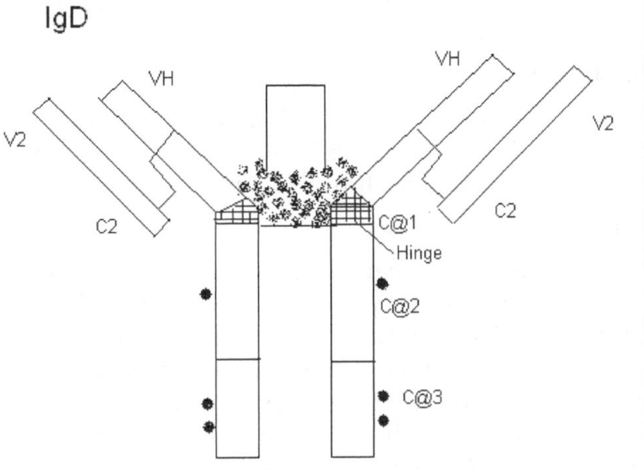

Fig. 6.8

In mammals it is in the form of cyclic pentamer link by disulphide bond and joining chain. A monomer form of Ig M is found on the B lymphocyte. Studies on the number of antigen binding site of immunoglobulin shows reacts has only 5 - 10 Ag binding sides are expected, but however human and rabbit Ig M.

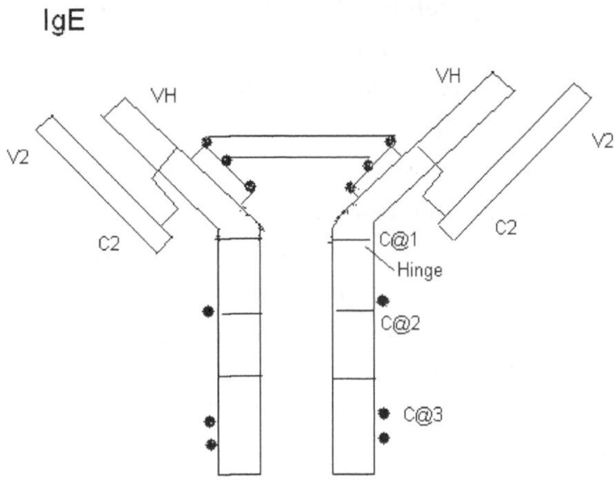

Fig. 6.9

6.4.3.1. Biological properties of IgM

It is the first antibody to appear in the primary but it does not persists for long time because of its short half life. It is therefore an useful indicator of recent infection.

Maternal IgM does not cross the placenta but the human fetus B cells due antigenically stimulated as in congenital infections. The IgM detection in new born is useful indication of intra uterine infection such as rubella toxoplasmosis etc.

Most natural anti bodies such as the natural classical ABO blood groups, is agglutinants many of the natural AB to microbes and also heterophile Ab all of the IgM classes.

By having multiple antigen binding sites (5 – 10).IgM has a high functional affinity four multivalent antigens. It also shows prospective such as optimization, complement function up to lysis etc.

Because of its macromolecular size, it is primary localized in blood & thus protects against the old serum infection.

9.4.4. IgD

This is normally present in minute concentration in blood and other body fluids. It occurs on the surface of B lymphocytes in function with IgM and serve as recognition receptors for antigens. It has 75% intravascular distribution.

6.4.5. IgE

This is a monomeric immunoglobulin. It has four constant domains (like IgM) and although it is the least plentiful of all immunoglobulins. Heat labile (inactivated at 56 ° C in one hour),has affinity for tissue cells particularly mast cells, is responsible for anaphylactic type of hypersensitivity. Its presence can be dramatically felt by its ability to bind to Fc receptors. On the surface of mast cells and basophils and when complexed with specific antigens, to trigger the release of inflammatory mediators. Its physiological role is with mast cells it function as gate keepers in regulating the exit of cells and plasma into extravascular sites.

6.5. ANTIGENIC DETERMINANTS ON IMMUNOGLOBULINS

Since Ab's are glycoproteins, they can them selves function as potent immunogens to induce an Ab response, Such anti Ig Ab's are powerful tools for the study of B cell development and humoral immune responses. The Antigenic determinants or episomes on I molecules fall into three major categories.

(*i*) Isotopic determinants
(*ii*) Allotypic determinants
(*iii*) Idiotypic determinants

which are located characteristic portion of the molecule.

6.5.1. Isotypic determinants

They are constant region determinants that collectively define each heavy chain class and subclass and each light chain type and sub type within a species.

Each isotope encoded by a seperate constant region gene and all members of a species carry the same constant region gene. With a species, each normal individual will express all isotypes is their serum. Different species infect different constant region gene and express different isotype.

Therefore when an Ab from one species is here into another species, the isotype determinants will be recognized as foreign inducing an Ab response to the isotopic determinants on the foreign Ab.

Anti type Ab is routinely used for research purposes to determine the class on subclass of serum Ab produced during an immune response (or) to characterize the class of membrane bound Ab present on B cell.

6.5.2. Allotypic determinants

Although all members of a species inherit the same set of isotype genes, multiple alleles exist from some of the genes. These alleles encode solute aminoacid differences called Allotypic determinants, that pure in some, but not all, members of a species.

In humans all types have been characterized for all four IG's sub classes, for 1 I A subclass and for the Kappa light chain. The r chain all types are referred to a Gm markers.

To date, 25 different Gm allotype have been determined, they all designated by class and subclass followed by the allele no for eg G1 M(1) G2m (23), G3m(11), G4m(4a). Of the two-type subclass, only IgA2 subclass all allo types, designated as A2 m (1), A2 m (2) and the kappa light chain has 3 allotypes designated Km (1), Km (2) and Km (3). Each of these Allotypic determinants are presents differences in one to four amino acids that are encoded by different alleles.

Anti body to allotypic determinants can be produced by injecting Ab's from one member of a species into another member of a same societies who carries different allotypic determinants.

Ab to allotypic determinants some times is produced by mother during pregnancy in response to maternal allotypic determinants on the further Ig. Ab's to allotypic determinants can also arise forming a blood transfusion.

6.5.3. Isotypic determinants

The unique aminoacid sequences of the Vh and Vc domins of a given Ab can function not only as an Ag binding site but also an Antigenic determinant.The idiotypic determinants are generated by the conformation of heavy and light chain variable region. Each individual Antigenic determinant of the variable region is referred to as idiotope.

In some class, an idotype may be the actual Ag binding site, and in some cases idotype may comprise vaccable region sequence outside of the Ag binding site, Each As will present multiple idiotopes the some of the individual idiotopes is called the idotype of the Ab.

Because the Ab's produced by individual B cells derived from the same lone have identical variable region sequences they all have the same idiotype.

Antidiotype Ab is produced by minimizing isotopic or Allotypic differences, so that the idiotypic difference can be recognized. Often, homogenous Ab's such as myeloma pertain (or) MCA are used. Injection of such as Ab into a signori recipient will result in the formation of Anti idiotype Ab to the idiotype determinant.

Avoiding to one the, anti idiotype Ab produced naturally during the coarse of a immune response plays a role in regulating the immune response.

6.6. DISEASES RELATED TO THE COMPLEMENT SYSTEM

The complement system is related to human disease in two general ways.

1. Deficiencies in anyone of the protein components usually due to abnormalities in gene structure can had to abnormal patterns of complement activation.If regulatory components are absent, too much complement activation may at the wrong time (or) wrong site, heading to excess inflammation and cell lysis and may also lead to lucle of plasma complement proteins. Absent of an integral component of the classical, alternative (or) terminal lytic pathways can result in too little complement activation & a back of complement medicated biologic function. Consequences in both cases are severe.

2. An intact, normally functioning complement system may be activated in response to abnormal stimuli Vies persistent micro organism, auto antibody responses to self antigens etc. In these infections auto immune diseases the inflammatory (or) lyric effect of complement may contribute significantly to the pathology of the disease.

6.7. COMPLEMENT DEFICIENCIES

Deficiencies can be categorized on the basis of the functional type of protein that is lating.Thus there are deficiencies of the components of the classical alternative and terminal pathways and deficiencies in either soluble membrane regulatory proteins.

(*i*) Genetic deficiencies in classical and alternative pathway components including C1q, C1r, C4, C2, C3 proper in and factor D have been identified. C2 deficiency in the most common one. The single most consistent clinical consequences of eminencies in the early classical pathway proteins is the development of systemic lupus erythrocytes (SLE). More than 50% of patents with C2 and C4 deviancies have Sleigh is suggested that the classical pathway is normally involved in regulating the immune system. It may neglect the requirement of classical pathway for the clearance and volatilization of cur collating immune complexes.

If normally generated immune complexes are not cleared from the limitation, they may deposit in blood vessel walls and tissues, where they activate the complement cascade and produce local inflammation. Such inflammatory reaction may promote the breakdown of peripheral toluene to self antigens leading to auto immunity. Other reasons including genetic linkage of defective complement alleles in the MHC with others.

C2 and C4 defections are often not associated with seed infections. This suggests that the alternative pathway may be adequate for the elimination of most material.

C3 deficiencies associated with frequented serious pyretic bacterial infections that may be fatal, thus shows the impotence of C3 for phonic suction, enhanced phagocytes and destruction of these organisms C3 deficiency leads to glamour nephritis.

(*ii*) Deficiency in members of the terminal complements via C5, C6, C7, C8 and C9 leads to inability to generate MHC. As a result there is an amiability to use foreign organisms.

(*iii*) Defiance's in soluble and membrane bound complement regulatory proteins are associated with abnormal complement activation and a variety of related clinical abnormalities.

6.7.1. Hereditary Angioneueodri OEDEMA (HENA).

Autosomal dominant deficiency of C1 interior accumulation of edema fluid in skin & mucus, lasting from 24 –72 hrs.

Serious symptoms resulting from edema in free, extremities etc include abdominal pain, nausea vomiting diarhoea fatal ait way obstruction.

The deficiency in C1 inhibitor in HANE proteins may be caused either by an absolute reduction in C1 inflator protein due to inherence of a non-functional gene (or) by synthesis of a normal amount of a dysfunctional protein due to mutations in the loading of the gene. Although these is usually one normal allele of the C1 inhibition gene in these patients the plasma levels of the C1 inhibitory proteins are sufficiently reduced, so that activation of C1 by immune complexes is not properly controlled. It is class that during attacks there are elevated plasma levels pf activated C1 and shed levels of the substrates of activated C1 namely C2 and C4.

The mediators of edema formation in HANE patients including a proteocytic fragment of C2.

C1 inhibitor is a major regulator of other plasma serum proteins besides C1. C11 inhibitor deficiency can lead to plasma generation which inturn can proctor Cytec ally generate a fragment from C2 with quince actively.

Deficiencies of the soluble alternative pathway regulators factors I and II/Plasma C3 is completely consumed in these patients as a result of in regulated formation of fluid phase C3 convertase. The clinical consequence in the said infection with progeny bacteria reflects it. Rare cases of factor H deficiency are characterized by excesses alternative pathway activation, consumption of C3 and glomerculoprits due to inadequated chance of immune complexes.

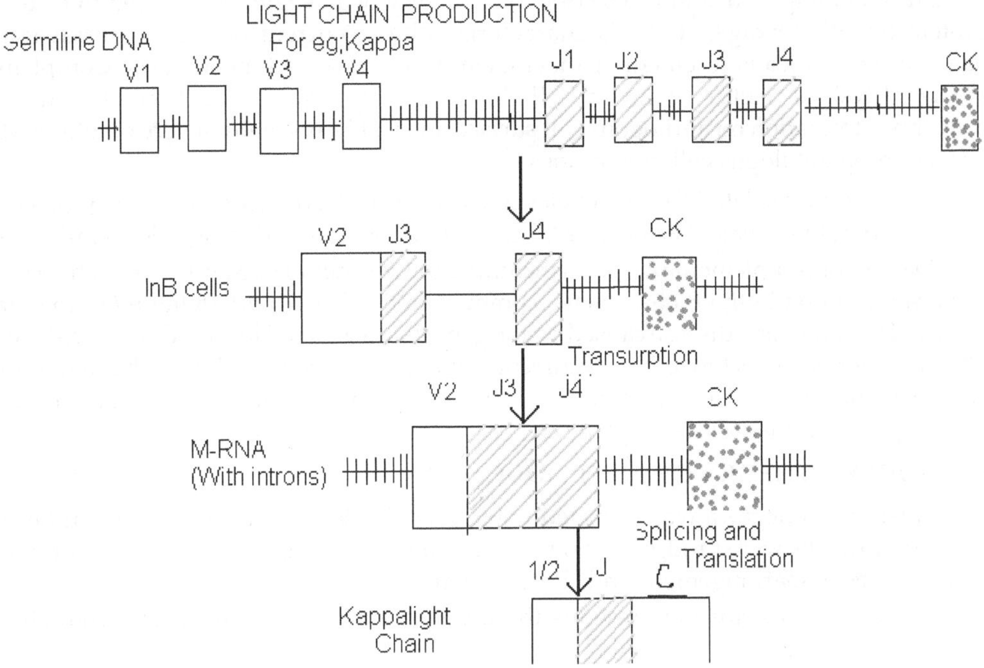

Fig. 6.10

The effect of a lack of factor J/H are similar to the effect of an auto Ab called C3 nephritic factor, which is specific for alternative pathway C3. This Ab stabilized C3 b BB and proteins the complexes from factor mediated dissociation. Causing charageted

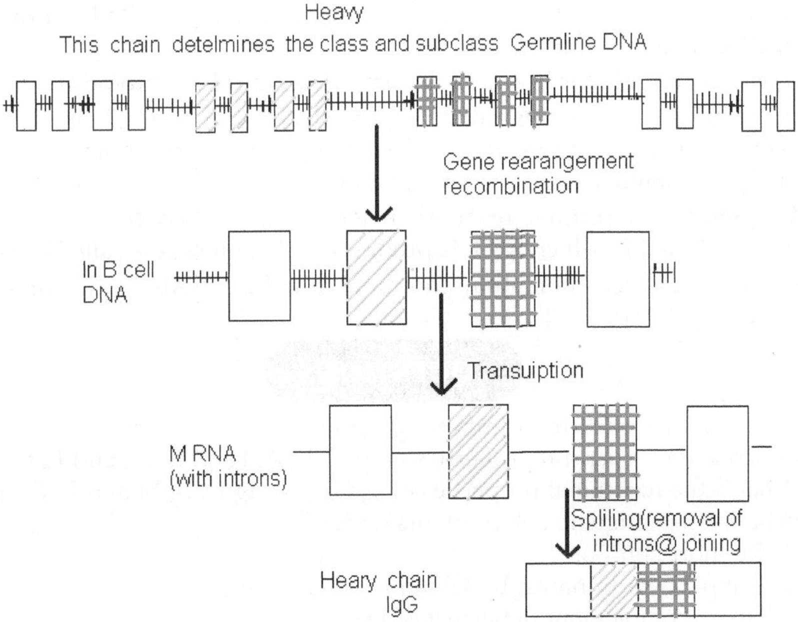

Fig. 6.11

comsumption of C3 patients with thus Ab often have glomerulonephritis. In maximal neural haemoglobinreas hells lack the ability to express phosphatydyl protein linked membrane proteins including DAF and CD 59 due to deficiency of an enzymes required to turn such protein hepatic linkage. PNH is characterized by recurrent bounds of intravascular haemolysis.This leads to cluenic pancytopeinca and various tuncomnosis. Complement mediated lysis of RBC's occurs in PNH because DAf normally functions to inhibit C3 convertase formation on antilogous cell surface and CD 5a normally severe of inhibit MAC formation on autologus cell membranes.

Rare cases of isolated CD 59 deficiency with normal DAF expression also have a DNA like syndrome but covalent isolated DAF deficiency does not lead to significant huimplusis.

Defines of complement receptors including the absence of Cr3 and CR4, both resulting from rare multion in the B chain gene, common to the CD 11 add 18 family of integrand deletes. The congenital disease caused by the gene defect is called locate adhesion deficiency. This disorder is characterized by recurrent phylogenic infections due to the inaugurated adherence of nuetrophiles to endothelisis tissue sites of infection and perhaps imparted iC3b dependent phagocytes of bacteria.

6.8. MODERN THEORY OF Ab PRODUCTION

Sutsuma, Tonegawa was awarded noble prize in 1987 for his molecules studies that showed how abs were produced. So that they could combine with a vast array of foreign molecules gene rearrangement and Recombination.

As abs are proteins has proteins the involvement of genes in their production is obvious.

An individual has the capacity to produce an estimate of 108 different Ab molecules., but Ig produced by clusters of genes present in different plates of the same as well as different chromosome of an immune competent cell. The Ag induction making a gene rearrangement or gene shufftin process. So that group of genes are selected and recombined the production of particular type of light or heavy chain of an Ig.

The gene responsible for Kappa light chain are located at 2nd chromosome and lambda at 22nd chromosome.

The gene responsible for heavy chain are located at 14th chromosome.

After the induction of an immunocompetent cell by an 'ag' 2 similar light chains and two similar heavy chains are produced separately and combined to form are monomeric Ig G. (number varies for dimer and pentamer)

Ig M is produced in priming response, where Cm is selected for the same Ag in the 2 response Cr is selected in which Ig G is produced. This process is called as **class switch**.

By shuffling and recombination enormous number of Abs with different specific antibodies are produced.

QUESTIONS

1. Describe the structure of Immunoglobulins
2. Describe the structural properties of Ig G, Ig A, Ig D, Ig M, and Ig E.
3. What is the functional properties of Ig G, Ig A, Ig D, Ig M and Ig E ?
4. What is the antigenic determinants on Igs ?
5. Write short notes on
   a. Isotypic determinants b. Allotypic determinants
6. What are complement deficiencies disease ?
7. Define Heriditary Angioneueodri Oedema.

CHAPTER 7

IMMUNO-DIAGNOSTICS ABO, Rh (D) Incompatibility, Widal Test, ELISA and RIA Techniques: Their Application

7.1. INTRODUCTION

Techniques in clinical immunology are enormously variable and it is important for a clinician to know something of the precision and reliability of the tests requested. This chapter illustrates the principles behind the techniques used and provides some idea of the difficulties in interpreting the results obtained.

Laboratory tests can be graded according to their value in the care of patients, sometests are essential for diagnosis or monitoring, some are optional but useful and others are of interest only (Research). Unfortunately, tests are useless of requested in inappropriate circumstances.These categories are used to define the clinical identity for each test.Examples of the use of individual tests are given in relevant chapters. Because methodology is not given as there are several textbooks and manuals dedicated to full detailed descriptions of the methods.

There are three groups of tests designed to evaluate immunological competence and the individual components of the immuune response. Humoral factors such as immunoglobulines, antibodies complement components and other specific proteins can be measured precisely. Normal ranges are well established, interpretation of results fairly straight forward. In contrast, tests of cellular components are more difficult to perform and to interpret. No standards are available and the methods used often vary enormously between laboratories. Valid interpretation demands close cooperation between clinician and immunologist. In vivo tests, which evaulate humoral or cellular factors, have some value in the investigation of immunodeficiency and allergy but are very difficult to standardize.

Despite differences in methods, all laboratories strive to ensure quality and accuracy of results. This is achieved by quality control and external quality assurance, often under the supervision of nationally organized schemes such as the National External Quality Assessment Schemes (NEQAS) in the UK.

7.2. ABO: Rh (D): INCOMPATIBILITY

The ABO red cell system is unique because naturally occurring antibodies (IgM) are found in the sera of some recipients, blood is therefore always matched for ABO antigens. To prevent sensitization to rhesus antigens, done and recipient are also tested for compatibity in relation to the immunologically strongest rhesus antigen (D antigen). Most other red cell authigens are poorly immunogenic mismatch for ABO and rjesus antigens alone should prevent major incompatibilities. However, reciepients sera are screened for rare pre existing antibodies which may destroy donor red cells. Transfusion reactions due to pre existing antibodies can be fatal. A test panel of red cells containing most antigens of known blood group systems is used as well as cells from selected donors. Sera are screened for complete (Ig M) antibodies, which will agglutinate red cells in saline and 'incomplete' antibodies (IgM), which are detectable by an indirect antiglobulin (coombs) test. It is not necessary to identify donor antibodies because these are rapidly diluted on transfusion.

Despite these precautious, red cells are sometimes destroyed by a haemolytic transfusion reaction.

POSSIBLE BUT RARE COMPLICATIONS OF TRANSFUSION

Outcome	*Cause*
Red cell destruction	* ABO mismatch
previously	* Antibodies to other red cell antigen undetected in recipients plasma
	* Boost of low-level antibody to sign level by transfusion
White cell destruction	HLA Antibodies, Antineurophil, Antiplatelet
platelet (or) cell destruction	antibodies
Serum protein reactions	Anti IgA antibodies in selective IGA deficiency
Infection	Hepatitis C
	Non A, B, C, hepatities viruses cytomegalo virus Malaria HIV

Unfortunately, the commonest cause of a transfusion reaction is a human error in handling blood samples or transfusion of the wrong unit of blood heamolytic transfusion reactions due to ABO incompatibility usually result in massive haemolysis, haenocoglobinuria, shock and disseminated intravascular coagulation. A mismatch due to rhesus incompatibility usually results in gradual haemolysis which does not interfere with renal function. Such reactions may be asymptomatic but the haemoglobin fails to rise.

Occasionally, transfusion reactions occur several days after a transfusion. This delay is due to the time lag required for the production of immune antibody by the patient. This is that the patients own cells are negative in a direct coomb's test but the transfused cells are positive for Ig G and complement. This type of delayed transfusions reaction occurs often multiple blood transfusions, such as those required in open heart surgery or severe gastrointestinal bleeding. Not all complications of blood transfusion are due to destruction of red cells. Immune reactions to white cells, platelets or serum proteins may occasionally result in acute febrile reactions urticaria or rarely anaphylexis.

A variety of blood borne infections can be transmitted by blood transfusion. Donors likely to carry or to have been exposed to such agents are discouraged from giving blood. Together with increased screening, these measures have dramatically reduced transmitted infection in the last 3 decades and have decreased the demand for authologous stored blood transfusion.

Transfusion from relatives (directed transfusion) is safe from infection risk but can lead to graft versus host disease unless the blood (in transfusions between close relatives) is irradiated first.

With the advent of concentration, leucocyte poor platelets is readily available, cell separators are based on the principle of differential centrifugation of cells of different density. The cells required can be removed from the centrifugal bowl while other blood elements are turned to the donor.

Platelets transfusions are particularly useful for patients with aplasic anaemia or a acute leukemia, provided the recipient does not have alloantibodies to platelet specific or HLA antigens. Such patients show no rise in platelets count after platelet transfusion. The problem can be overcome by HLA matching the platelet donor and recipient.

The commonest cause of haemolytic disease of the newborn (HDN) remains rhesus incompatibility between mother and fetus. Routine blood grouping of all antenatal patients and blood grouping of their spouses detect those rhesus D negative women who may be at risk. HDN is now preventable by administration of anti D antibodies immediately after delivery in the first pregnancy. These passive antibodies destroy any rhesus positive fetal cells and prevent the mother's immune system. from becoming sensitized. However rhesus (D) haemolytic disease has not been eradicated, possible due to sensitization during pregnancy (rather than at delivery) antenatal administration of anti D Ig to all mothers at risk is now advocated. Patients sensitized despite the injection of anti D antibodies, or inadvertently by blood transfusion, can still be successfully treated as in case 2.

7.3. WIDAL TEST

Widal test is used for the diagnosis of typhoid fever. In typhoid patients, the serum contains antibodies and Salmonella. In widal test two antigens are used. They are antigen H the flagellar antigen. When antiserum of the patients is added to the antigens, the antigens are clumped.

7.4. HAEMOLYTIC TRANSFUSION REACTION

The term transfusion reaction generally refers to the adverse consequences of incompatibility between patient and donor erythrocytes. Transfusion can include haemolytic reaction, (Red cell lesions) occurring during or shortly after a transfusion, shorten post transfusion survival of RPD an allergic response are disease transmission.

Transfusion reaction can be divided into haemolytic and non-haemolytic types.

Haemolytic reaction is associated with the infusion of incompatible erythrocytes. This reaction can be classified into actual (immediate) or delayed in their manifested types of transfusion reaction.

7.4.1. Intravascular haemolysis of RBC

Delayed haemolysis is intravascular haemolysis of erythrocytes.

7.4.2. Non-haemocytic

e.g.: Final reaction, anaphylaxis non cordial pulmonary edema, fever shoed vita conge trial heart failures, myocardial failure.

7.4.3. Non delayed hemocytic

Graft Vs Host disease.

Host transfusion purpose.

Iron over coat.

Allo immunisation to RBC, WBC or platelet Ags or plasma proteins.

7.4.4. Infection disease

Several factors influence whether the transfusion reaction will be actual or delayed. These features include.

(i) The number of incompatible erythrocyte inflexed.

The achievement of the optimal tough for Ab binding

7.4.5. Immediate hemocytic reaction

The most common cause of an actual haemolytic transfusion reaction in the transfusion of ABO incompatible blood, in patients with parenting Ab caused by prior transfusion or pregnancy, or other blood groups may be responsible.

7.5. EPIDEMIOLOGY

All the haemocytic reactions are the most serious and potentially lethal. Most catalytic resulting from acute haemocytic transfusion reactions occurs in areas of the tissue of unconscious patients, it heads to uncontrolled hypotension.

7.6. SIGNS AND SYMPTOMS

The reaction can own with infusion of as little as to 15ml of in compatible blood, most common initial symptoms are fever and chillers non haemolytic reactions caused by leucocyte in compatibility. Back pain and shoulder of backrest pain at the infusion site and hay potencies are additional symptoms. In addition to show, the residue of those blasts substance into collation cause dissimilated intravascular coagulation (DIC).

7.7. IMMUNOLOGY MAIN TITRATION

Auto hemolytic reaction our during or immediately after blood have been infuses in fusion of incompatible RBC in the presence foggier Abs, initiates the Ag Abs reaction with the activation of complement plasminogen, kinin and coagulation system. Other infectious of acute hemolytic reactions include bacterial contamination of blood or the infusion of demolished RBC.

Many reactions demonstrate both extra vascular and intravascular hemolysis. If an Abs is capable of activating complement and sufficiently activated invite.

Intravascular haemolysis, producing rapid increase of free Hb in the circulation. The immediate clinical symptoms may be due to pdt declease by the action of complements of the RBC's which triggers multiple shown mechanisms.

7.8. DELAYED HAEMOLYTIC REACTION

The DHR may not expressed itself until 7 to 10 days post transfusion. In contrast to an immediate reaction a delayed reaction occurs in the extravascular spread these reaction are associated with, decreased red cell survival become coating of the RBC, while promotes phages citrons and probative removal of RBC by the monocular phagocyte system. If an Ab does not activated complement or does. So very slowly extravascular haemolysis occurs.

Most immunoglobulin Ab wanted erythrocytes are destroyed mainly in the spleen. The delayed haemolytic transreaction may be of two types. It may represented an enormously in a pranumsly transfurred RBC Ag or it may resulted from alloimmunisation.

In an animistic response these are disease agents the After the reception has been previously by transfusion of programmy.

QUESTIONS

1. Explain Rh incompatibility.
2. Define Widal test.
3. Explain hemolytic transfusion reaction.
4. Explain immediate **haemolytic**reaction.
5. Explain delayed haemolytic reaction?

CHAPTER 8

HYPERSENSITIVITY

Hypersensitivity are immune reactions in which instead of giving protection to our body against exposed antigen of immune system, produces antibodies against self antigens which ends in fatal to normal cells.

Hypersensitivity reactions are divided into 4 classes by Coombs and Gell (1968).

8.1. TYPE I : ANAPHYLACTIC REACTIONS

These are mediated by Ig E antibodies which bind to receptors on most cells. When cross-linked by antigens, the Ig E antibodies triggers the mast cells to release several pharmacologically active agents that are responsible for the characteristic symptoms of anaphylaxis. Reactions are rapid occurring within minutes after exposure to antigen.

8.2. TYPE II : CYTOLYTIC (OR) CYTOTOXIC REACTIONS

It occurs when Ig G (or) Ig M antibodies bind to antigen on the surface of the cell and activate the complement cascade, which culiminates in destruction of the cells.

8.3. TYPE III : IMMUNE COMPLEX REACTIONS

It occurs when aggregates of antigen and Ig M or Ig G antibody accumulate in the circulation or in tissue and activate the complement cascade. Granulocytes are attracted to the site, and damage results from the release of lytic enzymes from their granules. Reactions occur within hours of exposure to antigen.

8.4. TYPE IV : CELL MEDIATED IMMUNITY (CMI) REACTION

Also called delayed type hypersensitivity (DTH) or the tuberculin reaction – is mediated by T cells rather than by antibody. Upon activation T cells release lymphokines that cause accumulation and activation of macrophages, which in turn cause local damage. This type of reaction has a delayed onset and may occur 1-2 days after exposure to antigen.

8.5. TYPE V : STIMULATORY HYPERSENSITIVITY

Here the antibody reacts with a key surface component such as hormone receptor and stretches on the cell.

Innate hypersensitivity

Most of the infections stimulate toxic shock syndrome involving excessive release of TNF, IL1 and IL6 and intramuscular activation of complement.

8.1. TYPE I : ANAPHYLACTIC REACTION

8.1.1. The sensitization phase

During which Ig E antibody is produced and binds to specific receptors on mast cell and basophils.

8.1.2. The activation phase

During which re-exposure to antigen triggers the mast cell to respond by release of the contents of their granules.

8.1.3. The effector phase

During which a complex response occurs as a result of the effects of many pharmacologically active agents released by mast cells and basophils.

8.1.1. Sensitization Phase

Approximately 50% of the population generates an Ig E response to antigens that are encountered only on mucosal surfaces, such as the lining of nose and lungs and the conjuctiva of the eyes. However 10% of the population develops hay fever after repeated exposure to a plethora of airborne antigens (called allergens) such as plant pollens mold spores and animal danders. Allergists frequently use the arochaic term atopy to refer the syndrome and the term atopic to describe affected patients. It has been suggested that the response is under the control of immune response (IR) genes.

Ig E antibody is made in very small amounts (its concentration in the blood is the lowest of all Ig's) and very rapidly becomes attached to mast cells as it circulates past them. Mast cells (and basophils) have specific receptors on their cell membranes that bind the Fc portion of Ig E molecules with very high affinity. Once they are bound to a mast cell Ig E molecules persist for weeks and the particular mast cell will remain "**sensitized**" as long as enough antibody remains attached to trigger the activation of the cell where it comes into contact with antigen.

Sensitization may also be effected passively by the transfer of serum, that contains IgL antibody to a specific antigen. A procedure known as the Prausnitz – kustner (P-K) test used to be performed as a test for antibodies responsible for anaphylactic reaction.

In the P-K (Prausnitz-kustner) test serum from an allergic individual was injected into the skin of a normal person. After a lag period of 1-2 days, during which the locally injected Ab diffused towards neighbouring mast cells and become bound to them, the site of injection was said to be sensitized, and it showed an urticial reaction after injection into it, of the antigen to which the donor was allergic. Such a reaction passively sensitized animals is called passive cutaneous anaphylaxis (PCA).

8.1.2. Activation Phase

The anaphylactic reaction itself may be triggered by injection of the specific antigen into the skin of a sensitized individual. The response called "wheal and flare" is characterized by erythema (redness due to dilation of blood vessels) and edema (swelling produced by release of serum into tissue). The anaphylactic reaction is the most rapid of all hypersensitivity reactions and reaches its peak with in 10-15 minutes, then it fades without leaving any residual damage.

Mast cells, the main effectors of type I responses are generally found around blood vessels in the connective tissue, in the lining of gut and in the lungs. They are large mononuclear cells, heavily granulated and deeply stained by basic dyes such as toludine blue. In some cases, a circulating polymorphonuclear granulated cells called a basophil also takes part in type I response.

Triggering of the mast cell to release its granules and their pharmacologically active contents require at least two of the receptors for the Fc portion of the Ig E molecules be bridged together in a stable configuration.

Among the rapid changes known to occur are

(*i*) Changes in membrane fluidity, which result from methylation of phospholipids and are followed.

(*ii*) By an influx of Ca^{2+} ions.

It requires energy generated by glycolysis and certain enzymes are activated that allow the movement of granules to cell surface.

(*iii*) The intracellular levels of cAMP and cGMP are responsible for mediating reaction as regulator subsequent events.

An increase in intracellular cAMP at this stage will slow or even stop the process of degranulation. Thus activation of adenylate cyclase, the enzyme that converts ATP to cAMP provides an important mechanism for controlling anaphylactic events.

The granules are moved, probably by microfilaments to the cell surface, where their membrane fuse with the cell membrane and all contents are released to exterior. This release of granules does not imply lysis and death of cell. Infact the degranulated cells regenerate and once the contents of granules have been synthesized, the cells are ready to resume their functions.

8.1.3. Effector Phase

It is categorized into two forms.

Preformed Mediators

Stored in granules by electrostatic attraction to a matrix protein and are released as a result of the influx of ions, primarily Na+.

8.1.4. Substance synthesized denova

They are released during the anaphylactic response. They are pharmacologically active substance.

8.4.5. Histamine

Histamine is formed in the cell by decarboxylation of histidine and it is stored bound by electrostatic interaction to an acid matrix protein called heparin. When released histamine binds rapidly to two major types of receptor H1 and H2.

When histamine binds to H1 receptor on smooth muscle it causes constriction; when it binds on endothelial cells it cause separations at their junctions resulting in vascular permeability. Hydrogen receptors are also involved in increased vascular permeability, as well as in the release of acid from stomach mucosa.

All these effects are responsible for systemic anaphylaxis – difficulty in breathing or asphyxiation due to constriction of smooth muscle around the bronchi in the lung, and a drop in blood pressure, that results from extravasation of fluid into tissue space as the permeability of blood vessel increases.

H1 receptor are blocked by antihistamines such as benadryl, by direct competition, and when these drugs are given they can counteract the effects of histamine. Blockage of the receptor require cimetidine. These antihistamines were infective in controlling constriction of smooth muscles that was slower in onset and more persistant than that produced by histamine. This observation lead to SRS-A, the flow reacting substance of anaphylaxis.

8.4.6. Serotonin

Serotonin is present in the mast cells of certain species such as rodents. Its effects are similar to those of histamine, in that it cause constriction of smooth muscle and increases vascular permeability.

8.1.7. Eosinophilic chemotactic factor of Anaphylasis (ECF-A)

It's a set of peptides that produces a chemotactic gradient capable of attracting Eosinophils to the site of release of these peptides. In anaphylactic reactions the eosinophils appear to serve as a belaved indicator of the presence of Ig E mediated reactions, they may also release arylsulfatase and histaminase, which affect mediators of hypersensitivity reaction.

8.1.8. SRS – A

It is potent and is present in such low concentration. It is a set of peptides which are coupled to a metabolite of arachidonic acid called leukotriens. The leukotriens have been named LTB4. LTC4 and LTD4 and in minute amounts, they cause prolonged constriction of smooth muscle. They are considered to be the cause of much antihistamine resistant asthma in humans.

8.1.9. Platelet Activation Factor

PAF induces platelets to aggregate and release their contents of mediators, which include histamine and serotonin. Activation of platelets may also include release of metabolites of arachidonic acid, thus augmenting effects generated by mast cells.

8.1.10. Localized Anaphylaxis (Atopy)

In localized anaphylaxis the reaction is limited to a specific target tissue of organs often involving epithelial surfaces at the site of allergen entry. The tendency to manifest localised anaphylactic reaction is inherited and is referred to as atopy. Atopic allergies, which affect atleast 20% of population include Ig E mediated disorders, including allergic rhinitis (hay fever) asthma, atopic dermatitis (eczema) and food allergies.

8.1.11. Allergic rhinitis

The most common atopic disorder is known as hay fever. This results from air borne allergens reacting with sensitized mast cells in the conjuctivae and nasal mucosa to induce the release of pharmacologically active mediators from mast cells. These mediators than cause localized vasodilation and increased capillary permeability. The symptoms include watery exudation of the conjuctivae, nasal mucosa and upper respiratory tract, as well as sneezing and coughing.

8.1.12. Allergic asthma

In some cases air borne or blood borne allergens such as pollens, dust, fumes, insect products, or viral antigens trigger an asthmatic attack.

8.1.13. Intrinsic asthma

In other cases an asthmatic attack can be induced by exercise or cold, apparently independent of allergen stimulation.

Like hay fever, asthma is triggered by degranulation of mast cells with release of mediators, but instead of occuring in the nasal mucosa, the reaction develops in the lower respiratory tract. The resulting contraction of the bronchial smooth muscle leads to broncho constriction. Airway edema, mucus secretion and inflammation contribute to the bronchial constriction and to air way obstruction. Asthmatic patients may have abnormal levels of receptors for neuropeptides.

The asthmatic response can be divided into two responses.

8.1.14. Early Response

It occurs with in minutes of allergen exposure and involves primarily histamine, leukotriens (LTC4) and prostaglandin D2 (PGD2). The effects of these mediators lead to bronchoconstriction, vasodilation and some buildup of mucus.

8.15. LATE RESPONSE

It occurs later and involves additional mediators including IL 4, IL 5, IL 6, TNF alpha, eosinophil chemotactic factors (ECF) and platelet activating factor. The overall effect of these mediators is to recruit inflammatory cells, including eosinophils and neutrophils into the bronchial tissue.

The neutrophils and eosinophils are capable of causing significant tissue injury by releasing toxic enzymes oxygen radicals and cytokines. These events lead to occlusion of the bronchial lumen with mucus, proteins and cellular debris, sloughing of the epithelium, thickening of basement membrane build up (edema) and hypertrophy and bronchial smooth muscle. A mucous plug often forms and adheres to bronchial wall. The mucous plug contains clusters of detached epithelial cell fragments, eosinophils, some neutrophils and spirals of bronchial tissue as aischmann's spirals.

8.1.16. Atopic dermatitis

Atopic dermatitis (allergic eczema) is an inflammatory disease of skin that is frequently associated with atopy. The disease is observed most frequently in young children, often developing during infancy. Serum Ig E levels are often elevated. The allergic individual develops skin eruptions that are erythromatous and filled with pus. Unlike a DTH reaction which involves TH cells, the skin lesions in atopic dermatitis have TH2 cells and an increased number of eosinophils.

8.1.17. Late Phase Reaction

As a type I hypersensitive reaction begin to subside, mediators released during the course of the reaction often induce a localized inflammatory reaction called *late phase reaction*. The late phase reaction begins to develop 4 6 hours following the initial type I reaction and persist for 1–2 days. The reaction is characterized by infiltration of neutrophils, eosinophils, macrophages, lymphocytes and basophils. The localised late phase response also may be partly mediated by cytokines released from mast cells. Both TNF alpha and I2 increases the expression of cell adhesion molecules on venular endothelial cells, thus facilitating the buildup of neutrophils, eosinophils and monocytes that characterize late phase reaction.

ECF (Eosinophil chemotactic factor) released by mast cell during the initial reaction, attracts large number of eosinophils to the affected site. Various cytokines released at the site, including IL 3, IL 5 and GM CSF contribute to growth and differentiation of eosinophils. Eosinophil express Fc receptors for Ig G and Ig E isotypes and bind directly to antibody coated antigen activates eosinophils, leading to their degranulation and release of inflammatory mediators, including leukotriens, major basic protein, PAF, cationic protein and eosinophil derived mediators may play a protective role in parasitic infections. However in response to allergens these mediators contribute to extensive tissue damage in late phase reaction. The influx of eosinophils in the late phase response contribute to chronic inflammation of the bronchial mucosa that characterizes persistent asthma.

Neutrophils are another major participant in late phase reactions also accounting for 30% of the inflammatory cells. Neutrophils are attracted to the area of a type I reaction by NCF, released from degranulating mast cells. In addition a variety of cytokines released at the site, including IL 8 have been shown to activate neutrophils resulting in release of their granule contents, including lytic enzymes, platelet activating factor and leukotriens.

8.2. ANTIBODY MEDIATED CYTOTOXIC (TYPE II) HYPERSENSITIVITY

Type II hypersensitive reactions involve antibody mediated destruction of cells. This type of reaction is best exemplified by blood transfusion reaction, in which host antibodies react with foreign antigens on the incompatible transfused blood cells and mediate destruction of these cells. Antibody can mediate cell destruction by activating the complement system to create pores in the membrane of the foreign cell. Antibody can also mediate cell destruction by antibody dependent cell mediated cytotoxicity (ADCC). In this process, cytotoxic cell with Fem receptors bind to the Fc region of antibodies on target cells and promote killing of the cells. Antibody bound to a foreign cell also can serve as an opsonin, enabling phagocytic cells with Fc or C3b receptors to bind and phagocytose the antibody coated cell .

This section examines three examples of type II hypersensitive reactions. Certain autoimmune diseases involve auto antibody mediated cellular destruction by type II mechanisms.

8.2.1. Transfusion Reactions

A large number of proteins and glycoproteins on the membrane of red blood cells are encoded by different genes, each of which has a number of alternative alleles. An individual possessing one **allelic** form of a blood group antigen can recognize other **allelic** forms on transfused blood as foreign and mount an antibody response. In some cases, the antibodies have already been induced by natural exposure to similar antigenic determinants on a variety of microorganism thought to be normal flora of the gut. This is the case with the ABO blood group antigens .

ABO blood group.(A) Structure of terminal sugars, which constitute the distinguishing epitypes, in the A, B, and O blood antigens. (B) ABO genotypes and corresponding phenotypes, agglutinins, and isohaemagglutinins.

Antibodies to the A,B, and O antigens, called isohaemagglutinins, are usually of the Ig M class. An individual with blood type A, for example, recognizes B like epitopes on intestinal microorganisms and produces isohaemagglutinins to the B like epitypes. This same individual does not respond to A like epitopes on the same intestinal microorganisms because these A like epitopes are too similar to self and a state of self tolerance to these epitopes should exist. If a type A individual is accidentally transfused with blood containing type B cells, the anti B isohaemagglutinir will bind to the B blood cells and mediate their destruction by means of complement mediated lysis. Antibodies to other blood group antigens may result from repeated blood transfusions because minor allelic differences in these antigens can stimulate antibody production. These antibodies are usually of the Ig G class.

Transfusion of blood into a recipient possessing antibodies to one of the blood group antigens can result in a transfusion reaction. The clinical manifestation of transfusion reactions result from massive intravascular haemolysis of the transfused red blood cells by antibody plus complement. These manifestations may be either immediate or delayed. Reactions that begin immediately are most commonly associated with ABO blood group incompatibilities, which lead to complement mediated lysis triggered by the Ig M isohaemagglutinins. Within hours, free haemoglobin can be detected in the plasma; it is filtered through the kidneys, resulting in haemoglobinuria. Some of the hemoglobin gets converted to bilirubin, which at high levels is toxic. Typical symptoms include fever, chills, nausea, clotting within blood vessels, pain in the lower back, and haemoglobin in the urine. Treatment involves prompt termination of the transfusion and maintenance of urine flow with a diuretic, because the accumulation of haemoglobin in the kidney can cause acute tubular necrosis.

Delayed haemolytic transfusion reactions generally occur in individuals who have received repeated transfusion of ABO compatible blood that is incompatible for other blood group antigens. The reactions develop between 2 and 6 days after transfusion and reflect the secondary nature of these reactions. The transfused blood induces clonal selection and production of Ig G against a variety of blood group membrane antigens. The most common blood group antigens that induce delayed transfusion reactions are Rh, Kidd, Kell, and Duffy. The predominant isotype involved in these reactions is Ig G, which is less effective than Ig M in activating complement. For this reason, complement mediated lysis of the transfused red blood cells is incomplete, and many of the transfused cells are destroyed at extravascular sites by agglutination, opsonization, and subsequent phagocytosis by macrophages. Symptoms include fever, low haemoglobin, increased bilirubin, mild jaundice, and anemia. Free haemoglobin is usually not detected in the plasma or urine in these reactions because RBC destruction occurs in extravascular sites.

Transfusion reactions can be prevented by proper cross matching between the donor's and the recipient's blood. Cross matching can reveal the presence of the antibodies in donor or recipient sera that can cause these reactions. '

8.2.2. Haemolytic Disease of the Newborn

Hemolytic disease of the new born develops when maternal Ig G antibodies specific for fetal blood group antigens cross the placenta and destroy fetal red blood cells. The consequences of such transfer can be minor serious, or lethal. Severe haemolytic disease of the newborn, called erythroblastosis foetalis, most commonly develops when an Rh+ fetus expresses an Rh antigen on its blood cells that the Rh mother does not express.

During pregnancy, fetal red blood cells are separated from the mother's circulation by a layer of cells in the placenta called the trophoblast. During her first pregnancy with an Rh+ fetus, an Rh woman is usually not exposed to enough fetal red blood cells to activate her Rh specific B cells. At the time of delivery, however, separation of the placenta from the uterine wall allows larger amounts of fetal umbilical cord blood to enter the mother's circulation. These fetal red blood cells activate Rh specific B cells, resulting in production of Rh specific plasma cells and memory B cells in the mother. The secreted Ig M antibody clears the Rh+ fetal red cells from the mother's circulation, but the memory cells remain, a threat to any subsequent pregnancy with an Rh+ fetus. Activation of these memory cells in a subsequent pregnancy results in the formation of Ig G anti Rh antibodies, which cross the placenta and damage the fetal red blood cells. Mild to severe anaemia can develop in the fetus, sometimes with fatal consequences. In addition, conversion of haemoglobin to bilirubin can present an additional threat to the newborn because the lipid soluble bilirubin may accumulate in the brain and cause brain damage.

Haemolytic disease of the newborn caused by Rh incompatibility in a subsequent pregnancy can be almost entirely prevented by administering antibodies against the Rh antigen to the mother within 24-48 hours after the first delivery. These antibodies, called rhogam, bind to any fetal red blood cells that enter the mother's circulation at the time of delivery and facilitate their clearance before B cell activation and ensuing memory cell production can take place. In a subsequent pregnancy with an Rh+ fetus, a mother who has been treated with Rhogam is unlikely to produce Ig G anti Rh antibodies; thus, the fetus is protected from the damage that occurs when these antibodies cross the placenta.

The development of haemolytic disease of the newborn caused by Rh incompatibility can be detected by testing maternal serum at intervals during pregnancy for antibodies to the Rh antigen. A rise in the titer of these antibodies as pregnancy progresses, indicates that the mother has been exposed to antigens and is producing increasing amounts of antibody. The presence of maternal Ig G on the surface of fetal red blood cells can be detected by a Coombs test. Isolated fetal red blood cells are incubated with the Coombs reagent, goat antibody to human Ig G antibody. If maternal, Ig G is bound to the fetal red blood cells, the cells agglutinate with the Coombs reagent.

If haemolytic disease caused by Rh incompatibility is detected during pregnancy, the treatment depends on the severity of the reaction. For a sever reaction, the fetus can be given an intrauterine blood exchange transfusion to replace fetal Rh^+ red blood cells with Rh cells. These transfusions are given every 10 to 21 days until delivery. In less severe cases, a blood exchange transfusion is not given until after birth, primarily to remove bilirubin; the infant is also exposed to allow levels of UV light to break down the bilirubin and prevent any cerebral damage. The mother can also be treated during the pregnancy by plasmapheresis. In this procedure, a cell separation machine is used to separate the mother's blood into two fraction, cells and plasma. The plasma containing the anti Rh antibody is discarded, and the cells are reinfused into the mother in an albumin or fresh plasma solution.

The majority of cases (65%), haemolytic disease of the newborn have minor consequences and are caused by ABO blood group incompatability between the mother and fetus. Type A or B fetus carried by type O mothers most commonly develop these reactions. A type O mother is most likely to develop Ig G antibody to the A or B blood group antigens either through natural exposure or through exposure to fetal blood group A or B antigens in successive pregnancies. Usually the fetal aneamia resulting from this incompatibility is mild; the major clinical manifestiation is a slight elevation of bilirubin, with jaundice. Depending on the severity of the aneamia and jaundice, a blood exchange transfusion may be required in these infants. In general the reaction is mild, however, and exposure of the infant to low levels of UV light is enough to breakdown the bilirubin and avoid any cerebral damage.

Drug induced Haemolytic Anaemia

Certain antibiotics (e.g.,penicillin, cephalosporin, and streptomycin) can absorb nonspecifically to proteins on RBC membranes, forming a complex similar to a hapten carrier complex. In some patients, such drug protein complexes induce forming a complex similar to a hapten carrier complex. In some patients, such drug protein complexes in duce formation of antibodies, which then bind to the absorbed drug on red blood cells, inducing complement mediated lysis and thus progressive aneamia. When the drug is withdrawn, the haemolytic aneamia disappears. Penicillin is noble in that it can induce all four types of hypersensitivity with various clinical manifestations.

The reaction of antibody with antigen generates immune complexes. Generally this complex in antigen with antibody facilitates the clearance of antigen by phagocytic cells. In some cases, however, large amounts of immune complexes can lead to tissue damaging type III hypersensitive reactions. The magnitude of the reaction depends on the quantity of immune complexes as well as their distribution within the body. Depending on where these complexes are carried, different tissue damaging reactions can be observed. When the complexes are deposited in tissue very near the site of antigen entry, a localized reaction develops. When the complexes are formed in the blood, a reaction can develop wherever the complexes are deposited in particular, complex deposition is frequently observed on blood vessel walls, in the synovial membrane of joints, on the glomerular basement membrane of the kidney, and on the choroid plexus of the brain. The deposition of these complexes initiates a reaction that results in the recruitment of neutrophils to the site. The tissue there is injured as a consequence of granular release from the neutrophil.

Type III hypersensitive reactions develop when immune complexes activate the complement system's array of immune effector molecules. C3a, C5a, and C5b67 are also chemotactic factors for neutroph's, which can accumulate in large number at the site of immune complex deposition. Larger immune complexes are deposited on the basement membrane of blood vessel walls or kidney glomeruli, whereas smaller complexes may pass through the basement membrane and be deposited in the sub epithelium. The type of lesion that results depends on the site of deposition of the complexes.

Much of the tissue damage in type III reactions stems from release of lytic enzymes by neutrophils as they attempt to phagocytose immune complexes. The C3b complement component acts as an opsonin, coating immune complexes. A neutrophil binds to a C3b coated immune complex by means of the type I complement receptor, which is specific for C3b because the complex is deposited on the basement membrane surface, phagocytosis is impeded, so that lytic enzymes are released during the unsuccessful attempts of the neutrophil to ingest the adhering immune complex. Further activation of the membrane attack mechanism of the complement system can also contribute to the destruction of tissue. In addition, the activation of complement can induce aggregation of platelets, and the resulting release of clotting factors can lead to formation of micro thrombi.

8.3. LOCALIZED TYPE III REACTIONS (IMMUNE COMPLEX MEDIATED DISEASE)

Injection of an antigen intradermally or subcutaneously into an animal that has high levels of circulating antibody specific for that antigen leads to formation of localized immune complexes, which mediate an acute Arthus reaction within 48 hours. Microscopic examination of the tissue reveals neutrophils adhering to the vascular endothelium and then migrating into the tissues at the site of immune complex deposition. As the reaction develops, localized tissue and vascular damage results in an accumulation of fluid (edema) and red blood cells, erythema, at the site.

After an insect bite, a sensitive individual may have a rapid, localized type I reaction at the site. Often, some 4 - 8 h later, A typical arthus reaction also develops the site, with pronounced erythema and edema. Intrapulmonary arthus type reactions induced by bacterial spores, fungi, or dried fecal proteins can also cause pneumonitis or alveolities. These reactions are known by a variety of common names reflecting the sources of the antigen. For example, "farmer`s lung" develops after inhalation of thermophilic actinomycetes from moldy hay, and "pigeon fancier`s disease" results from inhalation of a serum protein in dust derived from dried pigeon feces.

8.3.1. Generalized Type III Reaction

When large amounts of antigen enter the bloodstream and bind to antibody, circulating immune complexes can form. If antigen is in excess, small complexes form; because these are not easily cleared by the phagocytic cells, they can cause tissue damaging type III reactions at various sites. Historically, generalized type III reactions were often observed after the administration of antitoxins containing foreign serum, such as horse anti tetanus or anti diphtheria serum. In such cases, the recipient of a foreign antiserum develops antibodies specific for the foreign serum proteins; these antibodies then from circulating immune complexes with the foreign serum antigens. Typically, within days or weeks after exposure to foreign serum antigens, an individual begins to manifest a combination of symptoms that are called serum sickness. These symptoms include fever, weakness, generalized vascularitis (rashes) with edema and erythema, lymphadenopathy, arthrities, and sometimes glomerulonephritis. The precise manifestations of serum sickness depend on the quantity of immune complexes formed as well as the overall size of the complexes, which determine the site of their deposition. As mentioned above, the sites of deposition vary but, in general, complexes accumulate in tissues where filtration of plasma occurs. This explains the high incidence of glomerulonephritis (complex deposition in the kidney) and vasculitis (deposition in the arteries) and arthrities (deposition in the synovial joints) caused by serum sickness.

Formation of circulating immune complexes contributes to the pathogenesis of a number of conditions other than serum sickness. These include the following:

Autoimmune Diseases

1. Systemic lupus erythematosus
2. Rheumatod arthritis
3. Good pasture's syndrome

Drug Reactions

1. Allergies to penicillin and sulfonamides

Infectious Diseases

1. Poststreptococcal glomerulonephritis
2. Meningits
3. Hepatitis
4. Mononuicleosis
5. Malaria
6. Trypanosomiasis

Complexes of antibody with various bacterial, viral, and parasitic antigens have been shown to induce a variety of type III hypersensitive reactions, including skin rashes, arthritic symptoms, and glomerulonephritis. Poststreptococcal glomeruloneophritis, for example, develops when circulating complexes of antibody and streptococcal antigens are deposited in the kidney and damage the glomeruli. A number of autoimmune diseases stem for circulating complexes of antibody with self proteins, with glycoproteins, or even with DNA. In systemic lupus erythematosus, complexes of DNA and anti DNA antibodies accumulate in synovial membranes, causing arthritic symptoms, or accumulate on the basement membrane of the kidney, causing progressive kidney damage.

8.4. DTH MEDIATED (TYPE IV) HYPERSENSITIVITY

Type IV hypersensitive reactions develop when antigen activates sensitized TDTH cells by antigen on appropriate antigen presenting cells resulting in the secretion of various cytokines, including interleukin 2 (Il 2), interferon gamma (IFN r), macrophage inhibition factor (MIF), and tumor necrosis factor (TNF B) . The overall effect of these cytokines is to draw macrophages into the area and activate them, promoting increased phagocytic activity and increased concentration s of lytic enzymes for more effective killing. As lytic enzymes leak out of the activated macrophages into the surrounding tissue, localized tissue destruction can ensure. These reaction typically take 48 –72 hours to develop, the time required for initial T DTH cell activation and cytokine secretion to mediate accumulation of macrophages and the subsequent release of their lytic enzymes. The Hall marks of a type IV reaction are the delay in time required for the reaction to develop and the recruitment of macrophages as opposed to neutrophils, as found in a type III reaction. Macrophages are the major component of the infiltrate that surrounds the site of inflammation.

As explained, several lines of evidence suggest that the type IV (or DTH) reaction is important in host defence against parasites and bacteria that can live within cells. Once these organisms are inside the host's cells, circulating antibodies cannot reach them. However, the heightened phagocytic activity and the buildup of lytic enzymes from macrophages in the area of infection lead to nonspecific destruction of cells, and thus of the intracelluar pathogen. When this defence process is not entirely effective, the continued presence of the pathogen's antigens can provoke a chronic DTH reaction, which is characterized by excessive numbers of macrophages, continual release of lytic enzymes, and consequent tissue destruction. The granulomatous skin lesions seen with Mycobacterium leprae and the lung cavitation seen with Mycobacterium tuberculosis are examples of the tissue damage that can result when chronic delayed type hypersensitive reactions develop.

The reaction to an intradermal injection of an antigen can serve as a test for the presence of TDTH cells previously sensitized by that antigen. The use of the PPD antigen to detect previous exposure to *M. tuberculosis* was described. Similar skin tests to effect previous exposure to the bacterium that causes leprosy and the fungus that causes occidiomycosis use lepromin and coccidiodin, respectively, as test antigens. In these tests, the appearance of swelling and redness at the injection site within 48-72 h is a positive reaction. The depletion in CD4+ T cells associated with AIDS progresses, the decline in TDTH cells is reflected in decreased skin reactivity to such antigens.

Many contact dermatitis reactions including the responses to formaldehyde, trinitrophenol, nickel, turpentine, various cosmetics and hair dyes, poison oak, and poison ivy were mediated by TDTH cells. Most of these substances are small molecules that can complex with skin proteins. This complex is internalize by antigen presenting cells in the skin (e.g., Langerhans cells), then processed and presented together with class II MHC molecules, causing activation of sensitized TDTH cells. In the reaction to poison oak, for example, a pentadeacatechol compound from the leaves of the ant complexes with skin proteins. When TH cells react with this compound appropriately displayed by local antigen resenting cells, they differentiate into sensitized TDTH cells. A subsequent exposure to pentadecacatechol will elicit activation of TDTH cells and induce cytokine production. Approximately 48–72 hours after this secondary exposure, the secreted cytokines cause macrophages to accumulate at the site. Activation of these macrophages and release of their lytic enzymes result in the redness and pustules that characterize a reaction to poison oak.

Antibody dependent cell mediated cytotoxicity (ADCC)

ADCC is known as Type IV hypersensitivity. This is mediated through Natural killer (NK) cells, and the cytotoxic mechanism is independent of complement. Target cells are coated with antibody are killed by NK cells through an extracellular nonphagocytic mechanisms and bear T markers. Antigen when enters inside the body induce antibody production. Ag-Ab complex forms on target cells. Natural killer cells combine with Ag-Ab complex via Fc fragment of antibody and causes lysis of target cells.

Schwartz phenomenon

Schwartzmann (1928) proposed this phenomenon. He injected S.typhi culture filtrate to rabbit intradermally, some of was also intra venous injection in the animal after 24 hours. In the intradermal inoculation, a haemmorrhagic nuerobic lesion was produced whereas injected intravenously. This is superficial resemblance of hypersensitivity. It is probably a specialized type of disseminated intravascular coagulation (DIC) precipitated by endotoxin.

8.5. SUMMARY

1. Hypersensitive reactions are inflammatory reactions within the humoral or cell mediated branches of the immune system that lead to extensive tissue damage or even death. These reactions are classified into four main type according to the mechanism that induces them. Each type generates characteristic effector molecules and clinical manifestations.

2. A type I hypersensitive reaction is mediated by Ig E antibodies, whose Fc region binds to receptors on mast cells or blood basophils. Cross linkage of the fixed Ig E by allergen initiates a sequence of intracelluar elements leading to mast cell or basophil degranulation with release of pharmacologically active mediators (see Table 16 3). The mediators are the immune effector molecules in this reaction. The principal effect of these mediators are smooth muscle contraction and vasodilation. Clinical manifestations of type I reactions include potentially life threatening systemic anaphylaxis and localized responses such as hay fever and asthma

3. A type II hypersensitive reaction occurs when antibody reacts with antigenic determinants present on the surface leading to cell damage or death through complement mediated lysis or antibody dependent cell mediated cytotoxicity (ADCC). Transfusion reactions, resulting from mismatching of donor and recipient blood, and haemolotypic disease of the newborn are two clinical conditions caused by type II reactions. The latter is most likely when an Rh mother carries an Rh+ fetus.

4. A type III hypersensitive reaction is mediated by the formation of immune complexes and the ensuing activation of complement. Complement split products serve as immune effector molecules that elicit localized vasodilation and chemotactically attract neutrophils. Deposition of immune complexes near the site of antigen entry can induce an Arthus reaction, in which lytic enzymes released by the accumulated neutrophils and the complement membrane attack complex cause localized tissue damage. Generalized type III reactions occur when circulating immune complexes are deposited at various sites; the manifestations of these reactions vary depending on the site of tissue deposition.

5. A type IV hypersensitive reaction involves the cell mediated branch of the immune system. Antigen activation of sensitized TDTH cells induces release of various cytokines, which serve as the immune effector molecule in this reaction. The effect of these cytokines is to cause an accumulation and activation of macrophages, which release lytic enzymes that cause localized tissue damage.

QUESTIONS

1. Define Hypersensitivity.
2. Write the classification of Gell and combs on Hypersensitivity.
3. Outline briefly the Type I anaphylactic reaction.
4. Write an account on Type II Hypersensitivity cytotoxic reaction.
5. Explain Type III hypersensitivity immune complex reaction.
6. Enumerate on Type IV hypersensitivity, delayed reaction.
7. Enumerate on Type V hypersensitivity, delayed reaction.
8. Write an account on hypersensitivity stimulatory reaction.
9. Write an account on the factors influencing anaphylaxis.
10. Explain the mechanism of anaphylaxis.
11. Distinguish between immediate and delayed types of hypersensitivity.

12. Write a note on systemic anaphylaxis.
13. Explain about the localized anaphylaxis.
14. Write a note Prausnitz-Kustner (PK) reaction.
15. Write a note on Antibody-dependent cell mediated cytotoxicity (ADCC).
16. Explain Shwartzmann phenomenon.
17. Write short notes on:
    (*i*) Anaphylaxis.
    (*ii*) Prausnitz-Kustner (PK) reaction.
    (*iii*) Atopy.
    (*iv*) Cytotoxic reaction.
    (*v*) Arthus reaction.
    (*vi*) ADCC.
    (*vii*) Shwartz phenomenon.

CHAPTER 9

COMPLEMENT

The complement system consists of approximately 30 molecules constituting nearly 10% of the total serum protein and forming one of the major defence systems of the body.

The functions of the system include controls of inflammatory reactions, clearance of immunity complexes, cellular activation and anti-microbial defense.

The system also plays a role in the development of the immune response and is a major epsilon system in autoimmune and hypersensitivity reactions.

In plasma these are present in functionally inactive form.

9.1. COMPLEMENT CASCADE

Complement in cultivated sequentially in a cascading manner, each protein activating the protein that directly follows it in the sequence. Complement proteins are not antibody, but the cascade is activated by antibody.

(ii) Activation of the complement cascade has widespread physiologic and pathophysiologic effects. It causes lysis of erythrocytes in haemolytic anemias, sensitizes foreign particles for phagocytosis and causes release of histamine from meristem cells.

9.2. METABOLISM

9.2.1. Synthesis

The liver is the major site of synthesis of complement proteins. Tissue macrophages and fibroblasts can synthesize some complement protein as well.

(*a*) Complement proteins first appear in the fetus during the second month of pregnancy. The concentration of most components at births is approximately 50% of adult levels.

(*b*) Inflammation increases the synthesis of complement components, presumably through the action of interleukin –1 (K-1) and Gamma interferon.

(*c*) The catabolic rate of complement proteins is high 1% to 3% per hour.

The term complement refers to a system of functions which occur in normal serum that are activated characteristically by Ag – Ab interaction and subsequently mediate a number of biologically significant consequences. Some of its components are heat stable, complement as a whole heat labile.

9.3. PROTEINS THAT REGULATE COMPLEMENT SYSTEM

Because the complement system is non-specific and than capable of attacking host cells as well as microorganism elaborate regulatory mechanism are require to confine the complement activation to designate targets. Both the classical and alternative pathways include a number of extremely labile component which undergo spontaneous inactivation as they diffuse away from target cells for e.g,. The target binding site in C3b. Under spontaneous hydrolysis by time it has differed from 40nm away from C4b2a or b.

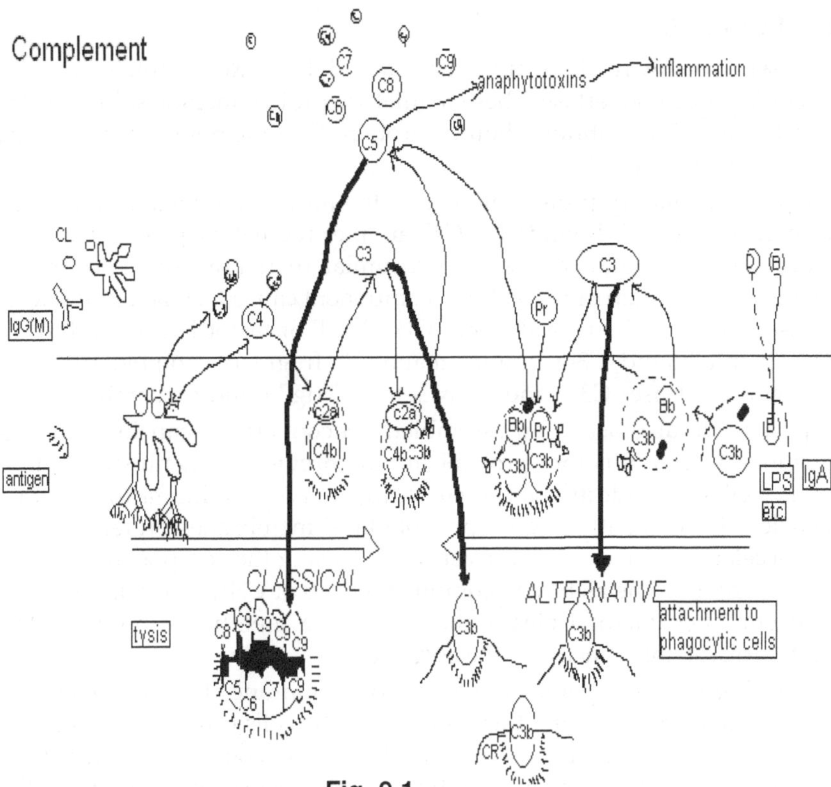

Fig. 9.1

convertase enzyme. This rapid hydrolysis limits binding of C3b to nearly host cell. In addition both pathways include a series of regulatory protein that inactivate various complement component e.g., glycoproteins called C, inhibition G can form a complex with C1r2 s2 causing it to dissociate from c1,q and preventing further activation of C4 or C2.

The reaction catalyzed by C3 convertase enzyme of classical and alternate pathway is the major amplification step in complement activation generates hundreds of molecules of C3b.The C3b generates by these enzyme can bind to near by cell mediating damage to healthy cells by opsonization to phagocytic cell bearing C3b receptor a by reduction of membrane attack complexes. It is estimated that irradiating red blood cells are exposed to thousands of C3b molecule each day. C3b mediated damage to healthy cell prevented by family of related proteins that regulate C3 convertase activity in the classical and alternate pathways. These C3 convertase regulatory protein all contain repeating amino acid sequences, containing about 60 residues, formed short consensus repeat. All these proteins are encoded at a single chromosomal location on chromosome 1 in lumen known as regulatory of complement activation.

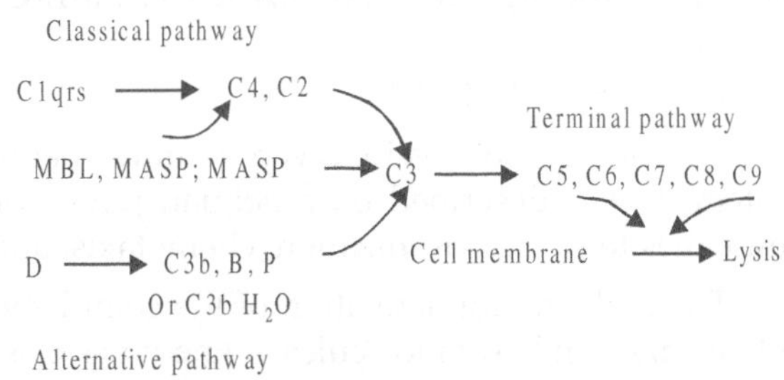

Fig. 9.2

9.4. GENE CLUSTURE

In the classical pathway, these are structurally defined RCA protein act similarly to prevent assembly of C3 convertase. These regulatory protein include soluble C45 binding protein (C4b Bp) and 2 membrane bound protein. Complement receptors type I and membrane cofactor proteins.

Each of these regulatory protein binds to C4b and prevent its association. with C2a. Once C4 bBpcR, or MCP is bounded to C4b another regulatory protein factor I Cleaves the C4b is to bound C4d and soluble C4C.A similar regulatory sequence in alternative pathway. In the case CR, map or regulatory component caused H factor bind to C3 b and prevents its association with factor B . Once CR1, MCP or factor H is bound to C3b factor I cleaves the C3b into bound C3bi fragment and C3f fragment (soluble) further cleavage of C3b1 by factor I releases C3C and leaves the C3 dg bound to membrane.

RCA protein do also act as an assembled C3 convertase, causing it to dissociate. Inoculated among these regulatory proteins are the previously mentioned C4 bBp, CRI & factor H as well as an additional protein (directly accelerating factor DAF) , a glyco protein is enclosed covalently to glycophospholipid membranes protein. Each of these RCA protein accelerates decay of C3 convertase, releasing the component with enzymatic activity (C2 a or Bb) from the all bound component (C4b or C3b). Once dissociate of the C3 convertase occur, then factor I cleans the remaining membrane bound C4b Or C3b component to irreversibly inactivate the convertase.

Regulatory Protein also operate at the level of membrane attack complex. The ability of C5b67 complex to be released and then bind to nearby cells poses a threat of innocent by stands lysis healthy cells. A number of serum protein can encounter these threat by binding to release C5bb7 and preventing its excretion into the membrane of nearby cells. Serum protein can bind to C5b6,7 including a hydrophilic transition and thereby preventing insertion of C5b67 into the membrane nearby cells.

The binding of ' s' protein to C56-67 also keep Cq from binding to C5b 67 and polymerizing there by preventing the consumption of Cq.

Complement mediated types of cells in effective of the complement in from different species than the cell begin lysed. This unusual phenomenon, which remained unexplained for several years, is now known to depend on two membrane proteins that block MHC formation.

These two proteins present in membrane of many cell types are homologous restriction factor. Membrane inhibitor of reactive lysis. Both MRF and MIR-C protect cell nonspecific complement mediated lysis by binding to C8 preventing assembly of poly-C9 and its insertion into plasma membrane.

However, the inhibition occur only if the complement components are from the same species of target cells. For this reasons, MIRC and MRF are said to play Homologous restriction.

COMPLEMENT SYSTEM

The eleven proteins involved in antigen antibody reaction is known as complement system. It was described by a belgian Jules Bordet in 1895. Complement works with antibodies to cause opsonization, chemotaxis, and lysis of bacterial cells .

The pathway for activation of the complement set into motion by the interaction of antigen and antibody molecules is known as classical pathway.

Louis Pillemes (1955) proposed an Alternate pathway or properdin system for activation of complement system.

9.5. ALTERNATIVE PATHWAY

It is activated in the presence of a variety of molecules and surfaces which provided an protected environment for reactions involved. This pathway is in a state of dynamic equilibrium. Since there is always a low level of turn over of the components involved which is rapidly unregulated in the presence of activation surfaces. The toxins are tightly controlled and limited by molecules present on mammalian cells involving DAF, MCD and CR1 and by factor H which is present in screen high concentrations.

The step for alternative pathway reaction in a molecule of activated surface bound C3 in this very be C3b generated by classical pathway or low level of spontaneous IC3 produced and deposited can acts as a focus on activator surfaces. C3b tends to bind B factor by N- terminal Ba domain analogous to binding of C2 by C4b.The C3b3 complex is acted on by the factor D, a serine protease which circulates at low levels, but in active form. Factor-D releases N- terminal Ba domain from factor B, leaving the catalytically active Bb fragment associated with C3b to form the alternative pathway acts as a positive feed back loop in the presence of activator surfaces and hence is sometimes called The amplification loop.

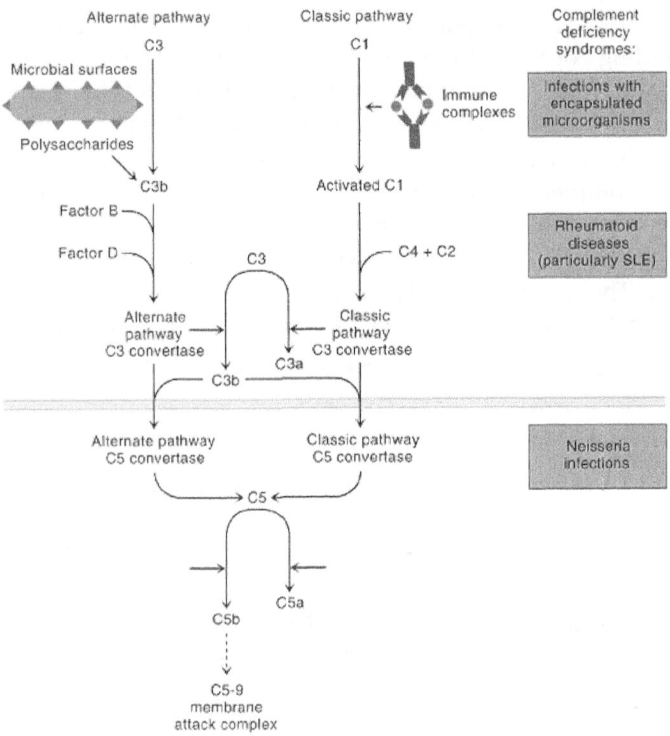

Fig. 9.3

This cycle is tightly controlled and a variety of factors hold these reactions. In follicle phase I the binding of factor H1leading to an inactive pathway in which C3b is cleaved in two phases by factor II releasing C3f and having IC3b which is degraded to C3C, C and C3dg by factor I and other serum proteases. C3dg contains thioester ring may remain covalently bound to the particle in which it was deposited.

Factor H is also promotes the decay of C3b Bb by release of Bb, complement C3b deposition. On mammalian cells, may induce death via lytic pathway and is damaging even when the cell is not killed. There are membrane proteins that is inhibit the assembly of C3bB and promotes the dissociation of C3b bB.

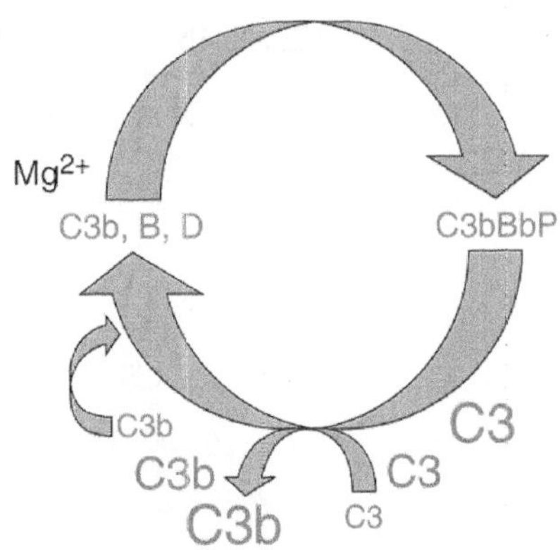

Fig. 9.4

DAF, present on erythrocytes, leucocytes and platelets accelerates the decay of classical and alternative pathway C3 convertase.

Several proteins controls C3B and C4b-CR1, CR2, DAF factor H and C4b are present on chromosome in man. This is called the "Regulation of Complement Activation" or RCA locus.

The molecule piperdin (factor P), the first component of alternative pathway was the activator of pathway and its role is in binding to and stabilizing the alternative pathway C3 convertase producing C3bBbP. Therefore it is not so much an activation as an up regulation of pathway. The structure of factor 'P' is of cyclic oligomers of 56 Kd unit.

9.6. CLASSICAL PATHWAY

The classical pathway of activation leading to the cleavage of C3 is initiated by the binding of two of more of the globular domains of the C1q component of C1, to its ligand. Immune complex II containing Ig G and Ig M and certain microorganisms and their products. This causes a conformational changes in the complex C1.

This causes a conformational change in C1 complex that leads to auto activation of C1r complement. The enzyme C1r converts C1s into an active serine esterase that acts on thioester containing molecule C4 to produce C4a and a reactive C4b. C4a is released less than 1y, C4b become activated to the surface.

The rest inactivated by reacting with H_2O. C_2 binds to the surface bound C4b1, became a substrate for the activated C1 complex and is split into C2a and C2b.

C1q → C1q (Active)

(IA) C1r → C1r (Active)

↓

(Inactive) C1S C1s (Active)

C1q, C1r1, C1S activates the following

C2 → C2a

↓

C4 → C4b

C2a+C4b → C2a4b

C3 → C3b

↓

C5 → C5b

↓

Membrane attack complex

The C2b is released leaving C4b2a, the classical pathway C3 convertase. This active enzyme then generates C3a and then unstable C3b from C3A. A small amount of C3b generated will bind to the activating surface and act as a bound for further complement activation.

The activation of the classical pathway is regulated by a inhibitor and by a number of molecule that limit the production of C3 convertase.

Membrane attack complex are lipophilic in nature. It is attracted towards the lipid part of the cell and leads to its lysis. C3a1, C5a acts as anaphylotoxins.

9.7. ALTERNATIVE PATHWAY (AMPLIFICATION LOOP)

The alternative pathway may act as a positive feedback loop for the generation of more C3b.This loop is controlled by a number of components. Properdin (p) can bind to the C3 convertase (C3bBb) to stabilize it, where as decay accelerating factor (DAF) and CR1 on cell surfaces factor 4 in solution promotes its decay by release of Bb. In the presence of activator surfaces, factor B binds to C3b to cause further alternative pathway activation but in the presence of inactivator surfaces, inactivated by factor ICR1, and membrane cofactor protein (MCP) on host cells also favour breakdown by this route.

9.7.1.Regulation

Protein	Type of protein	Pathway affected	Immunological function
1. C1 inhibitor	soluble	classical	serine protease inhibition causes C1r2s2 to dissociate from C1q.
2. C4 binding protein	soluble	classical	Blocks · formation of C3 convertase by binding C4b., cofactor for cleavage by factor I.
3. Factor H	soluble	Alternative	Blocks formation of C3 convertase by binding C3b cofactor for cleavage of C3b by factor
4. Complement receptor (CR1)	membrane bound	classical and alternative	It blocks the formation of C3 onvertase by binding C4b or . C3b1.,cofactor for factor 1 catalyzed cleavage of C4b or C3b
5. Decay accelerating factor (DAF))	membrane bound	classical & alternative	Accelerate dissociate C42A +C3bBb(C3 convertase).
6. Membrane cofactor protein(MCP)	membrane bound	classical & alternate	Accelerate dissociation of C4b2a + C3bBb (C3 convertase).
7. Factor I	soluble	Both	serine protease cleaves C4b/C3b using C4BP,CR1,factor H, DAF (or)MCP as cofactor.
8. S.protein	soluble	terminal	binds soluble C5b,6,7,8, prevents its insertion into cell membrane
9. Homologous			

9.8. BIOLOGICAL FUNCTION OF COMPLEMENT SYSTEM

Complement serves as a mediator of the humoral response by amplifying the response and converting it into effective defense mechanism to destroy invading microorganism and viruses.

(1) cytolysis (2) Opsonization of antigen
(3) Viral neutralization (4) Inflammatory response
(5) Cleavage of soluble immune complexes.

The membrane attack complex mediate the following functions.

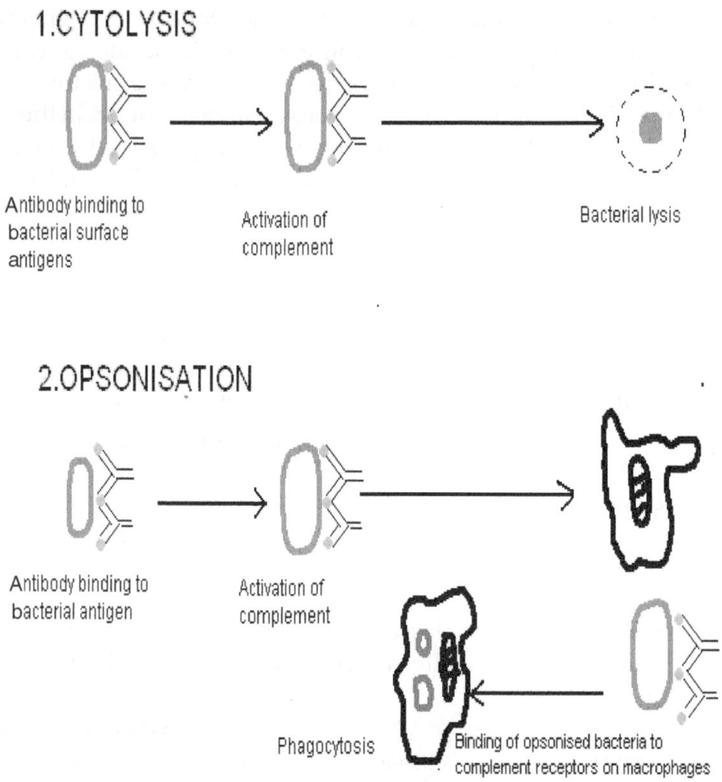

Fig. 9.5

9.8.1. Cytolysis

MAP or MAC formed by the complement activation is capable of lysing broad spectrum of microorganisms, viruses, erythrocytes nucleated cells (dendritic cells)

9.8.2. Opsonisation

C3b is the major opsonin of the complement system, although C4b & C3bi also have opsonin activity. C3 activation causes binding of C3b on immune complexes & particulate antigen (insoluble Ag).

Phagocytic cells express complement receptors Cr1, Cr3 & Cr4.

9.8.3. Viral neutralization

The complement system plays an important role by host defence by neutralizing the viral infectivity and Epstin bar virus.

These viruses activate alternative (or) classical pathway in the absence of Antibody. For most viruses the binding of antibody to the repealing units are virus surface Protein . It creates particular immune complex, then it stimulates classical pathway.

9.8.4. Inflammatory response

The complement split products C3a, C4a and C5a called anaphylotoxins. It binds to receptors on most cells and blood basophils and it induce degranulation with release of histamine and better pharmacalogically active mediators. These mediators induce smooth muscle contractions, increases vascular permeability.

C3a, C5a, C5b6, 7 act together to induce monocytes and neutrophils. These cells will adhere to vascular endothelial cells extravast through the Endothelial lining of the capillary and mild rate towards the site of complement activation in the tissues.

C5a is effective activation of complement causes influnce carries Antibody and phagocytic cells to the site if Ag entered.

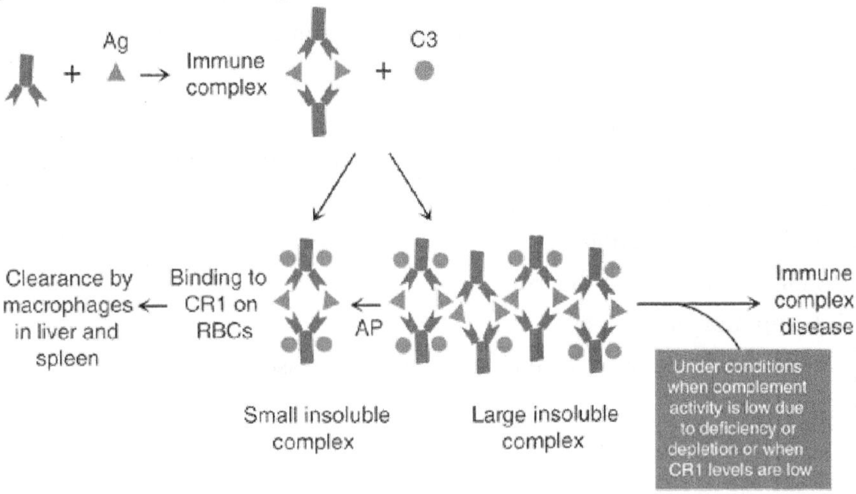

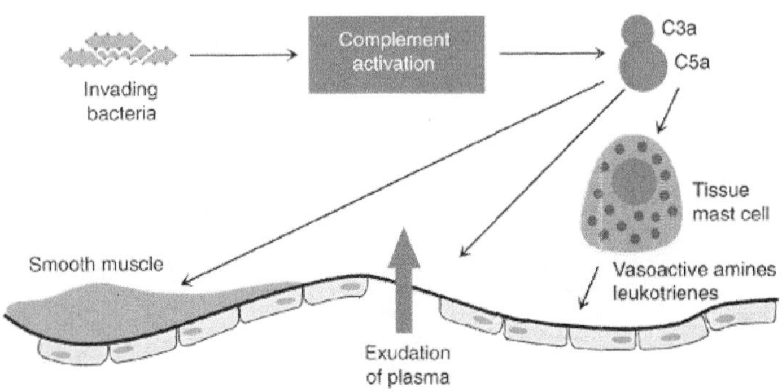

Fig. 9.6

9.8.5. Chavare of soluble immune complexes.

The role of complement system in clearing immune complexes can be seen in Patients with the auto immuno disease, systemic cubern's erythometors. These individuals produce large quantities of immune complement and suffer tissue damage as a result of Complement medical lysis and induction of type II and Type III Hyper sensitivity.

Enterocytes plays an important role in binding C3b caused immune Complexes and carrying these complexes to the liver and spleen. In these organs immune Complexes are strappied from RBC and are phagocytosed, there by preventing their disposition in the tissue.

In systemic lubes erythrometous (auto immune disease) patients deficiencies in C1, C2 and C4 each contribution to reduced leaves of C3b on immune. Complex and hence inhibit their clearance.

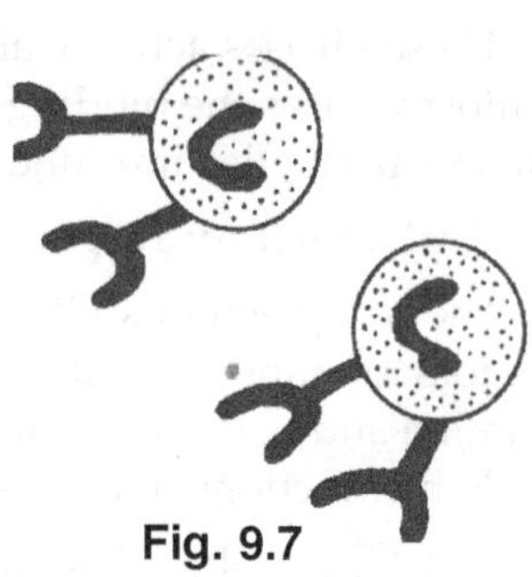

Fig. 9.7

QUESTIONS

1. Define complement.
2. What are complement components and how they help in biological function?
3. Explain the gene cluster of complement components.
4. Explain classical pathway.
5. Explain alternative pathway.
6. Explain the biological functions of complement components.
7. Write short notes on.

   (*a*) Viral neutralization;

   (*b*) Opsonization of antigen;

   Inflammatory response.
8. Mention two diagram for enzyme (e.g. convertase).

CHAPTER 10

AUTOIMMUNITY

10.1. INTRODUCTION

One of the cardinal properties of Immune system is its ability to recognize and response to foreign antigen but not to self antigen. This is called self (or) non self discrimination. The unresponsiveness of immune system to antigenic stimulation is called immunologic tolerance.

The necessity for maintaining tolerance to self antigen is defined as self tolerance. Self tolerance exist partly because of the absence of self recognizing lymphocytes from the repertoring (collection) of mature lymphocyte and partly because of the way the mature lymphocytes recognize and response to self antigens. Loss of self tolerance results in immune reaction against ones own or autologus antigen. Such reactions are called autoimmunity.

The disease they cause are called autoimmune disease (or) Disease caused by an immune response against own body tissue.

Autoimmunity results from a failure or breakdown of the mechanism normally responsible for maintaining self tolerance.

Loss of self tolerance may result from abnormal selection or regulation of self reacting lymphocytes and from abnormalities in the way self antigen are presented to the immune system.

Multiple interacting factors contribute to the development of autoimmunity. These factor include immunologic abnormalities affecting antigen presenting cells or lymphocytes, genetic backgrounds that predispose to Autoimmunity, tissue injury and microbial disease (or) infection.

Autoimmune disease may be systemic or organ specific and these may be caused by different types of antigen and different immunologic abnormalities. For example the formation of circulating immune complexes typically produces systemic disease. In contrast auto antibody (or) T cell responses against antigens with restricted tissue distribution leads to organ specific injury.

Various effector mechanism are responsible for autoimmunity in different auto immune disease. These mechanism include circulating auto antibodies, immune complexes and auto reactive T lymphocytes.

10.2. CLASSIFICATION OF AUTOIMMUNE DISEASE

Based on the site of the involvement and nature of autoimmune disease may be classified as

(*i*) Haemocytolytic

(*ii*) Localized or organ specific

(*iii*) Systemic (or) non organ specific disease

10.3. HAEMOCYTOLYTIC AUTOIMMUNE DISEASE

10.3.1. Autoimmune haemolytic anaemias

Autoantibodies against RBC are produced against RBC. Two groups of auto immune anaemias can be distinguished, characterized as cold and warm autoantibodies respectively.

The cold autoantibodies are generally complete agglutinating antibody belonging to the IgM type class and agglutinate RBC at 4° C and not to 37° C. Cold agglutinins are seen in primary typical anaemias trypanosomiasis, hay fever. The warm autoantibodies are generally incomplete non-agglutinating antibodies belonging usually to the IgG class. Warm erythrocyte autoantibodies are seen in patients taking drug alpha methyl DOPA. In autoimmune anaemias, the RBC coated with antibody are prematurely destroyed in spleen and liver. Complement dependent intravascular haemolysis appears to be a rare event.

10.3.2. Autoimmune thrombocytopenia

Autoantibodies derived against platelets occur in thrombocytopenic purpura. This condition is traditionally considered as antibody mediated hypersensitivity.

10.3. 3. Autoimmune Leukopenia

Non agglutinating antileucocyte antibody can be demonstrated in the serum of patients with systemic tuber erythromatosis and rheumatid arthritis.

10.3.4. Localized Autoimmune Disease

In an organ specific autoimmune disease, the immune response deviated to a target antigen unique to a single organ or gland, so that the manifestation is largely limited to that organ. The target organs may be subjected to direct cellular damage by humoral or cell mediated immunity.

Alternatively, the function of a target organ may be stimulated (or) blocked by autoantibodies.

10.3.5. Disease mediated by direct cellular damage

Autoimmune disease involving direct cellular damage, occur when lymphocytes or antibody's bind to cell membrane antigen, causing cellular lysis and an inflammatory response in the affected organ. Gradually, the cellular structure of an affected organ is replaced by connective tissues and function of organ destines.

10.3.6. Hashimoto's thyroiditis

It frequently affects the middle age women and individual produces autoantibodies and sensitized delayed type of its T cells, specific for thyroid antigen

The TDTH response is characterized by an intense infilteration of the thyroid gland by lymphocyte macrophages, plasma cell, which form lymphocytic follicules and germinal centers. The inflammatory response causes goitre or visible enlargement of thyroid gland. Antibodies are formed to a number of thyroid proteins including thyroglobulin and thyroperoxidase. Both of which are involved in the uptake of iodine.

Binding of autoantibodies to these proteins interferes with iodine uptake and leads to decreased production of thyroid hormone (hypothyroidism).

10.3.7. Autoimmune anaemias

Autoimmune antibodies includes pernecious, autoimmune haemolytic and drug induce haemolytic anaemia.

10.3.8. Pernicious Anaemia

Pernicious anaemia caused by auto antibodies to a membrane bound intestinal protein called intrinsic factors, this facilitates the uptake of vitamin B12 from the small intestine. Binding of the auto antibodies, to a intrinsic factor, blocks the intrinsic factor mediated by absorption of vitamin B12. In the absence of sufficient vitamin B12, which is necessary for proper haematopoiesis, the number of functional matured RBC's decreased below normal. Pernicious anaemia is treated with injection of vitamin B12, thus circumventing, the defect in its absorption.

An individual with autoimmune haemolytic anaemia makes autoantibodies, to RBC antigen, triggering complement mediated lysis or antibody mediated opsonisation and phagocytosis of the RBC. The immuno diagnostic test for auto immune haemolytic anaemias generally involves Coomb's test in which the RBC's are incubated with an anti human Ig antiserum. If Ig G autoantibodies are present on the red cells, the cells are agglutinated by the antiserum.

10.3.9. Good Pasteur's syndrome

In this auto antibodies specific for certain basement membrane antigen bind to the basement membrane of the kidney glomeruli and the alveoli of the lungs. Then subsequent complement activation inducing leads to direct cellular damage and an inflammatory response mediated by a complement split product. Damage to alveoli basement membrane leads to kidney damage and pulmonary haemorrhage with death often with several month of onset of symptoms staining of biopsy from patient with pasteur's syndrome fluroscent labeled Ig and anti C3b reveals linear deposits of C3b, Ig G along the basement membrane.

10.3.10. Insulin dependent Diabetes mellitus

It is caused by an autoimmune attack on the pancreas. The attack is derived against specialized insulin producing cells that are located in spherical clusters called the islets of pancreas scattered throughout the pancreas. The auto antibodies destroys the B-cells, resulting in decreased production of insulin and consequently increase level of blood glucose. This disease is characterized by insulitis. The condition in which a large number of delayed type of T-cells, infilterate the islets of langerhan. The cell mediated DTH response develops the following infilteration and activation of membranous macrophages. The subsequent B-cell destruction is thought to be mediated by cytokines released during the TDTH response and by lytic enzymes released from the activated macrophages. IFN gamma, Tissue IFN alpha, IL which has been implicated in destruction of B-cells.

Auto antibodies to B-cells may contribute to cell destruction by facilitating complement lysis or Antibody dependent cell mediated cytotoxicity (ADCC).

10.3.11. Diseases mediated by stimulating or blocking antibodies

In some autoimmune diseases antibodies act as antiagonists binding to hormone receptors and stimulating inappropriate activity. This usually leads to an overproduction of mediated or an increase in cell growth. In other auto immune conditions auto antibodies binds to hormone receptor but act as antagonist, blocking receptor function. This generally causes impaired secretion of mediators and gradually atrophy the affected organ.

10.3.12. Grave disease

The production of thyroid hormone is carefully regulated by thyroid stimulating hormone which is produced by the pituitary gland. Binding of TSH to a receptor on thyroid cells activate adenylate cyclase and stimulate the synthesis of two thyroid hormone, thyroxine and triiodothyronine. A patient with Grave's disease produces autoantibodies to the receptors for TSH. Binding of these auto antibodies mimics the hormonal action of TSH, activating adenylate cyclase and resulting in production of thyroid hormone. Unlike,

TSH, however, the auto antibodies are not regulated and consequently they over stimulate the thyroid. For these reasons the auto antibodies are called Long acting thyroid stimulating antibody (LATS).

10.3.13. Mysthemia Gravis

It is a prototype auto immune disease mediated by blocking antibody. A patient with these disease produces autoantibodies to acetylcholine receptor on the motor end plates. As a result, binding of these autoantibody to receptor present binding to acetylcholine, thereby activating muscle activation. The antibodies also induce complement mediated degradation of the receptor, resulting in progressive weakening of skeletal muscle.

10.3.14. Systemic Autoimmune diseases

In systemic autoimmune diseases the response is directed towards the broad range of target antigen and involves a number of organs and tissues. These diseases reflect a generalized effect in immune regulation that results in hyperactivation of T cells and B cells. Tissue damage is wide spread both from cell mediated immune responses and from the direct cellular damage caused by autoantibodies or by accumulation of immune complexes.

10.3.15. Systemic lupus erythromatosis

One of the best example of systemic autoimmune disease is the lupus erythromatosis which typically appear in women between 20 to 40 years of age with a female to male ratio of 10:1. Systemic lupus erythromatosis is characterized by fever, weakness, skin rashes and kidney dysfunction. Affected individuals may produce auto antibodies to a vast assay of tissue antigen such as DNA, histones, RBC's, platelets, leucocytes and clotting factors; interaction of these autoantibodies with a specific antigens produces various symptoms.

Autoantibody specific for RBC's and platelets, for example also lead to complement – mediated lysis resulting in haemolytic anaemia and thrombocytopenia respectively. When immune complex of autoantibodies with various nuclear antigens are deposited along the walls of small blood vessels, a type III hypersensitive reaction develops. The complexes activate the complement system and generate MAC's and complements split products that damage the blood vessel walls, resulting in vascularitis and glomerulonephritis.

Excessive complement activation in patients with severe systemic lupus erythromatoses leads to elevated serum levels of complement split products C3a and C5a which may be 3-4 times lighter than normal individuals. C5a induces increased expression of the type 3 complement receptor (CR3) on neutrophils facilitating neutrophil aggregation and attachment to vascular endothelium. As neutrophils attached to small blood vessels, the number of circulating neutrophils declines (neutrophenia) and various occlusions of small blood vessels develop (vascularitis), these occlusions can lead to widespread tissue damage.

Laboratory diagnosis of systemic lupus erythromatosis depends on the characteristic antinuclear antibodies which are directed against ds or ss DNA, nucleoprotein, histone and nuclear RNA.

Indirect immunofluroscent straining of serum from systemic lupus erythromatosis patients produce various characteristic straining patterns.

A clinical practice resembling systemic lupus erythromatosis has been observed in patient taking drugs such as diphenylhydantoin, and procaineamide. The symptoms subside an withdrawl of the drugs.

10.3.16. Rheumatid arthritis

It is a common acute immune disorder most oftenly affecting women from 40-60 years old. This is a symmetric polyarthritis with muscle wasting and subcutaneous nodules commonly associated with myocarditis, vascularitis etc.

The synovial membrane of the affected joints are swollen and oudematous with dense infilteration of lymphocytes and plasma cells.

A striking feature is the presence of a circulating autoantibodies called the Rheumatoid factor that are reactive with determinant in the Fc region of Ig G.

The classical Rheumatoid factor is an Ig M as reactive to the Fc of Ig G. Such auto antibodies binds to normal circulating Ig G, forming Ig M-Ig G complexes that are deposited in joints.

RF is detected by agglutination tests using an antigen, particle coated with globulins. In the rose-waaler test, the original technique for detection of RF, sheep erythrocytes coated with sub agglutinating dose of anti erythrocyte antibody are used as the antigen in an agglutinating test.

In modification of the test, letase and sentonite are used as the carrier particle for Ig G. antinuclear antibodies are frequently found in rheumatoid arthritis.

10.3.17. Multiple Sclerosis

Multiple sclerosis, an autoimmune disease affecting the CNS, is the most common cause of neurologic disability associated with disease.

Individual with this disease produce auto reactive T cells that participate in the formation of inflammatory lesion along the myelin sheath of nerve fibers.

CSF of patient with active multiple sclerosis contains activated T lymphocytes which infilterate the brain tissues and cause characteristic inflammatory lesion, destroying the myelin sheath.

Since myelin function to insulate the nerve fibers, a breakdown in the myelin sheath leads to numerous neurologic dysfunction.

Transitory autoimmune processes

These individuals conditions such as anaemia, thrombocytopenia or nephritis that follow certain infections and drug therapy. The infecting agents or drug induce antigenic alteration in some self antigen. The immune response set up causes tissue damage. The disease is transient and undergoes spontaneous cure when the infection is controlled or the drug withdrawn.

QUESTIONS

1. Define autoimmune diseases.
2. Write briefly on the spectrum of auto immune diseases.
3. Write various autoantibodies in human diseases.
4. What are the various environmental factors contribute the autoimmune disease.
5. Discuss whether autoimmunity is driven by antigen.
6. Explain the systemic autoimmunity.
7. Explain the following terms;
   (*i*) Autoantibodies;
   (*ii*) Autoimmune diseases;

(*iii*) Organ specific diseases;
(*iv*) Hashimato's diseases;
(*v*) Thyroid autoimmunity;
(*vi*) Multifactorial control;
(*vii*) Natural antibodies;
(*viii*) Idiotype disease;
(*ix*) Piggy-back T cell epitopes;
(*x*) Polyclonal activation;
(*xi*) Systemic Lupus Erythromatosus (SLE).

8. Explain the T cell mediated hypersensitivity as a pathogenic factor.
9. Write the diagnostic value of auto antibody test.
10. Enumerate various treatment of autoimmune disorders.

(*i*) Explain what is known as autoimmune diseases.
(*ii*) Write the genetic factors in autoimmune diseases.
(*iii*) Explain whether autoreactivity comes naturally
(*iv*) Explain whether autoimmunity is driven by antigen.
(*v*) Write the classification of autoimmune diseases
(*vi*) Write short notes on.

(*a*) Hashimotos thyroiditis;
(*b*) Pernicious anaemia;
(*c*) Good Pasteur syndrome;
(*d*) Graves disease;
(*e*) Multiple sclerosis.

CHAPTER 11

TRANSPLANTATION

Transplantation is the act of transferring cells, tissue or organs from one site to another.The degree of immune responds to a graft varies with type of graft. The following terms are used to denote different types of transplant.

11.1. AUTO GRAFT

The self tissue is transferred from one body site to another place in the same individual. These graft are often carried out to the patients with burns by transferring healthy skin to a burnt area.

11.2. ISO GRAFT

Isograft is known as if a tissue is transferred between genetically identical individual. In hundred strains of mice, an isograft can be performed between genetically identical tissue.

11.3. ALLO GRAFT

It is carried out in tissues which is transferred between genetically different members of serve species.

11.4. XENO GRAFT

Here the tissue is transferred between different species. Eg. The graft of baboon heart into a human.

Auto and isografts are usually accepted but the autograft which is genetically dissimilar to host is often recognised as foreign graft is rejected.

11.5. TRANSPLANTATION Ag

Tissues that are antigenically similar are said to be histocombatible, such tissue do not induce an immunological response that leads to tissue rejection.Tissue displaying significant antigenic differences are history on clopatriable, such tissue induce an immune response leading to tissue rejection

The various Ag's that determine histocompatibility are encoded by none than 40 different loci but the loci responsible for the most vigorous allograft reaction are located within major histocompatibility complex (complex of gene encoding cell surface molecules that are required for Ag presentation to T-cell and for rapid graft reaction. It is called the H-2 complex in the mouse and HLA complex in humans.

MHC molecules are proteins encoded by the MCH compared and classified as Class I, Class II and Class III MHC molecules.

Minor histocompatibility loci are the genes outside of the MCH that encode antigen contributing to graft rejection.

11.6. MECHANISM INVOLVED IN GRAFT REJECTION

Graft rejection is caused principally a cell medicated immune response to alloantigen (Primary MHC molecules) expressed on cells of the graft. Both day type hypersensitive and cell medicated cytotoxity reaction have been implicated. The process of graft rejection can be devided into two stages.

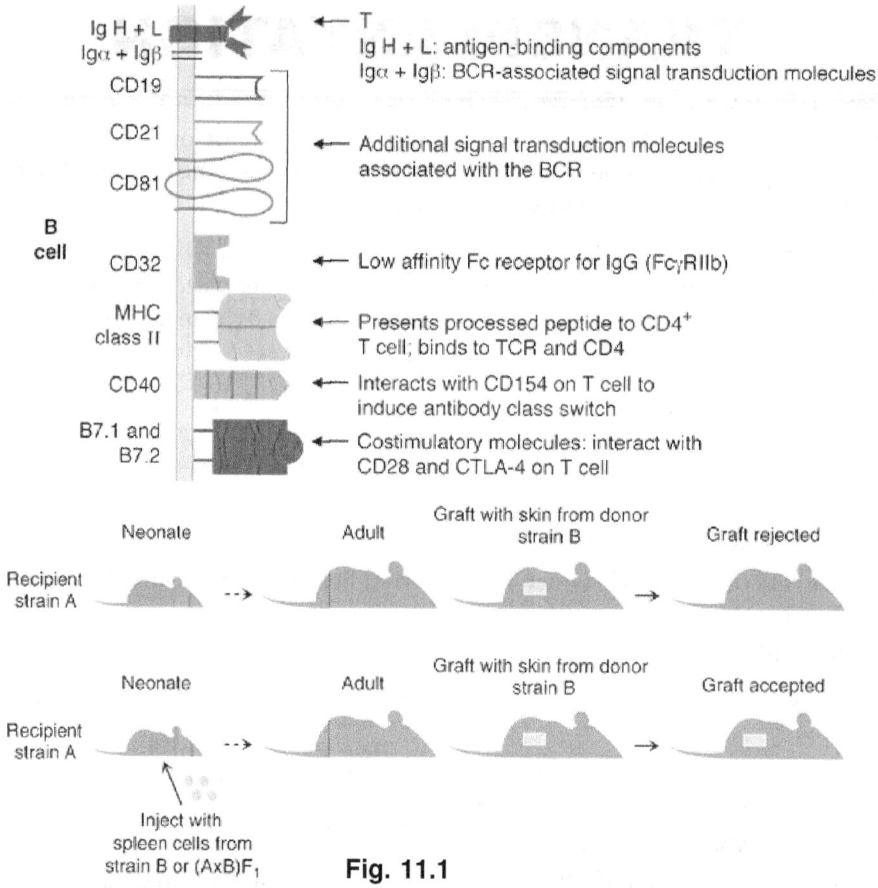

Fig. 11.1

A sensitive Place : in which antigen reactive lymphocytes of recipient proliferates in response to alloantigens on the graft. Effect on phase or stage in which immune destruction of the graft takes place.

11.7. SENSITIZATION STAGE

During this phase CD4+ and CD8 + cells recognise allo Ag expressed on cells of the foreign graft and proliferate in response. Both major and minor histocompatibility allo Ag can be recognised. In general the response to minor histocompatibility Ag is weak. Although the combined response to several minor difference can sometimes be quite vigorous, the response to major is involves recognition of both NHe molecule and an associated peptide ligand in the left in the MHC molecule.

The peptides present on groove of allogenic class I MHC molecules are derived from proteins synthesised within the allogenic cell. Peptides present in the groove of allogenic the peptides class II MHC molecule are generally proteins take up and precessed. Through the endocytic pathway of the allogenic Ag presenting cells. In some cases peptide fragments

Transplant rejection

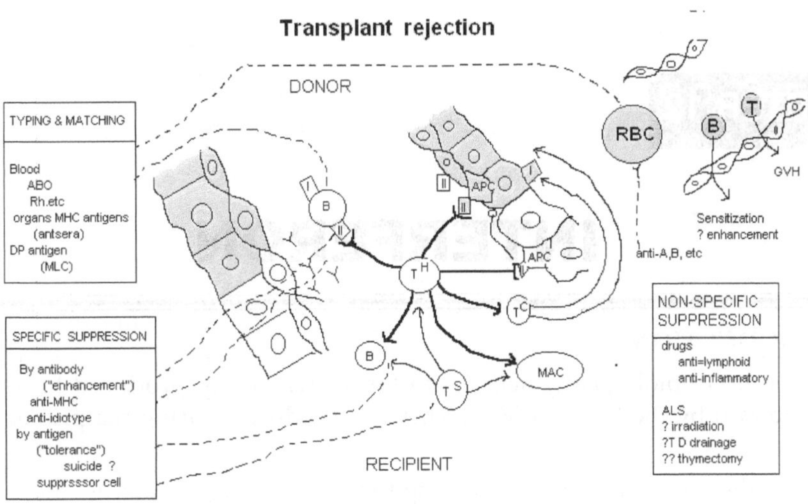

Fig. 11.2

of allogenic class I MHC molecules can be presented within the groove of a class II MHC molecule.

Activation of host TH cells required the initiation with an antigen presenting cells (APC) expressing an appropriate antigenic ligand MHC molecule complex and providing the requesite care – stimulatory signal.

QUESTIONS

1. Discuss whether graft rejection is immunological.
2. Explain the mechanisms of graft rejection.
3. Write the prevention of graft rejection.
4. Explain whether foetus as an allograft.
5. Write short notes on:
   (*i*) Autograft;
   (*ii*) Isograft;
   (*iii*) Allograft;
   (*iv*) Xenograft;
   (*v*) Cell mediated lympholysis;
   (*vi*) Graft Vs Host reaction (GVH).
6. Explain the role of humoral antibody.
7. Explain various immunosuppressive drugs.
8. What are the clinical experiences in grafting.
   (*a*) Methods of Tissue typing;
   (*b*) Kidney grafts;
   (*c*) Heart transplants;
   (*d*) Liver transplants;
   (*e*) Bone marrow grafting.

CHAPTER 12

INTERFERON

12.1. INTRODUCTION

Interferon are biologically active substances which are produced by the body of animals and also by the body of Man. They are produced during the primary infection with virus.

Though they are produced during the first infection by the virus, they protect the organism from the virus.

Thus, interferon protect organism from the second infection by the virus as infected previously.

Interferons are produced not only against living viral particles, but also against beat killed particles and unactivated virus particles.

Once interferous are produced in the body of an animal, the animal maintains the immunity against virus for protecting itself during second attack. Interferons are also produced against common cold, hepatitis and herpes virus.

Interferon also inhibit the rapid multiplication of some abnormal cells. Thus interferons possess an anti tumourogenic effect for controlling the multiplication of cells.

Interferons are proteins. They are sensitive to typisin which destroys the activity of interferon. The molecular weight is about 25,000 daltons. Each interferon has its own specificity.

12.2. INTERFERONS

The best established anti microbial cytokines are interferons.

In 1957, interferons was discovered by Alec Isaac and Jean Lindenmann

12.3. DEFINITION

Viral interference is a process in which the presence of one virus, in an infected cell interferes with replication of the other. This viral interference are 'Interferons'. In 1980, Interferons were successfully produced from recombinant DNA technique.

12.4. ISAAC AND LINDEMANN EXPERIMENT

Isaac and Lindenmann (1957) while they were investigating viral interference, they found that cells infected with influenza virus secreted a soluble protein into the surrounding fluid that protected the cells from infection with the same or other viruses.

12.5. CHARACTERISTICS OF INTERFERONS

Interferons are naturally occurring non-immune, non-antigenic, soluble, diffusible, cellular glycoproteins, with a broad spectrum antiviral action. They are produced by a living cell infected with viruses, and block further viral infections of the cell by inhibiting viral replication. They are more or less cell-specific, and species – specific, but not virus specific.

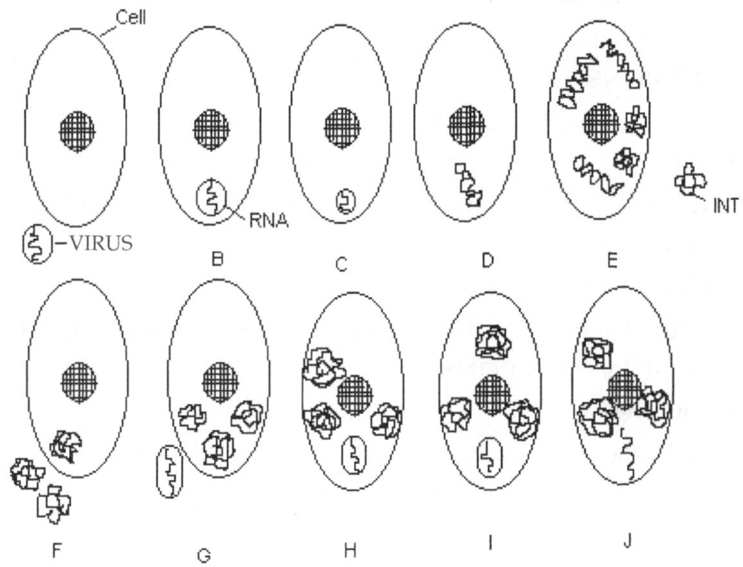

12.6. TYPES OF INTERFERON

According to origin of cells, they are typed into three major group α IFN, β IFN, γ IFN.

	α IFN	β IFN	γ IFN
Alternative name	Leucocyte IFN	Fibroblast IFN	Immune IFN
Sub types	14 (α1- a14)	2 (β1-β2)	—
Principle source	all cells	all cells	Lymphocyte
Inducing agents	Viral infection (ds DNA)	Viral infection	Antigen (mitogen)
Chromosomal Location of genes	9	9	12
Antiviral activity	+++	+++	+
No activation	—	—	+
MHC regulation	—	—	+

12.7. ALPHA INTERFERONS

- It is found is Leucocytes.
- More than 14 sub-types.
- Induced by sendai virus.
- Molecular weight 15,000 to 20,000.
- Not glycosylated.

12.8. β- INTERFERONS

- Derived from fibroblast.
- INF-b genes are located in the chromosomes
- This genes have no introns
- Induced by ds RNAs.

- Molecular weight 20,000.
- Glycosylated.

12.9. γ – INTERFERONS

- Derived from blood buffy coat (or) spleen.
- Induced by lectins, mitogens and some antigens.
- This INF differs from other two by its antigenecity, acid liability, Host range and subunit structure.
- Molecular weight 17,000.
- Natural and- g INF have high antiviral activity than g-g INF.

12.10. CLASSES OF INTERFERONS

12.10.1. Alpha Interferon

They represent the major antigenic species in human leucocyte cuctivars. More than 14 sub types are there. They are induced by *sendai virus* and other *paramyxo viruses*. The molecular weight of the different species varies from 15,000-20,000. They are not glycosylated.

12.10.2. Beta Interferon

It is derived from immunologically stimulated fibroblast. INF–b genes are located in three different chromosomes (2,5,9). This genes have no introns. The INF has been induced with poly (Ic) or other ds RNA's. Its cloning and expression have been carried out in bacteria. It is a glycosylated protein with a molecular weight of about 20,000.

12.10.3. Gamma Interferon

Its cell sources are blood buffly coat or spleen. It is induced by lectins, mitogens and some antigens. INF-g differs from INF –a INF-b by its antigenicity, acid lability, host range and possible subunit structure. Molecular weight 17,000. Natural Interferon – g having high antiviral activity when compare with recombinant–g interferons.

12.11. INTERFERON AS HORMONES

Interferons can be regarded as hormones, resembling polypeptide hormones or growth factors, Eg- insulin, epidermal growth factor, prolactin or vasopressin in some respects. The similarities are follows.

(*i*) IF N with specific receptors in the cell membrane.

(*ii*) Effect on cell wall proliferation of a wide variety of cell types.

(*iii*) Effect on a range of cell functions. Growth factors

(*i*) They are glycoproteins.

(*ii*) Inhibit multiplication of viruses.

Virulent infected cell	IFN binds to Uninfected cell Cell remains uninfected,	IFN induces synthesis of Antiviral molecules Antiviral State develops in few hours, lasts For a day

IFN induces synthesis of

↓

2′ 5′ oligo adenylate synthetase → Adenine trinucleotide Synthetase → Activates endonuclease → Aeglades viral mRNA

Protein kinases → Phosphorylation and inactivation of elongation factor-2(ELF-2) → inhibits protein synthesis

12.12. INDUCTION OF INTERFERON

Many substances induce the formation of interferon. Two main types of inducers involved in INF production. They are **type I** production and **type II** production.

12.12.1.Type-I Production

Which follows stimulation with specific intron inducers. This production is brought about by two types of inducers,

Class-A inducers (ds RNA, good inducers)

Class–B inducers (Non-Nucleic acid inducers, Moderate inducers)

12.12.1.1. Class- A Inducers

They frequently produce more than 1,000 units/ml of interferon in cell cultures or the blood stream of animals after injection. The different types of Class-A inducers are as follows: RNA animal viruses, DNA animal viruses. Plant viruses, insect viruses, fungal viruses.

12.12.1.2. Class-B Inducers

They have poor inducing ability (less than 1,000 units). They are frequently produced in virus injected animals and not in cell cultures. Class-B inducers consists of a heterogenous group of macromolecular substances. They are Intracellular microorganisms, Bacterial products (exo and Endo) polymers.

12.12.2. Type-II Production

Is by specific and non-specific immunologic activation of lymphocytes and macrophages. It is of three types.

(1) Immune–specific induction.

(2) Mitogen stimulation of lymphocytes.

(3) 'spontaneous' or autogenous production.

I-S Induction is the production of INF by sensitized lymphocytes on exposure to specific antigens.

Mitogens : Include phytohaemagglutinin, concanavalin, pokeweedmitogen and bacterial lipopolysaccarides.

12.13. BIOLOGICAL PRODUCTION OF INTERFERON

1. It is produced by separating WBC's from human blood and exposing them to sendai viruses.
2. The viral capsid stimulates a human cell chromosome (presumably numbers) to produce INF.

3. That produced INF can't affect produced cell.
4. Those INF are absorbed by adjacent cells, where they induce chromosome number 21 to produce an unknown protein that binds to mRNA molecules specified by the viruses.
5. INF is finally released in the culture medium and is purified using monoclonal antibodies.

INF a, INF b are produced rapidly within 24 hrs of insertion and constitute a major part of the early response to viruses. IFN a is mainly a T cell product and is therefore produced later, although in some cases an early

- IFN g response may be mounted by NK cells. Interferons can also inhibit virus assembly at a later stage. (Eg, Retro virus)
- IFN is active against ricketsia, Mycobacteria and several protozoa.
- IFN normally inhibits without damaging host cell. Eg. Leishmania can survive MHC expression with the help of INF a.

12.14. PRODUCTION OF INTERFERONS BY USING RDNA TECHNOLOGY

DNA technology used to prepare large level of interferons.

Cloned Interferon's genes are being expressed in large amounts in bacteria and yeast.

12.14.1.Step 1: cDNA Preparation

1. Size-fractionated mRNA was isolated from human leukocytes, reverse transcribed and inserted into the Pst I site of plasmid pBR 322.

2. The approximately 6,000 clones that were produced following transformation of *E.coli*.

3. Crude interferon mRNA were prepared. They are protein synthesis system.

4. DNA – mRNA hybrids were transferred to cell free Protein synthesis system.

5. This translation mixture assayed for IFN antiviral activity.

6. Sub grouping process was repeated until a clone the complete cDNA for a human. IFN identified.

12.14.2. Step 2 Large scale production

Purified human IFN cDNA was inserted into *E.coli* expression vector. These vectors transformed to *E.coli* or yeast cells and produced in large amount.

12.15. DISORDERS OF HUMAN INTERFERON

	Interferon	Disorders
1.	α2 a	HCV, hairy cell leukemia.
2.	α 2 b	Bladder cancer, Head and neck cancer, malignant melonama, multiple myeloma, non-Hodgkins lymphoma, renal cell carcinoma and HIV infection.
3.	α n3	AIDS, cervical dysplasia, papilloma virus infection and genital warts.
4.	β1a	Multiple sclerosis.
5.	β1b	Chronic progressive multiple sclerosis.
6.	γ 1b	Renal cell carcinoma chronic granulomatous disease.

Some recombinant proteins that have been approved for human use by the U.S Food and Drug Administration

Compound	Company	Disorder
1. IFN – α 2a	Hoffman-La Roche	Hairy cell leukemia, genital warts.
2. IFN – α 2b	Schering – plough	Hairy cell leukemia, kaposi's Sarcoma hepatitis B and C.
3. IFN - α n3	Interferon Science	Genital Warts.
4. IFN – β 16	Berlax laboratories And clinron.	Relapsing multiple sclerosis.
5. IFN – γ 1b	Gene tech	Chonic granulomatous disease.

QUESTIONS

1. Define Interferons.
2. Types of interferons and their characteristics.
3. Define alpha, beta and gamma interferons.
4. Production of Interferon by using different types of inducers.
5. Production of interferon by using rDNA technology.
6. Biological functions of Interferon.

CHAPTER 13

CYTOKINES

The development of an effective immune response involves lymphoid cells in inflammatory cells and haemopoietic cells. The complex interaction among these cells are mediated by a group by several low molecular weight proteins that are collectively designated as cytokines, to denote their molecule in cell to cell communication.

Cytokines consist the regulation of development of immune effectors cells and some cytokines posses direct effector functions of their own. So cytokines serve as messengers of immune system. Unlike endocrine hormones which exert the effects over long distances, cytokines act locally.

13.1. PROPERTIES OF CYTOKINES

They are a group of low molecular weight regulatory proteins secreted by WBC and a variety of other cells in the body is response to a number of inducing stimulate.

Cytokines binds to specific developers on the membrane of largest cell triggering signal transduction, apart that ultimately alter expression in the target cells. The nature of largest cell for a particular cytokines is predominate in the preserve of specific membrane receptors.

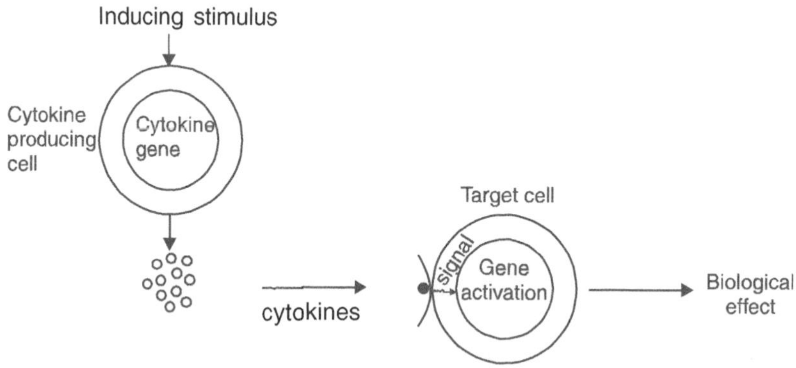

Fig. 13.1

A particular cytokine may bind to receptors on the membrane of the same cells that secreted by exerting autocrat action. It may bind to receptor on a largest cell in close proximity to the product or cell, exerting peregrine action, in a few lakhs it may bind to target cell in distant parts of the body exerting into peregrine action. Cytokines regulate the intensity and duration of the immune response by stimulating or inhibiting the proliferation and or differentiation of various cells and by regulating the secretion of antibodies of other cytokines.

Binding of a given cytokines to responsive taught cells generally stimulates the expression of cytokines receptors as well as of other cytokines which inturn affect other target cells.

Thus, cytokines secreted by a single lymphocytes following Ag specific activation can influence the act with various cells involved in the immune response for eg. cytokines produced by activated T- cells can influence the activity of B cells. T-cells are natural killer cells marveling granulocytes and haematopoietic cells activating an entire network of interacting cells.

Cytokines exhibit the attributes of plerotrophy residency synergy actively in a coordinated interactive way. A given cytokines that has different biological effects on different target cells has a plerotropic action.

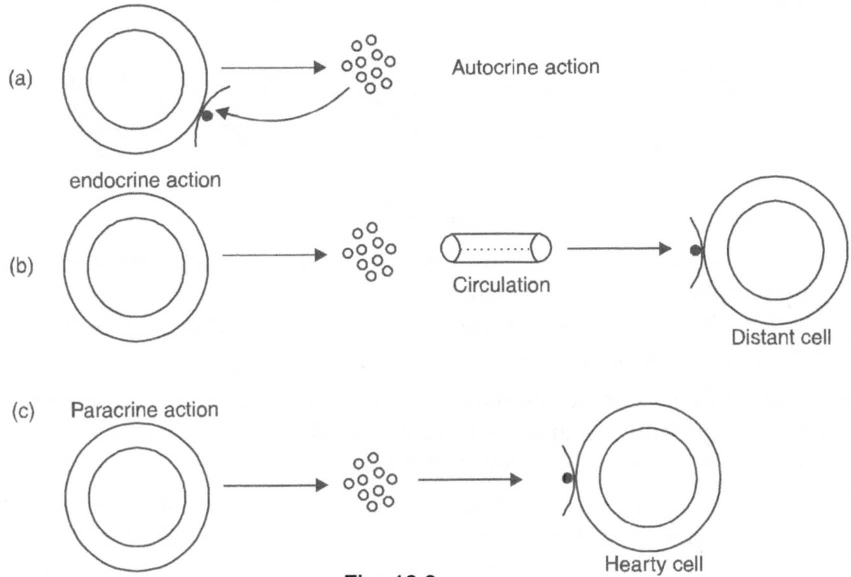

Fig. 13.2

Two or more cytokines that mediate similar function are said to be reductant. This property makes it difficult to ascribe a particular way to signal cytokines.

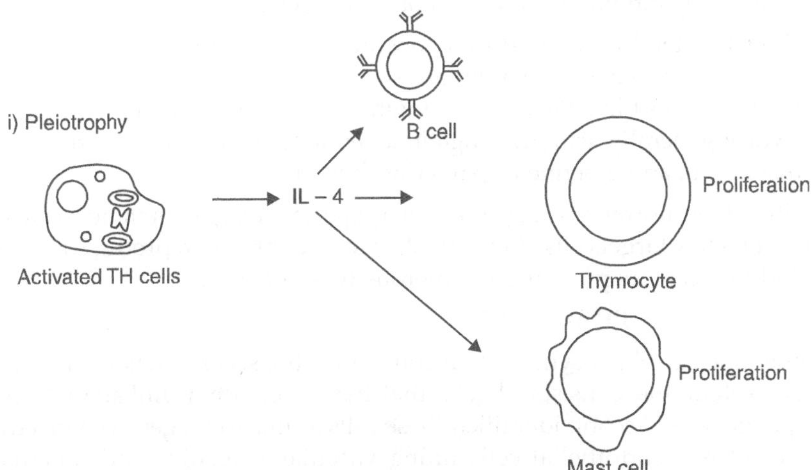

Fig. 13.2

Cytokines synergism occurs when the B cell combined effect of two cytokines on cellular activity is greater than the additive effect of the individual's cytokines.

In some cases cytokines excitant antagonism that is the effect of one cytokine inhibit or affect the effects of another cytokines.

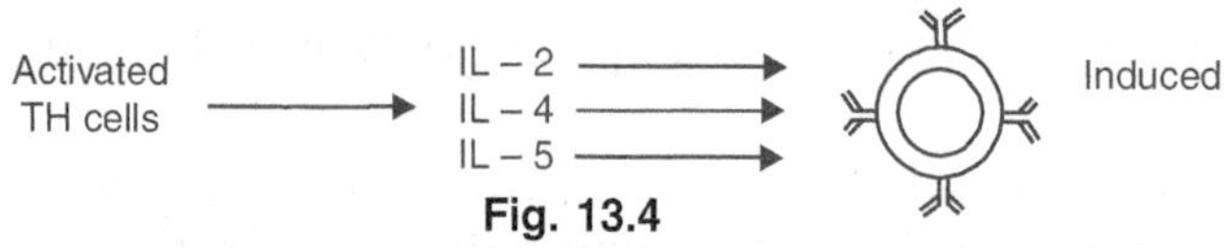

Fig. 13.4

The term cytokines encomposses those cytokines secreted by lymphocytes formerly known as lymphokines and those cytokines secreted by monocytes and macrophages known as monklines.

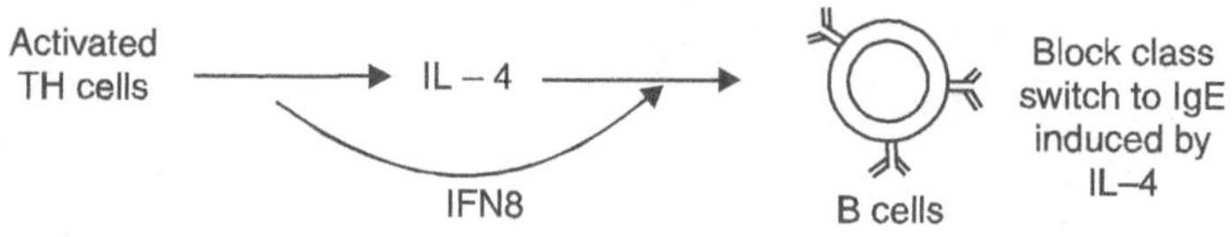

Fig. 13.5

Although these two terms contain to be used in the literature they are misleading because many lymphokines and monokines are secreted by a broad spectrum of cells and not supply by lymphocytes and monocytes as their name would imply. For this reason the more inclusive term cytokine is preferred.

Many of the cytokines are referred to as interleukins a name in denoting that they are referred to as inter cytokines a name indicating that they are secreted by some leukocytes and act upon other leucocytes. Presently interleukins 1-17 have been identified.

Other cytokines are known by common names. These include the IFNs and INF s. A group of low molecular weight cytokines, including IL 8 is classified in the chemokine family. These molecules play an important role in the inflammatory response.

Because cytokines share many properties with hormones and growth factors, the distinction between these three classes of mediators is often blurred. All these classes of mediators are secreted by soluble factors that elicit their biological effects at pico molar pico concentration by binding to receptors on target cells.

Growth factors tend to be produced constitutively, whereas cytokines production is carefully regulated. Cytokines generally are secreted following activation of a particular cell and secretion is short lived generally ranging from a few hours to a few days. Unlike hormones, which generally act long range in an endocrine fashion. Mostly cytokines, act as a short distance in an autocrine as paracrine fashion.

In addition most hormones are produced by specialized glands stand to have a unique action on one or a few target cells. In contrast, cytokines are often produced by a variety of cells and bind to receptors present on numerous types of cells.

13.2. MONONUCLEAR PHAGOCYTES

The mononuclear phagocytic system constitutes the second major cell population of the immune system and consists of cells that have a common linkage whose primary function is phagocytes. Aschoff identified these cells as macrophages in connective tissues, mitoganglia in CNS, endothelial cells lining vascular sinusoids and reticular cells of lymphoid organs. He suggested that these varied cell types fractioned in host defence by phagocytes of foreign invaders such as microbes and grouped them collectively into reticulo endothelial system. It is now clear that the pinocytosis of which the endothelial and reticular cells are capable and fundamentally different from the active phagocytes is of macrophages.

It is more appropriate of macrophages. It is appropriate to classify monocytes and macrophages as members of the mononuclear phagocytes system.

13.2.1. Development

All the cells of mononuclear phagocytic system originate in the bone marrow and after maturation and subsequent activation can achieve varied monophlogic forms.

The first cell type that enters the peripheral blood after leaching the marrow is completely differentiated and is called the monocyte. Monocyte is 10–20 Mg in diameter and they have been shaped nuclei and finally granular cytoplasm containing lysosome phagocytic vacuoles and cytsheleated filaments. Once they settle in tissues, these cells mature become macrophages, which are also called histocytes.

Monoplanes can be activated by a variety of stimuli and may assume different forms. Some develop abundant cytoplasm and are called epithelial cells because of their resemblance to epithelia cells of the skin. Macrophages can fused to form polykaryons also termed multinucleate giant cells.

13.2.2. Avidity

Some use this term to desirable the increased affinity that occurs as a result of multiple Ag binding sites on an Ig. Others use it to desirable the strength of binding of serum with a complete antigen, while still others use it to indicate the rate of reaction between Ag and Ab. The term is probably best avoided.

Some of the biological consequences of the continuation of antigen and antibody on the surface of the cell.

13.2.3. Phyto haemeagglutination (PHA)

A lectin derived from the red kidney bean and it acts as a 7 cell mitogen.

13.2.4. Advantages of inactivated vaccines

Produce humoral immunity.

May be incorporated into other vaccines.

will not react to virulent form.

May be used in immune deficient individuals.

May be used in tropical countries.

will not spread.

13.2.5. Advantages of live vaccines

Produces humoral and intestinal immunity.

Offers required onset of protection.

Easy to administer

Blocks interfinal spread.

Cheaper than inactivated

- More people get the full course since few boosters are required.

QUESTIONS

1. Properties of cytokines.
2. Explain mononuclear phagocytes and their development.
3. Define avidity.
4. Define phytohaemoagglutination.
5. Difference between Live vaccines and inactivated vaccines.
6. Discuss about attenuated and inactivated Vaccine give two eg for each.
7. What are genetically engineered vaccines, describe their preparation and advantages ?
8. Structure of Ig and elucidation.

CHAPTER 14

TISSUE TYPING

Testing for histocompatability between an organ donor and the selected recipient is based on the presumption that tissue compatibility will promote graft acceptance and avoid immuno rejection.

Comparison is the tissue typing of a donor and the recipient usually some degree of Ag mismatching because of the extensive polymorphism of HLA Ag System.

The basic approach as termed " HLA typing". It is carried out at the.

Phenotypic level (i.e) What types of protein are expressed on the cell surface

Genotypic level (i.e) What DNA sequence is present in the genes.

Generally HLA typing was carries out almost exclusively at the "phenotypic level" using Ab's to different HLA molecule. Since it is based on the use of serum derived Abs, the approach is known as "serological typing". This serology remains the main technique for class I HLA molecules and although used DNA based approaches.

Class II HLA typing, serological technique are not as applicable class II serological typing was to analyze the alleles at the "genotypic Level".

14.1. TISSUE DISTRIBUTION OF HLA MOLECULES

Class I molecules are expressed on the surface of virtually all cells, except mature erythrocytes and thromboblast cells.

Class II molecules are only constitutively expressed on the surface of a small number of cell types, including macrophages, monocytes, follicular dendritic cells and b lymphocytes. On some cells, such as macrophages only DR molecules are expressed, where as on other, such as is lymphocytes are the three subclasses are expressed.

14.2. STRUCTURE OF HLA MOLECULES

The class I molecules HLA-A, HLA-B, HLA-C are formed from polymorphic heavy chain glycoproteins that binds non covalently to b2 microglobulin. The heavy or a chain of class I molecules carriers three distinct extra cellular regions or domains formed by disulfide binding. The polymorphic and membrane anchoring heavy chain contains both peptide binding domains (a1 and a2) and are Ig domain (a3).

The class II molecules HLA-DP, HLA-DQ, HLA–DR are formed from 2 glycoproteins, a and b chain class II molecules are heterodimer consisting of a non –covalently linked peptides (a and b).

14.3. HLA POLYMORPHISM

Polymorphism is the phenomenon whereby numerous different alleles (allotypes) can occur at a single locus. Gene polymorphism in the HLA region is used in the HLA typing.

Polymorphism of HLA genes has 2 important consequences for the individual, unselected individual have a diverse susceptibility to disease and also prompting reject organ transplants between each other.

14.4. MICROCYTOTOXICITY

In this test WBC from the potential donor and recipient are distributed into a separate series of wells on a microtitre plate and then monoclonal Abs specific for various class I and class II MHC alleles are added to different wells.

After incubation, complement is added to the wells, and cytotoxicity assessed by the uptake or exclusion of various dyes (trytan blue or eosin Y) by the cells.

If the WBC express the MHC allele for which a particular monoclonal Ab is specific, then the cells will be lysed on the complement and these dead cells will take up by a dye such as tryptan blue.

HLA typing based on Ab mediated microtoxicity can indicate the presence of absence of various MHC alleles.

14.5. LYMPHOCYTOTOXICITY

When a fully HLA–compatible donor is not available transplantation may be successful. In this condition 'mixed lymphocyte reaction' can be used to assess the degree of class II MHC compatibility between potential donor and a recipient.

Lymphocytes from a potential donor, that have been X–ray irradiated or treated with mitomycin which serves as a stimulator cells, and lymphocytes from the recipient serves as a responder cells.

14.5.1. Method

Individual HLA Ag antisera are predisposed in ml quantities into the micro test wells of plastic trays composed of 60 or 72 wells of 15 ml capacity. 2000 isolated lymphocytes are dispersed per well, and the tray is incubated for 30 min at room temperature allow anti HLA antibodies to bind their specific target HLA Antigens.

Complement is added, and the tray is incubated for 60 minutes. To visualize the dead and live cells under microscopy, a vital dry eosin Y is added. Live cells are exclude the dye and appear bright and refractile, while dead cells takes up the dye and are swollen and dark.

% Dead cell lymphocytes	Interpretation
0-10	- ve
11-20	Doubtful +ve
21-50	Weak +ve
51-80	+ ve
81-100	Strong +ve.

14.5.1.1. Mixed lymphocyte reaction

Mixed lymphocyte reaction can be used to assess quantitatively the degree of class II mixed histocompatiblity between potential donors and a recipient.

Lymphocytes from a potential donors, that have been X–ray irradiated or treated with mitomycin – C, which serves as stimulator cells and lymphocytes from the recipient serves as a responders cells.

Proliferation is indicated by the uptake of [^{3}H] thymidine. The greater difference between the donor and recipient cells the more [H^3] thymidine uptake will be observed in MLR assay.

14.5.1.2. Advantage

MLR over microtoxicity typing is that it gives a better indication of the degree of TH cell activation generated in response to the MHC II Ag of the potential graft.

14.5.1.3. Disadvantages

MLR takes about 66 days to run assay. If the potential donor is a cadaver eg., it is not possible to wait 6 days for the results of the MLR.

14.5.2. RFLP method (Genotyping)

RFLP is currently used by the basic serologically defined class II HLA antigen. The technique of RFLP typing is based on the knowledge of the germ line nucleotide sequence. The resulting band patterns are diagnostics for most of the genetic Class II antigens. Locus specific probes have been designed for the detection of the products of the different "subloci of DR, DQ, DP" Eg: DR, DR4, DR5.

14.5.2.1. Method

The isolated DNA sequence is digested with a particular restriction endonuclease eg. The DNA of the HLA haplotype will be cut into fragments of different lengths depending on the location of restriction side out. Subjecting the fragments to gel electrophoresis by their size.

These different sized fragments are visualized by probing with cDNA's that hybridizes with the consensus sequences. This digested enzyme fragments transfer to the nitrocellulose membrane, by blotting of the gel, and hybridization by labeled probes is known as southern blotting.

14.5.2.2. Uses and limitations

RFLP typing cannot identify the many alleles. Its main use has been to confirm and classify serologically assigned DR and DR types and to identify genetic DP alleles.

The main advantages which is time consuming technique, usually requiring from 2-3 week and at least of 5-10 mg of DNA.

14.5.3. PCR method

PCR allows complication of specific targeted region of DNA by providing multiple copies of selected DNA sequences PCR provides a useful amount of selected portions of HLA genes. This method is now applied only in VNTR (Variable number tandem repeats) probing.

14.5.3.1. Method

PCR is an automated method for amplification of a specific known DNA sequence. Two oligonucleotide primers are synthesized that are complementary to the 2 flanking regions of the DNA segments to be amplified.

The primers will hybridizes to the flanking regions of the genes and 'prime' the replication of the gene by 'thermostable polymerases' PCR involves the repetition of 3 steps.

1. Denaturation of the sample DNA to single strand.
2. Annealing of the sequence- specific oligonucleotide primers to the boundaries of the segment.
3. Extension of the primers by the polymerase to form new ds DNA across the segments.

It is a chain reaction since reaction polls becomes the template for the next cycle of the process.

This reaction requires template DNA, a mixture of nucleotides, the 2 PCR primers. Tag polymerase and buffer.

There are 2 types of PCR can be used to amplification of the DNA sequences of both donor and recipient.

1. Genetic or group specific amplification.
2. Allele specific amplification.

14.5.4. SSOP (Sequence specific oligonucleotide probe)

The amplified portion of HLA genes obtained from PCR can be further tested to reveal specific alleles within the amplified segments such as SSOP method.

Oligonucleotide probes are short synthetic segments of single stranded DNA (18-24 nucleotide long) complementary to known polymorphic sequences in HLA alleles. These probes hybridized only to their exactly complementary sequences under stringent hybridization conditions. Even a single-base difference cause the probes to break away from the pairing with the sample DNA. This property gives this method of class II HLA typing.

For this techniques multiple probes are required to identify an HLA allele because of the sharing of sequence polymorphism primer pairs are selected that amplify the DNA in the region which contain sequences present in the group of Antigens of interest.

Dots of amplified DNA plot are allow to adhere to a membrane in a "slot blot" The blot are then cut in to strips, are rearranged and the patterns of hybridization signal is compared with the pattern of known alleles.

14.5.5.Attenuated DNA sequencing

In this method PCR plot derived from polymorphism region of HLA class I and class II genes are directly sequence to determine HLA specificity.

Typing strategy involves preliminary PCR step in which a sample of genomic DNA is amplified with primers selected to produce a group of locus specific product.

Primers pairs are designed to amplify shorter regions within the first PCR product, so these smaller segments can be sequence in both 3′ – 5′ and 5′ – 3′ directions This directional sequence provides confirmation of accuracy of sequences. Separate sequences reaction tubes are prepared in which one of the bases A T G C is labeled with an unique colored fluorescent dye.

Generated DNA fragment are then pooled and subjected to gel electrophoresis to produce a ladder chromatogram. Ladders are anatomically scanned for position and color by laser beam and an electrophorogram is produced showing position *vs* relative intensity of fluorescence.

The intensities of peaks are examined with particulate attention to position of known polymorphism.

The nucleotide sequences are then compared with library of HLA sequences and antigens are assigned based on sequences identity.

14.6. IMMUNOSUPPRESSIVE THERAPY

Growth of clinical immunology has uncovered covering number of diseases that are due to abberant immune response. This as result in a search for drugs and capable of inhibiting these unwanted responses. Particularly organ transplant may created a strong impetus for the development of safe and effective immune suppressive reaction.

14.7. CYTOTOXIC DRUGS

It is a group of chemicals with pharmcologic property of killing cells capable of self-replication. These drugs are for anticancer therapy, it may possess immuno suppressive activities.

This drugs may extended to treatment of disease caused by abberant immune response and to inhibition of transplant reaction.

Four days are required for the action of these drugs currently in general clinical use for immunosuppression. Cyctophosphasmide azothioprine, methotrexate and chloranbusil other drugs like corticosteroids, cyclosporine, tacrolinus, papamygum, total lymphoid irradiation.

14.8. SPECIFIC IMMUNO SUPPRESSIVE TEST

Monoclonal antibodies are assuming fine clinical importance in treatment of organ graft rejection reactions. Because of their potency trials have been initiated to assess their activities are prophylactically inhibiting organ rejection and therapeutic agents for immunologically related diseases. These antibodies believed to immuno suppresant by virtue of their ability to bind to inactive cell surface receptors on T lymphocytes or APCS. An important T cell interaction is that of adhesion molecules. Leucocyte function antigen (LFA-1) on lymphocyte with its ligand intercellular adhesion molecule-1 ICAM-1 or CD54 on APCs. Mc Abs directed at ICAM –1 have been successfully used in initial clinical trials to inhibit transplant rejection reactions.

14.9. GENETIC BARRIER TO TRANSPLANTATION

It is termed as genetic disparity between the donor and recipient.

Autograft : From one part of the body to another part eg., Trunk to arm.

Isograft : Between genetically identified individual eg., monographic twins or within inbred strain.

14.9.1. Auto graft

Between different members of same species eg, man to man.

14.9.2. Xeno graft

Between members of different species eg: monkey to man.

14.9.3. Rejection or Incompatible

Cell mediated immunity is more effective than, other because HLA system is recognize the T cell receptors. There are 2 sets rejection.

1st set of rejection - cell mediated immunity acts as Ag and it is rejected.

2nd set of rejection - Role of memory cells which is developed.

Rejection is of 3 types:-

1. Hyperacute. It will take minus of hours to cause preformed anti-donor Abs and complement.

2. Acute Days to week and to cause primary activation of T cells.

3. Chronic Months to years to cause unclear Abs immune complexes, slow cellular reactions, recurrence of disease.

14.10. CROSS REACTIVITY

Identical or similar epitopes are sometimes encountered in unselected molecules. This is known as " **Cross-reaction**". These are 2 types.

1. Cross-reaction between totally unrelated molecules.
2. Cross- reaction between the structurally similar ties.

Eg. For Type 1

Many bacteria possess cell wall polysaccharide that are also found on mammalian RBC. For eg. some intestinal bacteria have blood group glycoproteins A and B on their cell walls. When these glycoproteins are absorbed through the intestinal wall into the blood stream, so the animals mounts an immune response against the foreign epitopes.

Thus the blood group Glycoprotein A is foreign to a person of blood group. B. This person produce anti-A antibodies in response to the bacterial Ag cross reaction of Abs this types called "heterophilic antibodies".

Eg., for type 2

This method is used to detect the relationships between animal species. Thus the antisera to human sera cross-react with chimpange and gorilla but react much more weakly with sera from other monkeys. This reflex the presence of shared epitopes on these proteins. The more closely related the proteins are the more epitopes they share. This is a useful tool in determining evolutionary relationships cross reactivity with micro organisms.

14.11. MOLECULAR MIMICRY

Molecular mimicry is a term used to describe the sharing of epitopes between an infectious agent or parasite and its host. In this case the immune response directed vice the normal self-Ags cause disease.

The shared epitopes should have to be sufficiently similar for tolerance to be broken. Once initiated the infections agent could be removed while the autoimmune damage continued a sort of "hit and run" process.

Eg. Persons infected with the protozoan parasite *"Trypanosoma cruzi"* the cause of chaga's disease". This organism possess antigens that cross-react with mammalian neurons and cardiac muscle. In infected individual these antigens provoke auto Ab formation and auto Abs provoke nervous disorders and heart disease.

Auto Abs to some microbial heat shock proteins seem to be associated with autoimmune arthritis in human and rats.

14.12. HEAT SHOCK PROTEINS

These are up to 25 new proteins are fold by stressed cells. These serves include heat, starvation exposure to O_2 radicals, toxins such as heavy metals, protein inhibitors and viral infections.

Heat shock proteins (HSP) are present in all organisms at very low levels at normal temperature. Mild heat treatment such as a low grade fever induces HSP in cells and rises their levels significantly.

There are 3 major heat shock proteins.

HSP-90

HSP-70

HSP-60

The gene for HSP-70 is located with in the MHC-III reaction in mammals which have been detected in patient with malaria, *Schistosomiasis filariasis*. HSP 60 is the immunodominant Ag by mycobacterium *Trypnosoma* and *Borrelia* infections. These HSP are highly antigenic for several reasons.

1. They are produced in abundance.

2. They are readily possessed by APCS.

3. These may be usually large numbers of lymphocytes capable of responding to HSP.

QUESTIONS

Define the following :

1. HLA polymorphism;
2. Microcyto toxicity;
3. Lymphocyto toxicity;
4. Cross reactivity;
5. Molecular mimicry;
6. Heat shock proteins.

CHAPTER 15

ADJUVANTS

A substances that, when gives with Ag, enhances the immune response of that antigen. Under some circumstances, as in vaccination, it may be desirable to enhance the normal immune response by administrating an adjuvants with an Ag. A large variety of compounds have been employed as adjuvant. Many adjuvants act by slowing the release of Ag into the body being Ag given response in the presence of Ag and terminates the response. Once Ag is eliminated, it is possible to flow the rate of Ag release into the tissue by first mixing it with an insoluble adjuvant injected into an animal, thus forms a four. Eg, of depount forming adjuvants include isolated aluminum salts such as Al_2 OH_3, Aluminum phosphate and $AlKSO_4$ (Alum). When an Ag is mixed with one of the salts and injected into the animal a macrophage rich granulosa forms in the tissues. The Ag within thus granulosa slowly release into the body and so provides a prolonged antigenic stimulies. Antigens that normally persist for only a few days may be retain in the body for several weeks by means of this peptide.

These depot adjuvants influence only the primary immune response and have little effect on 1° immune response and adjuvant with similar mode of action in Berillium sulfate which also forms local granulosa and so enhances the Ab formation. This adjuvants has no effect in thymectomized animal, it does not influence cell mediated immunity. Silica and carbon also promote antibody formation when given with Ag and probably act in a similar fashion.

An alternative method of forming a depot is to incorporate the Ag in a water in oil emulsion known as Freund's incomplete adjuvants. This oil stimulates a coal, chronic inflammatory response as a result a granulosa form around the site of innoculum. The antigen is slowly reached from the aqueous phase of the emulsion. Oil droplets may also be carried to other sites though the lymphatic system. If killed tuberculin bacilli (*Mycobacterium tuberculosis*) are incorporated into the water in oil emulsion. The mixture is known as Freund's complete adjuvants (FCA) is extremely potent not only does FCA from a depot but a tuberculi bacilli have a nurmide diputide (NDP) (N-actetamyl neuromyl–L

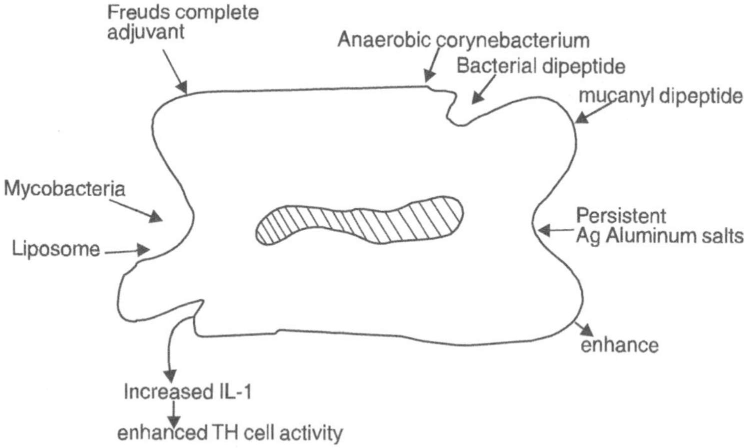

Fig. 15.1

alaine D-isoglutamine) NDP acts on macrophages to produce inter leubin which stimulates the cells of the immune system and so enhances the immunity.

Because IL-1 also causes side effects including fever, muscle wastage and depression, non biogenic derivative of NDP have now been synthesized. NDP can be covalently linked with synthetic Ag to create potent, synthetic chemically defined Ag's.

Freuds complete adjuvants works best when gives subcutaneously or intra- dermally and when the Ag doses relatively. It acts specifically to stimulate T-lympho- cyte function and therefore enhances responses.

Only to thymus dependent Ag FCA promotes Ig G production over Ig M. It inhibits tolerance induction, causes delayed hypersensitivity reaction, accelerate graft rejection and promote resistant of tumors. FCA is required to induce some experimental autoimmune diseases such as experimental allergic encephalities and thyroiditis. It also stimulates macrophages, promoting the phagocytosis and cytotoxic activity.

Other bacterial pelts besides MDP possess adjuvant activity. For eg. endotoxins enhances Ab formation if given about the same tissue as the Ag. They have no effect on delayed hypersensitivity, but they can break tolerance and they have a general immuno stimulatory activity, which is reflected in a non-specific resistance to bacterial infections. Endotoxins act by stimulating macrophage production IL-1. Similar is their effects are anaerobic cornybacteria, especially propionic bacterium acnes. They promote the release of IL-1 leading to B-lymphocyte and T-lymphocyte activation. As a result their general immuno stimulating action leads to enhanced antibacterial and anti tumor activity.

Bordetella pertusis, the cause of whooping cough, also has endotoxins like activity but it causes a rise in blood lymphocytes in some species and render redents highly susceptible to the toxic effects of histamine. It prolongs and enhances immunological memory and stimulate macrophage activity. B-pertusis may selectively enhance Ig E production.

Poly ribonucleotides consisting of double stranded nucleic acid such as the copolymer (or) polyinosinic acid and poly cytidilic acid.(poly IC) stimulate T-lymphocytes, probably by functioning as IFN inducers.

Certain surface activated agents, such as Sodium alginate, Lanolin, lysolecithin, vit-A, Saponin and phospholipid, liposomes act on cell membranes to enhance immuno reactivity. Their mode of action is unknown, but they may reduce Ag destruction within cells. Polyanions like dextran sulfate (or) suramin operate similarly. Micells constructed using protein antigen and a mature of quick A appear to be highly effective adjuvants with, so far, no observed side effects.

Compounds that act like adjuvants are the immune enhancing drugs of which the most widely used is levamisole, used originally to kill parasite worms in animals, levamisole appears to function like the thymic hormone thymopoietin, i.e. it stimulates T- lymphocytes differentiation and the response of T-lymphocyte to antigen. It also stimulates phagocytic activities of macrophage and neutrophil. The effects of levamisole are maximal in animal with depressed T-lymphocyte function and it has little (or) no effect on the immune system of normal animals. Levamisole may therefore assist in the treatment of chronic infections and neoplastic diseases, but it may rise diseases caused by excessive T-lymphocyte function. Other immuno enhancing drug include iso purinosine and audine and macrophage activity. It probably acts by stimulating IL-2 production. Antigen is a lipoidal amine that enhances neutrophil function when depressed.

QUESTIONS

1. Define Adjuvants.
2. Define Frauds complete adjuvant.

CHAPTER 16

ANTIBODY FORMING THEORY

There are two theories proposed for the formation of the antibodies. They are

1. Selective theory (side chain selective theory).

2. Directive or template theory constructive theory.

These two theories, that form the base assume that antibody producing cells are already programmed to produce antibody specific to an antigenic determinant and the production of specific antibody is controlled by genes

16.1. SELECTIVE THEORY

In 1880, Paul Ehrlich put forth the theory about the Ab formation i.e known as Ehrlich's side chain theory. It was thought that Ab forming cells have, antibody molecules as side chains. On their surface the Ag selects its corresponding side chain, after to this, and finally knock it down .This act as a trigger and when the cell starts to synthesis similar side chain, repeatedly which are dislodged to freely circulate and unit with the Ag specifically. This concept was based on lot of speculations and as such was nearly forgotten until recent past when it was been shown that Ab forming cells do posses immunoglobulin molecule on the surface. But the cell posses one type of Ig present knowledge of immunologically competent cells has review that information for the synthesis of all possible configuration of Ab exist in the genome of antibody forming cells and that specific Ig molecules are normally present on these cells. The presence of Ag on lymphocyte, it reviewed by the fact that these lymphocytes can be activated by antibiotic Ab's to produce Ab's similar to that induced by Ag.

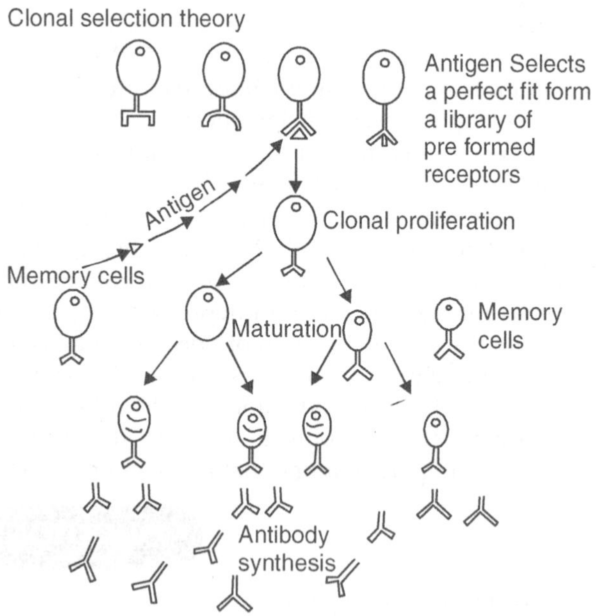

16.2. INSTRUCTIVE THEORY

Two theories have been seen put forth to explain the synthesis of Ig in a normal human. These are template theory of Ehrlish which is now known as the instructive theory and the selective type of theory with clonal selection theory as its model name.

16.3. CLONAL SELECTION THEORY

A clonal is a population of cells arising from a single parent cell as per this theory in a mature animal.

The lymphocyte is genetically endowed with the capability of synthesis of rising Ig. At rest, unstimulated by Ag, only small amounts of Ig are formed. These are found as Ig M and Ig D on the surface of the lymphocyte. On context with corresponding Ag lymphocyte capping heralds a cleavage of that lymphocyte to reproduce and differentiate into a clone of Ig secreting plasma cells. The resulting clone of cell would consists of a large enough population that the Ab produced would become measurable into the blood.

Such clone have been identified under fluorescent microscope and this theory is hormonic to the kinetics of immune response in both its chymeric and anaemic response. Recent source of support for clonal theory has been the discovery that embryonic cells do contain structural genes for Ig prior to exposure to Ag, and that the mature cells are the same biochemical process to synthesis Ig's which they are for synthesis of other proteins.

16.4. TEMPLATE OR DIRECTIVE OR INSTRUCTIVE THEORY

According to the template theory the Ag is the Ab forming cells act as a template for the synthesis of Ab having complementary configuration. The Ab so formed dissociate from the Ag molecule which further acts as template for next molecule of Ab. However persistent of Ag molecule for as long as Ab synthesis goes is not acceptable. As an alternative, it has been suggested that, Ag brings about genetic change in the cell, so that its daughter cell continued to produce specific Ab. However , not all Ag's are capable of influencing the genetic information.

16.5. INSTRUCTIVE THEORY

This theory assumes, the Ag can inform or cell in some say to made an Ab against that Ag. As per this theory, the Ag's enters a cells, i.e. routinely engaged in normal γ-globulin synthesis. The Ag's interferes with the process by complexing with mRNA in polysome .This results in a change in g-globulin synthesis and now take a shape of Ab. The Antibody dissociates itself from Ag and is excreted into the blood. The Ag is retained and acts as the template for continuous Ab synthesis. However, this theory is incompatible with known biochemical knowledge regarding protein synthesis. Polysaccharides acting as Ag, and phenomenon of immunological tolerance.

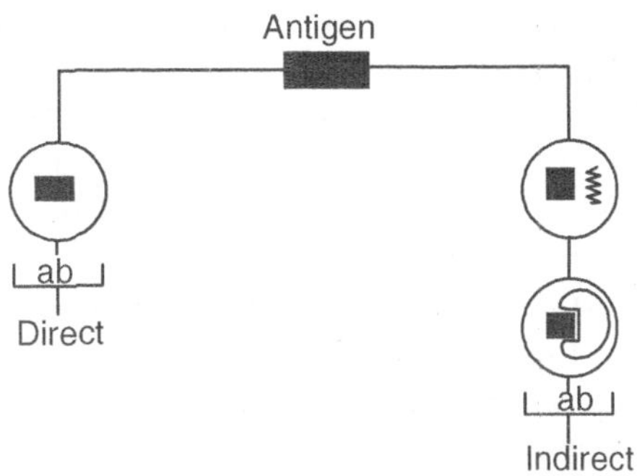

Instructive theory that includes direct and indirect template theories

16.6. POLYCLONAL ANTIBODIES

Different Ab's are produced against a single Ag.

An antiserum passively to a single Ag containing a number of of Ab's to that Ag from a different plasma cell clone (all antisera are infant polyclonal)

16.6.1. Production

It is most important to understand the even a small hapten of a relative molecular mass as low as 200 can induce a mammal to produce upto 6 different Ig each differencing in aminoacid composition and capable of combining different parts of the surface of the haptens.

Therefore it is not surprising that large Ag such as proteins induce the production of a multitude of different Ig, each reacting to a slightly different epitope on the surface of the Ag. This in turn can lead to slight cross reactions between an antiserum and Ags and having only a minimal common structural feature.The different Ig have produced as a ensure the interaction between the epitopes in the Ag surface and different clones of B-lymphocytes, each carrying a specific receptors for that epitopes. Hence the name polyclonal antiserum. The relative amounts of each different Ig in such a polyclonal serum depends on the route of inoculation of Ag and species and strain inoculation as well as adjuvant used and frequency of inoculation. Thus it is not surprising of the standardisation of so called Ag specific polyclonal antisera is very difficult task because most animals contain many different clones of Ig each capable of reacting of the different Epitope or the Ag, which may or may not produce Ig G. In practice it is impossible to obtain homogenous population of Ig molecules (MCAS) from poly clonal antiserum, despite intensive efforts by immunochemists for many decades.

However modern cell biological techniques, includes function of B lymphocyte with immortal cell surface, have facilitated the production of monoclonal anti bodies.

16.7. CELL AND CELL COLLABORATION

16.7.1. Three types of cells are involved in the immune response

Many studies in vivo and in vitro indicate that the central role in the induction of the immune system response is played by accessory cells of the monocyte macrophage series. After infection of antigen, macrophage in draining lymph nodes and in the spleen trap and concentrate the antigen.

Studies have shown that production of antibody involves the formation of clusters of lymphocytes around central macrophages with subsequent, intimate contact. If the macrophage are removed, the immune response is abrogated.

An experiment was demonstrated by Claman, that two types of lymphocytes are involved in the production of antibody. The study of immune response was carried out in lethally irradiated Mice.

In such mice the immune functions are destroyed by irradiation, can be restored by an infusion by spleen cells. When lethally irradiated were given cells from the thymus, the mice failed to produce an immune response to antigen.

They also failed to produce a response when graft with bone marrow cells. But when they given a mixture of bone marrow and thymus cells, the irradiated mice produce an immune response comparable to the response made possible by infusion with spleen cells. Restoration of immune responsiveness by administration of the mixture of bone marrow and thymus cells, showed that at least two distinct lymphocytes, both which are present in the spleen are necessary for co-operation.

By the use of appropriate cell markers, it was determined that B-cell from the Bone marrow become the actual producers of antibody, while T-cells from the thymus are the co-operation "helper" cell type.

16.7.2. T-B cell co-operation

Antigens having multiple repeating epitopes, have the ability to cross link the BCR and directly activate the B cell. Most protein antigens do not contain multiple repeating units. They contain small single epitopes that contain cross link the BCR by itself. For this reason, B cell responses to nearly all protein antigens require additional signals, provided by CD4+ T Cells.

The major class of antigens that require CD4+ T Cells is referred to as thymus dependent (TD) antigens, and the set of CD4+ T Cells that participate in TD antigen response is referred to as has helper T Cells (TH). The production of antibody is response to a TD antigen required that both B and T cells to be activated and interact.

In addition for T-B co-operation to occur, the B and T cells must respond to epitopes that are physically linked in order to produce antibody. Thus the phenomenon of T-B cell cooperativity is also known as linked recognition.

When antigen binding causes secretion of cytokines by T cells, the B cells is activated to proliferate and to differentiate to an antibody producing cell. The nature of the cytokines secreted by T cells determine to which Ig isotype the B-cell switches.

The importance of T cell involvement in B cell antibody production can be found with antigens that do not use T cell help, so called T independent (Ti) antigens. These Ti antigens do not induce memory B cell and B cell do not undergo Ig class switching from IgM to other Ig isotypes.

In response T cells are most effectively activated by Ag processed and presented by dendritic cells. Activated T cells interact with and activate B cells.

In 2° response, both the relevant B and T cells have been previously activated and expanded by antigen and is contrast to 1° response, the T-B co-operation is achieved without the dendritic cells.

16.7.3. Mechanism for T-B cell co-operation

The B-cell captures antigen by binding it to its specific Ig receptors. After antigen binding at the B cell surface the complex of antigen and Ig is taken into the cell (internalized) and degrade in acid compartments, which also contain MHC class II markers.

The peptide formed by the degradation of antigen selectively bind to MHC class II molecules with in the cells. The antigen peptide MHC class II complex then goes to cell surface, where it is recognized by an antigen specific activated helper T cell. The B cell, having displayed antigen in conjugation with the MHC class II molecules has now moved from the initial to B1 stage into cycle.

As a result of activated T helper cell, B cell interaction, the T cell produce IL-4 and IL-6, formerly B cell growth and differentiation factors (BCGF, BCDF). B cells move into the B2 stage of activation and display receptors for IL4 and IL5.

When the receptors and the corresponding interleukin molecules interact, B cell grow and proliferate yielding several clones of B3 cells. B3 cells in turn display receptors for IL6 interaction of IL6 within receptors on the B-cell pushes the B cell into the final differentiation stage-the plasma cell and antibody production ensures. A set of B3 cell remain un stimulated by IL6 .This pool of B3 cells sub populate the memory cells.

Subsequent exposure of B cells to antigen will yield a much faster and created antibody response, since the terminal differentiation stages before antibody production. To go into production, these B cell only need renewed swelling with IL6.

The nature and essential features of this T cell-B cell cooperation were elucidated by a classical experiment conducted by Mitchson in 1969, who used a hapten carrier system.

16.7.4. Carrier effect

A small hapten, such as DNP (dinitrophenyl) group can be used to tool to examine the production of antibody directly against a particular epitope. Thus it is possible to immunize an animal with attracted to one protein carrier and then look for production of antibody to only the DNP, when it attached to a different carrier.

The experiment that showed this hapten carrier effect is below:

A mouse primed with an infection of DNP'S-BSA molecule of DNP coupled at multiple positions to one molecule of BSA carrier,which make a good secondary anti DNP and response if given second injection of DNPs-BSA.

If the same mouse is primed with DNPs-BSA, but given a second injection DNPs-OVA it fails to make a good anti-DNP antibody response. In this instance the second-carrier OVA is immunologically unrelated to and non cross reactive with BSA.

Its inability to boost the anti-DNP response shows that a secondary response to the hapten requires that the carrier molecule be the same as the one used in the primary infection. The requirement here for identical carrier molecule is an example of the carrier effect.

The critical experiment, showed that of the same animal was primed with DNP on the BSA carrier and separately primed with OVA alone, it was then able to mount a secondary anti DNP antibody response when injected with DNPs-OVA. Thus the carrier effect could be overcome by priming with the second carrier and then immunizing with the hapten linked to the second carrier.

Primary Immunogen	Secondary Immunogen	Secondary anti DNP antibody response
← DNP_8BSA	← DNP_8 BSA	← ++++
← DNP_8BSA	← DNP_8OVA	———
← DNP_8BSA+OVA	← DNP_8OVA	++++

From this experiment, it was concluded that antibody mediated responses to antigens such as proteins, require the participation of thymus-derived cells. Hence the term thymus-dependent antigens associated with protein in general, since they have multiple but different epitopes on one molecule.

Upon 1° immunization, T-cell clones, recognizing some of the epitopes on the protein carrier expand. Similar clones of B-cells recognizing other epitopes on the antigen also expand.

When the animal is boosted, then the expanded clones of T and B cells cooperate to give a 2° response as an increased production of anti DNP antibody. If the animal is given a booster injection with the DNP on a different carrier to which there has been no priming then there are no expanded clones of T-cells specific for epitopes on the heterologous carrier. Therefore no co-operation with B cells specific for the DNP epitope can occur and no production of anti-DNP antigen ensures.

If the animal is separately primed with the second carrier, the expanded clones of T cell specific for epitopes on this carrier are available for co-operation with the B cells that are making an anti-DNP response.

An important point is that, while the T cells and B cells could be separately primed by infection with different antigens , a successful secondary response require that epitopes, for which each cell type was specific, to be given on the same molecule.

A protein antigen even with added haptenic determinant has a variety of epitopes towards with T cell and B cell clones respond. Since the epitopes to which the T cells are responding and those to which the B cells are responding to the same molecule cellular co-operation between B and T cells is arised by physical bridging between them.

QUESTIONS

1. Write a short notes on selective theory.
2. Explain instructive theory.
3. Write a note on clonal selection theory.
4. Outline polyclonal antibodies.
5. Write an essay on Antibody forming theory.

CHAPTER 17

ANTISERA

Antigens are more immunogenic when presented in an insoluble form or with an adjuvant. The most commonly used adjuvant is Freund's complete adjuvant into which soluble antigen is combined as a stable water in oil emulsion. The mode of adjuvant action is unclear but it is probable that the slow release of the antigen from the emulsion 'depot' acts as a prolonged series of small injections. In addition a proportion of the subsequent antibody production occurs within the granuloma induced by the *Mycobacterium tuberculosis* in the adjuvant.

17.1. FREUND'S ADJUVANT

Freund's complete adjuvant is a mixture of oil (Bayol F) and detergent (mannide mono oleate) containing *Mycobacterium tuberculosis*. The incomplete adjuvant is a mixture of oil and detergent alone. The complete and incomplete adjuvant may be purchased commercially.

For obvious reasons, this adjuvant is not clinically acceptable for human use and indeed great care should be exercised to avoid personal contamination during its experimental use. Although Freund's adjuvant was first described almost 40 years ago, it is still the most effective non specific immuno potentiator known. The relative abilities of some common adjuvants to stimulate humoral or cell mediated immunity have been extensively investigated by Bomford (see references at end of chapter).

Every laboratory has its own method of raising antisera. Those we describe below work and are rapid.

17.2. RABBIT ANTI MOUSE IMMUNOGLOBULIN

17.2.1. Materials and equipments

Mouse, globulin, Freund's complete adjuvant, Glass syringe with lever lock, Large rabbits.

17.2.2. Method

1. Dissolve 500 mg of mouse, globulin in 1 ml of saline.
2. Add protein drop wise to 1 ml of Freund's complete adjuvant. Homogenize with syringe and needle to a white cream after adding each drop.
3. Continue homogenizing until a stable water in oil emulsion is obtained. Check this by gently placing one drop from the syringe onto saline. If the emulsion is stable the first or second drop will not disperse.
4. Inject approximately 1 ml of the emulsion intramuscularly into each hindquarter of the rabbit.) Many investigators favor a foot pad injection regime. Theoretically this is advantageous as there is good lymphatic drainage to the local nodes, in practice, however, it is not advisable especially if the rabbits are housed over wire grilles).

5. Two weeks later repeat the injections. (If the granuloma ulcerates after the second injection of antigen in Freund's complete adjuvant, we suggest you omit the M. tuberculosis from this second injection in your general immunization schedule),
6. After two further weeks take a 5 ml sample bleed from the central ear artery.
7. Transfer the blood to a glass tube and allow it to clot at 37° for one hour. Loosen the clot from the tube to aid retraction.
8. After incubation, store at 4°C until serum is expressed.
9. Test the antiserum in an interfacial ring test.

17.3. AMMONIUM SULFATE PRECIPITATION

17.3.1. Methods

1. Dissolve 1000 g ammonium sulfate in 1000 ml distilled water at 50 percent, allow to stand overnight at room temperature and adjust the pH to 7.2 with dilute ammonia solution or sulfuric acid.
2. Bleed chicken by cardiac puncture and separate the serum from the clotted whole blood.
3. Dilute serum 1:2 with saline and add saturated ammonium sulfate solution (prepared in 1) to a final concentration of 45 percent v/v.
4. Stir at room temperature for 30 minutes.
5. Spin off precipitate (1000 g for 15 minutes at four degree).
6. Wash precipitate with 45 percent saturated ammonium sulfate and re centrifuge.
7. Re dissolve the precipitate in the same volume of PBS as the original serum.
8. Centrifuge to remove any insoluble material.
9. Re precipitate the 7 globulin using a final concentration of 40 percent saturated ammonium sulfate.
10. Spin off the precipitate and was with 40 percent saturated ammonium sulfate.
11. After centrifuging the washed precipitate, redissolve in a minimum volume of PBS.
12. Dialyze the F7G against 5 changes of PBS at 4°C. Centrifuge off any precipitate.
13. Prepare a 1:20 dilution of the F7G and determine the absorbance at 280 nm using a UV spectrophotometer.

17.3.2. Calculation of protein content

At 280 nm, an od of 1.0 (1 cm cuvette) is equivalent to an F7G concentration of 0.74 mg/ml.

Example: if od at 1:20 = 0.95

F7G concentration = 0.95X 0.74 X 20

= 14.1 mg/ml

17.4. PURIFICATION OF ANTIBODIES

As we know from chapter 5 an antibody reacts specifically with its own antigenic determinant to form an antigen antibody complex. If an animal is immunized with an antigen it will respond with antibodies all reacting with the antigen to some degree. Serum from this animal will have the usual range of immunoglobulins but those reacting with this antigen will be at a relatively higher concentration, compared to normal serum.

To study a particular antibody in detail it is of great advantage to be able to separate it from the surrounding, non specific antibody molecules. The precipitated antigen antibody complex has already done this for us. Unfortunately, the antibody in combination with its antigen has already completed the interesting reactions before we could follow them.

We must separate the complex, remove the antigen and we will then have purified antibody in a reactive state.

The forces binding antibody to antigen are those involved in any protein-protein interaction.

1. Columbic
2. Dipole
3. Hydrogen bonding
4. Vander waals
5. Hydrophobic bonding

All these forces depend on the charge of the molecules taking part in the reaction the net charge of the molecules inturn depends on the pH of the medium. If the pH of the medium is lowered sufficiently the protein molecules change conformation, gain H+ ions and so repel each other. We are now faced with the problem of physically removing the antigen or the antibody, because when the pH is returned to neutrality the complexes would reform.

If the antigen is insoluble it can be removed easily.

There are many methods available for rendering either the antigen or antibody insoluble, some of which are described in the following sections. The antigen in its insoluble form is known as the Immunoadsorbant and the whole purification process is affinity chromatography.

17.5. PREPARATION OF IMMUNOADSORBENT

In this experiment antibodies to mouse immunoglobulin are purified but the identical method can be used for other proteins.

17.5.1. Materials and equipments

Sepharose 4B, Cyanogen bromide (this chemical is very toxic and must be handled in a fume cupboard), 2.0 M sodium hydroxide, Phosphate buffered saline (PBS), Borate saline buffer, pH 8.3, ionic strength 0.1,mouse immunoglobulin.

17.5.2. Method

1. Pipette 14 ml of sepharose (about 200 mg) into a 50 ml glass beaker and add 10 ml of distilled water.

   All procedures must be carried out in fume cupboard.
2. Weight a stoppered tube, add some solid cyanogen bromide, replace the stopper and reweight the tube.
3. Dissolve the cyanogen bromide in distilled water to a final concentration of 50 mg/ml.
4. Place the sepharose beads on a magnetic stirrer and titrate the pH to 11.0 –11.5 with 2.0 M sodium hydroxide.
5. Add 10 ml of the cyanogen bromide solution (use a pipette safety bulb).
6. Maintain the pH at 11.0 – 11.5 by drop wise addition of sodium hydroxide for 5 –10 minute until the pH becomes stable.

7. Wash the activated beads on a sintered glass funnel with 100 ml of water and then 100 ml of borate buffered saline.
8. Wash the beads into a glass beaker, allow them to settle and remove the supernatant.
9. Add 100 mg of mouse immunoglobulin to 10 mg/ml (initial concentration).
10. Leave the beads stirring with the protein overnight at 4 degree (most of the uptake occurs within the first 4 hours and so this stage can be abbreviated).
11. Wash the beads on a sintered glass funnel with 10 ml PBS and collect the washings. (Use negative pressure and collect washings in a tube standing in a side arm flask).
12. Wash the beads thoroughly with PBS to remove the rest of the unbound immunoglobulin.
13. A uv spectrophotometer reading of the washings will give the amount of unbound protein and so the approximate quantity of protein bound to the column can be calculated.

The immunoadsorbent is now ready. Store in PBS containing azide (0.1 *M*).

Technical note: In step 6, wash the gel with borate saline buffer as soon as the pH becomes stable. The rate of inactivation by hydrolysis is highly pH dependent.

17.5.3. Use of immunoadsorbent for antibody purification

17.5.3.1. Materials and equipments

Rabbit anti mouse immunoglobulin, Immunoadsorbent, mouse 1 g on sepharose 4B, Glycine HCl buffer, 0.1 M, pH 2.5, Trichloroaceticacid, 10 percent aqueous solution(TCA), Tris (Hydroxymethyl) amino methane, Phosphate buffered saline (PBS), chromatography column.

17.5.3.2. Method

1. Pour the immuno adsorbent into the column and equilibrate with 20 ml PBS. Close the column.
2. Run 20 ml of antiserum through the column, do not use positive pressure allow it to run under 1 g.
3. Wash the unbound protein from the column. If a UV cord or similar flow through UV cell is available, wash the column until the absorbance is < 0.1. Otherwise wash with 200 ml PBS. Close the column.

We now have the antigen antibody complex.

17.5.4. Dissociation of complex

1. Pipette out twenty 0.5 ml aliquots of TCA into small glass tubes. (Use this to sample the effluent for protein during elution if a flow through UV cells is not available).
2. Add glycine HCl buffer to the top of the column and collect the effluent when protein is first detected.
3. Stop collecting the effluent when protein is no longer detectable.

The first stage of the elution is now complete and part of the antibody has been recovered. The low pH will, however, eventually denature the antibody so we must raise the pH.

The unbound protein was washed away with phosphate buffered saline until the absorbance at 280 nm was below 0.1°.A first population of antibody molecules was eluted with 0.1 M, pH 2.5° glycine HCl buffer (first arrow).When elution was complete a second population of antibody was eluted off with the same buffer containing 10 degree dioxane (this second elution was started at the second arrow).To reduce the total sample volume only the volumes A and B were collected.

4. Titrate the protein solution to pH 8.5 with solid Tris. Mix thoroughly and monitor with a pH meter or indicator papers. We are now going to alter the elution conditions to recover a second batch of antibody.
5. Add glycine HCl + 10% dioxane to the column. Monitor the effluent and collect the second batch of antibody.
6. Adjust the pH to 8.5 with solid Tris.
7. Read the absorbance of each protein solution at 280 nm and calculate the recovered protein.(remember to use the correct reference solution for the spectrophotometer).
8. Concentrate the samples in dialysis tubing against either sucrose of polyethylene glycol 40,000 or by negative pressure dialysis.
9. When the sample volume has been reduced to 3.5 ml, dialyse against 5 × 1.1 PBS.
10. Spin off the precipitate and determine the protein content of each sample.

This method of antibody purification is highly reproducible and so it is not necessary to calculate the antibody content of the sample routinely. However, a specimen calculation is given below.

17.5.5. Calculation of recovery from Immunoadsorbent

Total weight of immunoglobulin on column = 92.0 mg on 200 mg of sepharose 4B.

Volume of antiserum for antibody purification = 10 ml

Antibody content of serum calculated from fig 8.3 at equivalence condition as in section 5.3.1.

Antibody content of serum = 5.2 mg/ml

Percentage yield of antibody from serum: immediately 81.5 %. after concentration and dialysis 47.8%

17.5.6. Eluates from immunoadsorbent total protein concentration in eluate immediately after concentration and dialysis

Glycine HCl	36.4 mg	23.0 mg	42.4 mg	25.0 mg
Glycine HCl + 10% dioxane	6.0 mg	2.0 mg		

17.5.7. Calculation of Antibody content of eluate

From weight of antibody in 200 mg of eluted protein

= 490 160

= 230 mg.

Hence all the recovered protein has retained antibody activity. (In general, at least 90% of the recovered protein should be antibody).

QUESTIONS

1. Define Freund's adjuvant.
2. Explain the purification of antibodies.
3. Explain the preparation of immunoadsorbent.
4. Explain the purification of antibodies using immunoadsorbent.

CHAPTER 18

ANTIBODY – RECEPTOR STRUCTURE

18.1. INTRODUCTION

People or animals who recover from certain kinds of infection are resistant to reinfection with the same pathogen. This "immunity" is highly specific the resistance does not extend to pathogens other than the immunizing one and may last for a very long time. i.e., decades. Only after the development of methods for the measurement of the state of immunity of an individual did immunization become an important medical and veterinary practice. Today we are able to follow a particular immune reaction by directly monitoring either the response of the cells involved in that reaction, or the secreted products of these cells. Undoubtedly the simplest and most rewarding method has been to measure the concentration of circulating antibodies.

Although immunology evolved as a branch of medicine, it soon became clear that animals could be immunized against a wide variety of substances (antigens) unrelated to pathogens. The methods for measuring antibody concentration have been adapted in various ways in order to measure antigen concentration. In this way the techniques of immunology acquired a practical importance for investigators in a wide variety of disciplines other than medicine. The special features of the immunological (serological) methods which make them so generally useful are the facility with which the antisera can be raised against widely different antigenic substances and the ability of the antibodies to pick out one molecular structure in a complex mixture.

Since the introduction of methods for the production of monoclonal antibodies (McAb) the immunological approach has been extended in two important ways. First, specific antibody reagents can be derived without the need to achieve high purity of the antigen used for immunization. Second, a given antibody can now be produced in large quantities at acceptable economic cost. In order to exploit the advantages of the immunological approach, a minimal knowledge of fundamental immunology would benefit even those researchers in fields remote from medical science.

18.2. CELLULAR ASPECTS OF THE IMMUNE RESPONSE

In fulfilling its role of defending the organism against infection the immune system must be able to recognize a wide variety of pathogens, at the same time, however, since the immune response is destructive, it must avoid recognizing and reacting against itself. Since both the host and the potential invaders are composed of the same kinds of chemical structure protein, carbohydrate, lipid, nucleic acids. The immune recognition must be finely tuned, a breakdown in the system will result in either fatal infection, or self destruction.

The first requirement of an immune system which is not to be capable of responding to a huge variety of antigens, is a set of receptors with the capacity of discriminating precisely between all of them.

The evolution of a small family of molecular structures with these properties, although attractive in terms of genetic economy, may be impossible to achieve. The alternative solution, which has been adopted by the immune system of higher vertebrates, is the continuous random production of thousands of different receptor structures able to bind, more or less efficiently, any possible antigen. In the absence of the antigen each of these receptors is present only in small amounts. When antigen enters the system the receptor able to bind to it are selected for expansion. Each antigen binding receptor is carried by a cell and each cell carries only one receptor. For example, in human being, with as many as 2×10 lymphocytes, a very wide spectrum of receptors can be represented at any one time. In this way the system is ready in advance to tackle any antigenic challenge. At the same time mechanisms are ready to suppress cells bearing receptors which bind to self molecules. This latter process of self tolerance is not yet well understood and we will not go into a detailed account of it here.

The population of lymphocytes can be divided into two main lineages, the T cells, which depend for their normal development on the thymus gland, and the B cells, which do not. Both these cell types are descendants of pluripotent stem cells present in the bone marrow in adults. These two types of lymphocytes can be distinguished in practice by several differences in protein expression, particularly a number of cell surface molecules. The most important characteristic of B cells is the expression of receptors on their surface classically identified as immunoglobulin (Ig), that can also be secreted in large quantities in soluble form by the terminally differentiated B cell, the plasma cell. The T cell does not express immunoglobulin but carries some easily identifiable specific marker proteins of course the most important, but historically very elusive, feature of the T cell surface is the presence of an antigen specific receptor similar to but distinct from the Ig receptor of the B cell.

A very important physiological difference between T and B cells is the manner in which they are activated by antigen. Whereas B cells will respond under the appropriate circumstances to antigens which are in solution T cells seem only to be activated by antigen presented on the surface of specialized antigen presenting cells. Whereas B cells seems to have but one function, i.e., the ultimate production of specific antibody, the T cell population contains cells which can perform a number of physiological roles e.g., the killing of infected or aberrant cells and the regulation of the immune response, either by the production of pharmacologically active factors (lymphokines) or by direct cell to cell contact. In this chapter we will not be discussing the autonomous activities of T lymphocytes but rather will confine ourselves to a description of the process involved in the production of antibody by B cells. Although some antigens can induce antibody production with little or no involvement of T cells (in particular, substances with a repetitive polysaccharide based structure), for most antigens the presence of T cell is an absolute requirement.

The majority of immune responses are initiated in solid lymphoid tissue (viz, the lymph nodes and the spleen), where antigen is presented to immunocompetent lymphocytes. Foreign antigen reaches these sites via lymphatic vessels (in the case of lymph nodes) or via the bloodstream (in the case of the spleen). Both prior to and after reaching, say a lymphnode, antigen is taken up and processed by specialized lymphomyeloid cells (eg., macrophages, "dendritic" cells). In a T lymphocyte dependent response, a T "helper" cell bearing appropriate receptors is activated when it recognizes antigen on the surface of one of these antigen presenting cells. In the last decade a lot of effort has gone into the identification of the form in which the antigen is recognized by the T cell. Two features of this recognition of protein or glycoprotein antigens have been revealed. First, the T helper cell does not "see" antigen per se but sees antigen in some form of association with a cell surface molecule, a class II histocompatibility molecule; second, an individual T cell is usually activated by a relatively short peptide derivative of the complete antigenic protein associated with this "self molecule" and not by the native form of the antigen. The binding of a T cell (in the phase of the cell cycle), carrying an appropriate receptor, to antigen in the form presented by a dendritic cell triggers differentiation and proliferation of the T cell to generate a clone of active helper cells. It is these cells which "help" antigen specific B cells to make antibody. In parallel with T cell activation, B cells carrying surface receptors able to bind to the antigen in solution undergo the first steps in their differentiation sequence. This includes internalization of receptor bound antigen followed by its processing. Although the intracellular events in antigen processing are probably very similar in B cells and dendritic cells, the uptake of antigen is specific in the case of B cells (being receptor mediated), but non specific for dendritic cells. In this way the same associations of antigen derived peptides and class II molecules can be present on both the dendritic cells and the B cell. Specific helper T cells, which have previously been activated by binding to dendritic cells can "help" B cells when they recognize and bind to the same combination of peptide and class II molecule on the B lymphocyte surface. This cell cell contact permits, by mechanisms unknown, the further differentiation of the B cell towards the terminal plasma cell. In consequence, a clone of T cells specific for one particular part of the antigen is able to help all the different B cells which can bind to any part of the antigen. This process is a satisfying explanation of the classical observation that one can obtain antibody response to small, intrinsically non immunogenic molecules (haptens) by conjugating them to larger carrier proteins. B Cells, which are able to bind to a free hapten cannot, however, proceed to terminal differentiation and antibody secretion, unless they receive the second, obligatory stimulus from a T cell. In the case of small hapten molecules, however, it seems that no appropriate form of the antigen can be presented, either by dendritic cells (to stimulate the resting T cell) or by a B cell (to receive help from the activated T cell). By coupling the hapten to a large protein, one is providing several T cell activating structures.

18.3. RECEPTOR STRUCTURE - IMMUNOGLOBULIN

The antigen binding receptor of a B lymphocyte is one of a family of proteins called immunoglobulins (Ig). These proteins can be inserted into the plasma membrane of the lymphocyte where they function as classical receptors to mediate the cellular response when binding to specific ligands, i.e., antigen. One result of this response is the differentiation of B cells into plasma cells which are highly specialized secretors of large quantities of the same receptor in a soluble form (serum Ig). Immunoglobulins are formed by the covalent linking of larger heavy polypeptide chains, (M.W. 50 65,000) with smaller light chains (M.W. ca.25,000). The basic unit of Ig is a Y shaped molecule consisting of two identical heavy light (H, L) heterodimers in covalent association (M.W. 150 180,000). This unit is a bi-functional structure; antigen recognition is achieved by two identical antigen binding sites situated at the ends of the arms of the Y. Both H and L chains contribute to the formation of the binding site. Interactions with other components of the immune system are mediated through sites on the tail portion of the molecule. This part of the molecule is formed by H chain sequences. In higher vertebrates the immunoglobulins can be categorized in five different classes based on their biochemical and biological properties. These differences are due only to the structure of their heavy chains.

Light chains can also be classified into two types, and these are free to associate with any class of heavy chain. The most common immunoglobulins in serum are IgG. In most mammals Ig G can be found in a number of forms called subclasses, which again are a

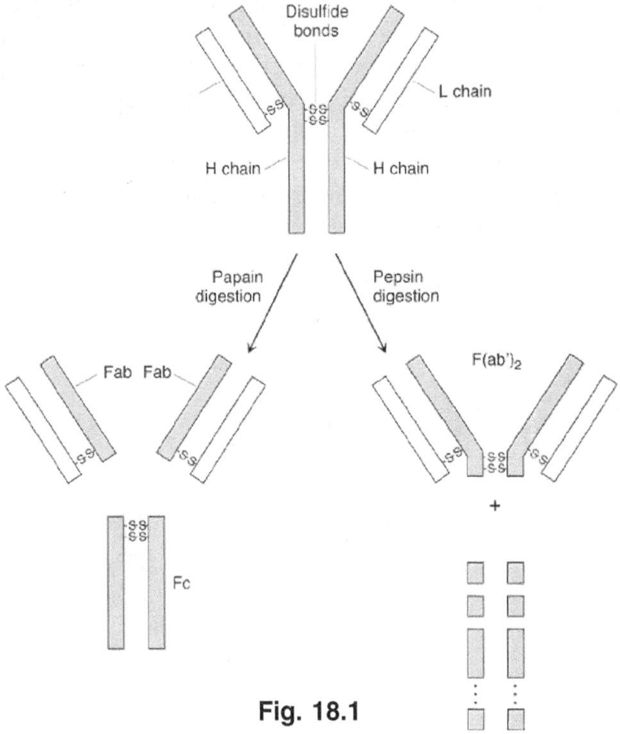

Fig. 18.1

consequence of differences in heavy chain structure. A notable exception to this is the rabbit which appears to possess only one form of Ig G.

As we have just said, the bi functionality of the Ig structure is a consequence of a basic asymmetry of all Ig polypeptide chains. In both heavy and light chains the N terminal part will contribute to the formation of the antigen binding site. Amino acid sequence analysis of a large number of immunoglobulins has shown that the N terminal region is very variable; the differences between Ig chains of the same type are almost exclusively located within the first ca.115 residues. This region is generally referred to as a the variable or V region, while the rest of the polypeptide is known as the constant or C region.

18.4. IMMUNOGLOBULIN GENES

The work of protein chemists in analysis of Ig chains made it clear that an individual is capable of generating a very large number of different B cell receptors. This number is so high that, if each molecule had to be coded for individually in the genome, the genetic load would be enormous. On the other hand, the alternative possibility of a small set of genes, highly unstable in somatic cells, has not been substantiated by studies of mutation rate in Ig producing cell lines in vitro. Only recently, with the application of gene cloning methods, the unravelling of the genetic complexity of the Ig genes has provided an adequate explanation for the generation of receptor diversity. In the mammalian genome there are three separate sets of genes for Ig chains. One coding for kappa light chains, another for lambda light chains and a third for all the types of heavy chains, each set being located on a separate chromosome. The genetic organisation of these three Ig gene regions is essentially similar and we will therefore discuss only that for the heavy chains.

Ig chains were the first example contradicting the old dogma one gene one polypeptide chain in that each Ig chain is encoded by at least three different genes. It is now clear that the configuration of Ig genes in B cells is different from that in any other cell type. In non

B cells there is a panel of several hundred different genes coding for part of the variable region (V genes) located several kilo bases (kb) upstream of a smaller number (at least 10) diversity (D) genes. Several Kb downstream of D is a set of a few (1 7) joining (J) genes which is in turn separated by a smaller distance from the regions coding for different heavy chain constant genes. In the course of differentiation of every individual B cell, one V gene is joined to one D segment and one J segment and the intervening DNA is excised. Any of the V, D and J genes can be chosen for rearrangement, but once a successful VDJ combination has been produced, the cell seems to lose the ability for further V region

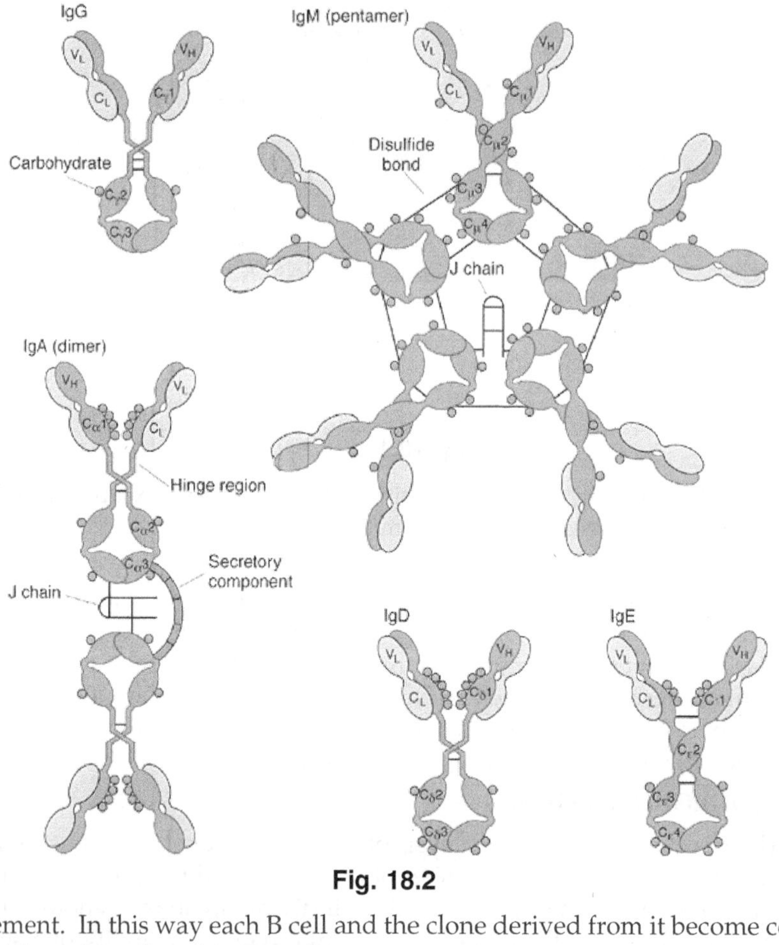

Fig. 18.2

rearrangement. In this way each B cell and the clone derived from it become committed to the expression of a unique binding site.

During B cell differentiation every successful VDJ combination is joined to the first (5') constant region gene, Cu. In later stages of differentiation of the cell the same VDJ combination can be moved next to the other heavy chain constant region genes e.g. C delta, C gamma, C E. Therefore within a single clone of B cells the commitment to the Ig binding site is greater than that of the effector portion of the molecule.

This complex genetic arrangement is a very economical way to create the large heterogeneity of receptors required by the immune system. There are four different contributions to the overall receptor diversity.

1. Germ line diversity : Each of the V, D and J segments are represented by multiple, different copies in the genome.

2. Combinatorial diversity : The different V, D and J segments can be permutated for the production of V regions. Two different V regions (of the heavy and the light chain) combine to make a functional binding site.

3. Junctional diversity : The position of joining of the different gene segments can vary over a few nucleotides.

4. Somatic mutation : V regions are subject to somatic mutation, probably at a normal rate. Under the influence of selection by antigen this can be important in the maturation of receptor affinity during an immune response

18.5. THE T CELL RECEPTOR

Recent advances have uncovered some of the features of the membrane bound antigen specific receptor of T cells. This receptor seems also to possess a two chain structure, and although the protein and DNA sequence information indicates that both chains are non immunoglobulin, they are nevertheless clearly of the Ig family of proteins, which includes other cell surface molecules such as Thy 1 and the class I and class II histocompatibility antigens (Williams, 1984). Genetic and molecular biological studies further indicate that the T cell receptor genes, while being distinct from the Ig genes, also undergo the kinds of somatic rearrangement which we have described above for B cells and so probably employ similar strategies for the generation of receptor diversity. It is, however, too early to predict how close the parallel between B and T cell differentiation will turn out to be.

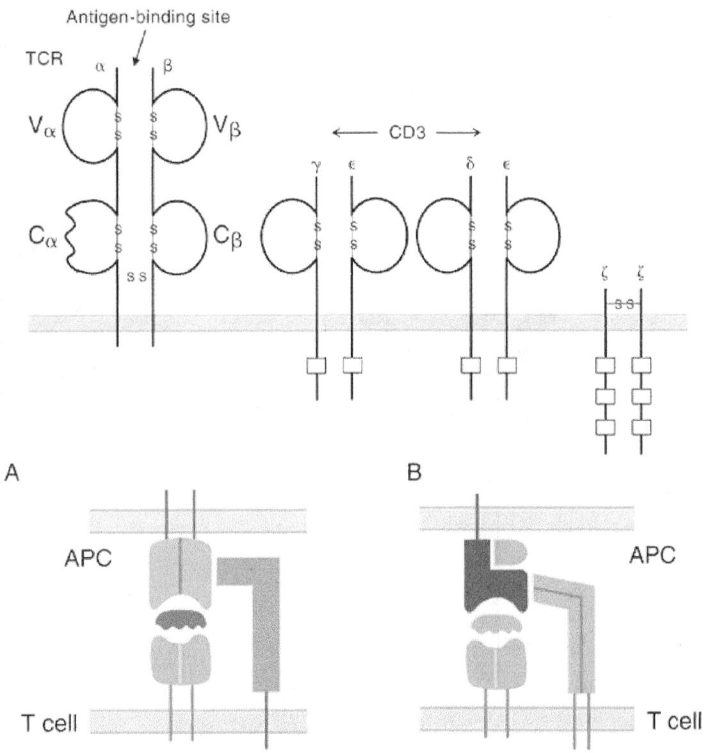

Fig. 18.3

18.6. MATURATION OF THE IMMUNE RESPONSE

During clonal expansion of an antigen specific B cell a series of morphological events take place to transform the resting lymphocyte, which has little cytoplasmic activity into the antibody secreting plasma cell, with a high ratio of cytoplasmic to nuclear material. These events are necessary to build up the protein synthetic machinery of the cell, e.g., development of the rough endoplasmic reticulumn and Golgi apparatus. More subtly, there are frequently also changes in the immunoglobulin expressed by the cells as a result of changes in the genomic regions coding for the immunoglobulins. The phenomenon concerned is known as class switching, because it involves the movement of the previously rearranged VDJ combination from its position near the Cu gene to an equivalent position

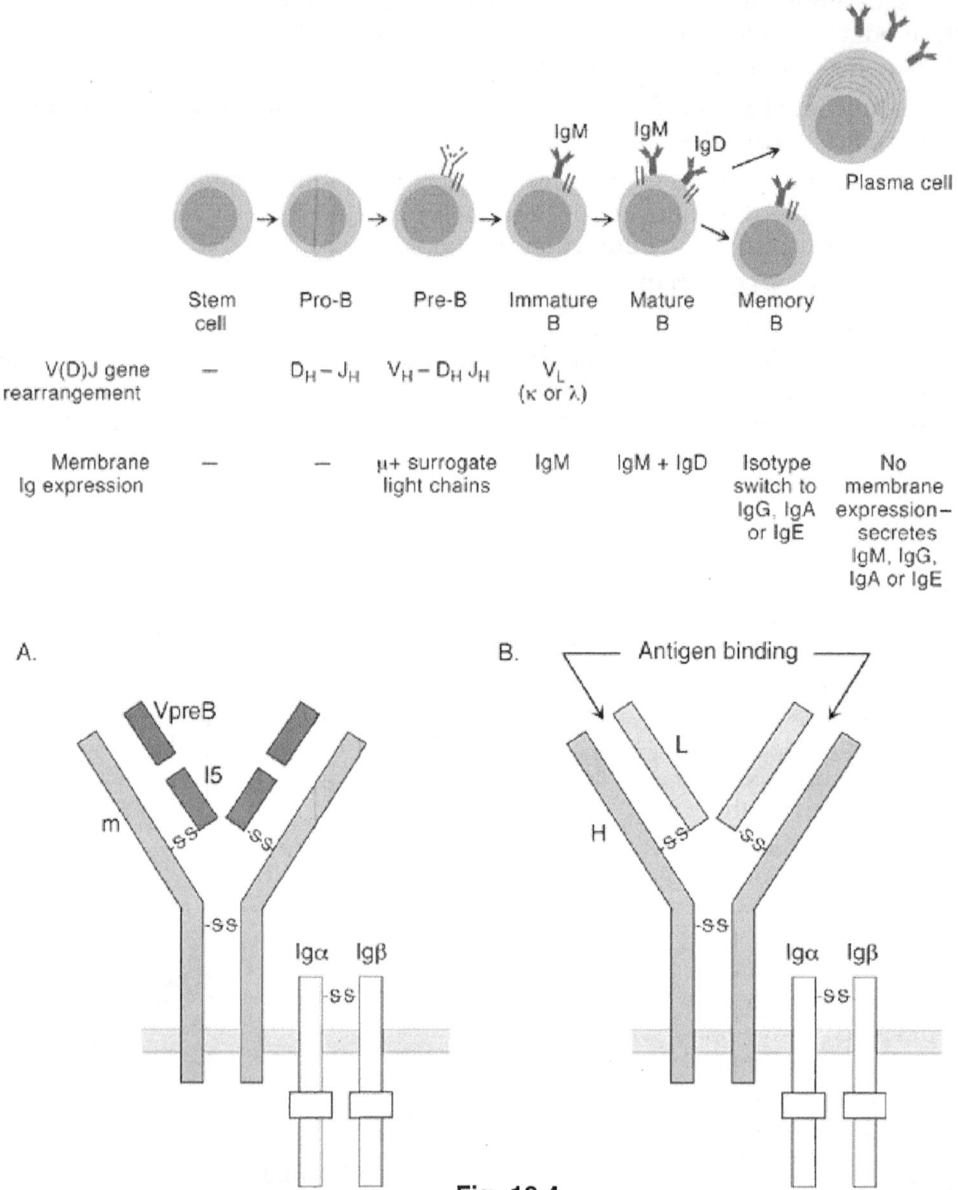

Fig. 18.4

upstream of one of the other heavy chain region genes. Thus, in a common case, a B cell producing antibody of the IgM class against a given antigen will generate progeny producing antibody of the IgG class with the identical combining site.

A property of the immune response which we have not so far considered is that of immunological memory. When an animal is challenged with an antigen against which it has previously reacted, its response is generally more rapid and more potent than on the first occasion. This phenomenon is specific the animal can somehow remember the first antigenic challenge and is primed for subsequent contact with the same antigen. The initial encounter with antigen will usually lead to the expansion of specific T and B cell clones. Simple clonal expansion, however, is probably not adequate to explain the longevity and specificity of immune memory instead the generation of a separate class of specialized memory lymphocytes seems most likely. Although the mechanism of memory cell generation is not fully understood it is presumed that the cells arise as side lineage(s) during antigen dependent B and T cell clonal expansion.

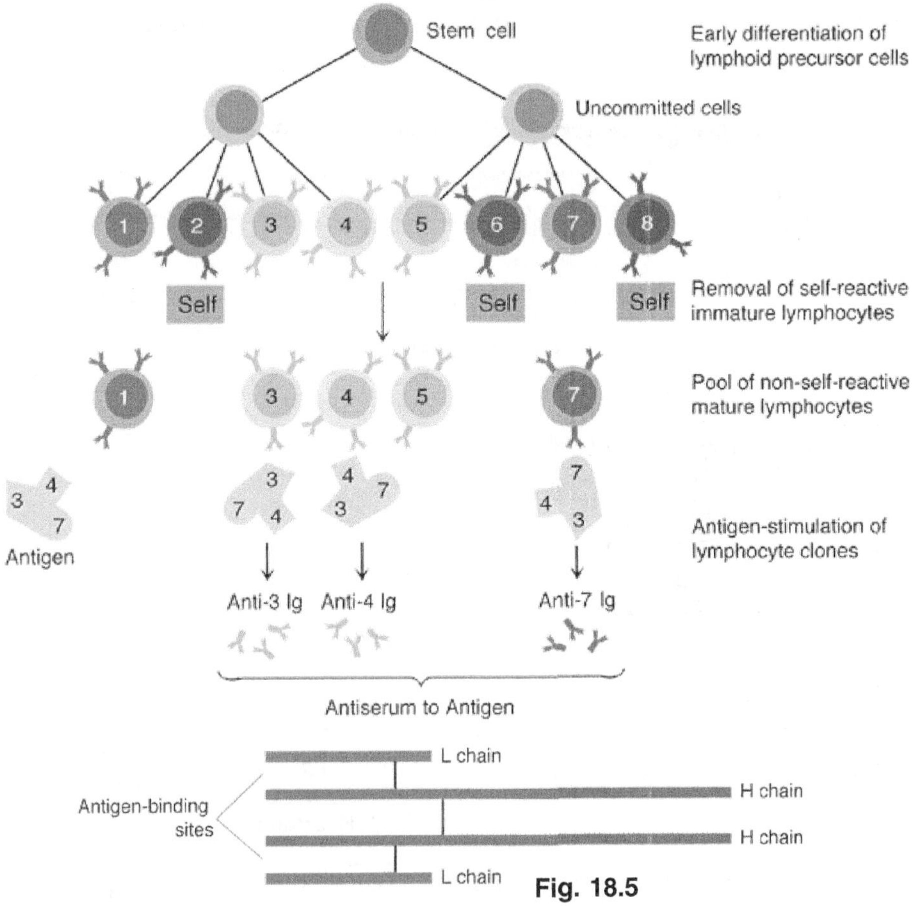

Fig. 18.5

In the course of specific immune response, during clonal expansion and differentiation in both primary and secondary responses, there is a continuous maturation of the antibody response towards high affinity receptors. Since B cell activation is antigen dependent, affinity maturation is likely to result from competition between different B cell clones for limiting quantities of antigen. Occasional somatic variants, if they express receptors of higher affinity than their progenitors can also be selected by this same process. The production of high affinity antisera by long term immunization schedules takes advantage of this process.

18.7. THE USE OF ANTIBODIES IN MOLECULAR BIOLOGY

"Molecular Biology" is a diffuse descriptive term which covers a variety of disciplines. However, the aims of molecular biologists may be succinetyl stated as the attempt to define biological phenomena by a description of the underlying molecular events to understand the reification of genetic information in structures which participate in the maintenance of cellular functions. These aims are being achieved through studies of the molecular structure of cellular components and through studies of the process of information transfer itself: the process of gene expression.

The mammalian immune system is a highly developed self defence mechanism. The introduction of foreign matter (antigens) to animals elicits a vigorous response directed towards the inactivation of these antigens. This response includes the production of a class of proteins, the immunoglobulins (antibodies), which recognize individual determinants on the surface of the antigen and react with them to bring about their immobilization. Subsequently, they are destroyed by other participants in the immune

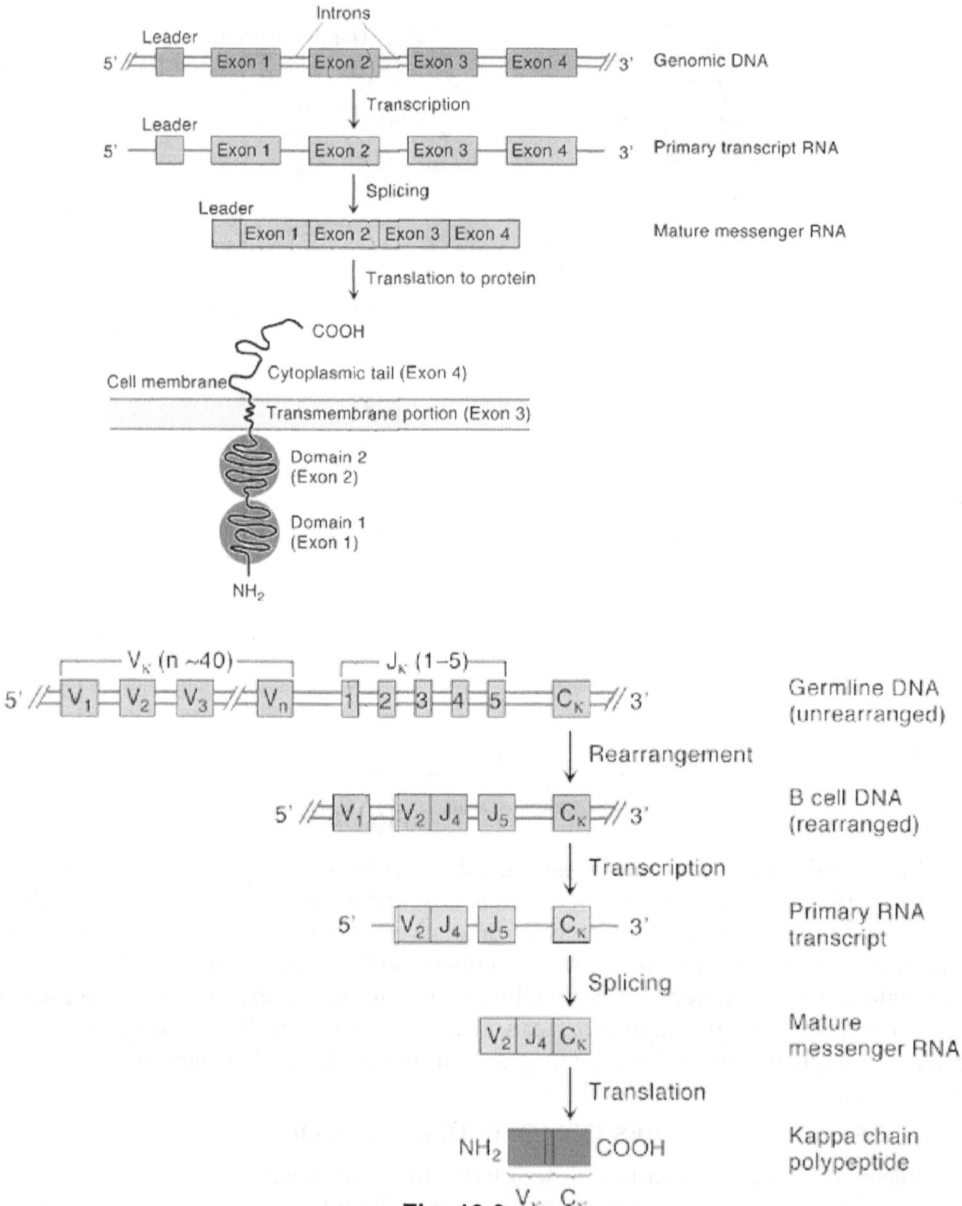

Fig. 18.6

response. The exquisite specificity with which antibodies recognize particular determinants makes them ideal tools for the molecular biologist. Their specificity permits them to be used as probes for molecular structure and function and also for the recognition and isolation of individual molecular species from the cell inclusion.

During the expression of eukaryotic genes. It is clear that the cascade of information transfer is complex, with opportunities for regulatory events to occur at each level in the pathway. Prior to transcription, the state of the DNA may dictate its activity. The expression of a gene may first require changes in the chromatin conformation, resulting from altered associations between DNA and DNA binding proteins, which permit access to polymerases. Subsequently, DNA sequence amplification, sequence rearrangement or base modification (e.g. methylation/demethylation) may be necessary for expression. In addition to regulation of this nature, it has become increasingly apparent that the transcription of individual genes at specific stages of development, or in response to environmental stimuli can be a consequence of the association of the structural sequence with a regulatory promoter sequence "upstream" of the transcriptional initiation site, or with an "enhancer" sequence, which may be located either 3' or 5' to the coding sequence.

Following transcription, processing of the primary transcript occurs to produce the translatable messenger RNA (mRNA). Processing events can include the removal of non coding intervening sequences ("splicing"), base modification (e.g. by methylation) and polyadenylation, both of which enhance the stability of the messenger, and 5' terminal "capping" which confers resistance to attack by exonucleases and increases the efficiency of the initiation of translation. During the production of a messenger, the transcript also becomes associated with proteins, to form the mature messenger ribonucleoprotein particle (mRNP). These proteins contribute to the stability of the mRNA and may also regulate its recruitment into polyribosomes. Thus the translational activity of a messenger is not simply a function of its abundance, but is also a consequence of its degree of base modification, secondary structure, protein associations and resistance to degradation within the cell.

Following translation, the initial polypeptide product may not be identical with its active form. For the gene product to become functionally active, it may have to be transported to a particular sub cellular destination, eg. with in an organelle, or to an extracellular location. Frequently, such polypeptides contain the information for this targeting in the form of aminoterminal "signal" or "transit" peptides. These amino terminal extensions are removed by the action of endonucleases during the act of transport. The activity of individual polypeptide gene products may further be modulated by secondary modifications, such as phosphorylation, glycosylation, ribosylation, etc., or by posttranslational proteolytic cleavage, such as occurs in the synthesis and packaging of seed storage proteins. Finally, the gene product may interact with the products of other genes, as in the assembly of multimeric enzymes. Between DNA sequence and functional gene product lie a myriad of possible intermediate forms. To trace the path by which the expression of an individual gene is achieved, if is therefore necessary to identify each of these intermediates and to do so among the 5 "background" expression of up to perhaps 10 other genes. For this task, specific probes are required for the recognition of the polynucleotide or polypeptide of interest. Cloned DNA sequence provide probes for DNA and RNA sequences and mono specific antibodies are used to identify individual polypeptides.

18.8. THE USE OF ANTIBODIES TO IDENTIFY POLYPEPTIDES

The analysis of complex mixtures of proteins is most conveniently achieved by polyacrylamide gel electrophoresis (PAGE). When a unidimensional separation is performed (e.g. by sodium dodecyl sulfate (SDS) PAGE), the number of polypeptides 2 resolved is of the order of 10. Two dimensional separations achieve far higher resolution, up to 10 polypeptides being resolved. The identification of individual polypeptides within these complex electrophoretic patterns may be achieved by the use of an antibody to precipitate, with high specificity, a radio-labeled antigen. For such a procedure to succeed, it is necessary for the mixture of polypeptides being assayed to be radio labeled to high specific activity. This is frequently difficult to achieve by the application of radioactive isotopes to plant tissue in vivo although, as we shall demonstrate, the technique of immuno precipitation is well suited to the analysis of polypeptide mixtures labeled in vitro. An alternative strategy has recently been developed to identify unlabelled polypeptides in electrophoretic separations by immunoassay and is based on the ability of proteins to bind to inert matrices such as nitrocellulose or nylon membranes. The procedure involves the electrophoretic transfer of polypeptides, resolved by PAGE, from a polyacrylamide gel to the membrane, where they become immobilized. The membrane thus forms a "filter replica" of the gel, with the bound polypeptides in a conformation accessible to interaction with specific antibodies. Incubation of the filter with a specific antiserum will result in the antibody in that serum binding to the antigen in its position on the filter. When unbound antibody is removed by washing, the position of the antibody antigen complex can be located on the filter by locating the position of the bound antibody. This can be achieved either by binding a peroxidase linked second antibody (e.g. goat anti rabbit immunoglobulin (Ig G) or by binding 125 I protein A. (Protein A isolated from *Staphylococcus aureus* cells specifically binds Ig G molecules, the main component of antiserum). This "immunoblotting" technique is analogous to the blot transfer methods used for the analysis of nucleic acids and is therefore frequently (if frivolously) referred to as "Western" blotting.

In the diagram it is shows how such an assay has been used to screen a series of monoclonal antibody raised against polypeptides contained in photosynthetic membranes. The photo system I (PS I) particle of pea chloroplasts is a chlorophyll lipoprotein complex containing 16 stainable polypeptides. Entire PS I particles were used to immunize mice, whose spleen cells were fused with myeloma cells to generate a collection of hybridoma cultures which secreted antibodies to the individual polypeptides of the complex. Track 2 illustrates the binding of antiserum to each polypeptide of the PSI particle. In this case, the serum used for the immune reaction was obtained by bleeding the mouse whose spleen was used to generate the hybridomas. Track 3 and 4 demonstrate the binding of the antibodies secreted by two of the hybridoma lines. In track 3, the antibody specifically recognized polypeptide P 2 of the PS I associated light harvesting complex (LHC 1) and in track 4, the antibody recognized polypeptide P 3 of the LHC 1 and its dimeric form. The protocol by which the hybridoma lines were screened by the immunoblot method is illustrated in cartoon form. Immunoblot identification of polypeptides of the pea chloroplast photo system I complex.

Track 1 Electrophoretic separation of PS I polypeptides; Coomassie blue stain.

Track 2 Filter replica of PS I polypeptides reacted with polyclonal anti PS I serum.

Track 3 Filter replica of PS I polypeptides reacted with monoclonal antibody to polypeptide P2 of LHC1 (the PS I associated light harvesting complex).

Track 4 Filter replica of PS I polypeptides reacted with monoclonal antibodies to polypeptide P3 of LHC 1. In this gel system the slower migrating band corresponds to a dimer of P3, which is incompletely dissociated. The polypeptide at 110kd corresponds to the undenatured PS I reaction center polypeptide, and that at 68kb to the fully denatured form. LHC 2 indicates the position of polypeptides of the light harvesting chlorophyll alb protein complex, which is associated with both PS I and PS II.

It is important to note that the use of the immunoblot technique coupled with the production of monoclonal antibody obviates the need to isolate a single antigenic polypeptide in pure form for the immunization of animals. This makes the procedure particularly useful for the study of polypeptides which comprise only a small proportion of the cellular protein content, or which are otherwise difficult to purify.

18.9. THE USE OF ANTIBODIES TO ASSAY MESSENGER RNA

Although immunoblotting is a powerful technique for the identification of small quantities of individual polypeptides in complex mixtures, it is not suitable for the identification of transient forms of these polypeptides such as precursors. Proteins which pass through membranes or those migrated into the structure of membranes are usually synthesized in the form of a precursor. These precursors are difficult to identify, in vivo, because they are usually rapidly transported and processed to the mature form. However, when they are synthesized in vitro in the absence of their "target membrane", the precursors are not processed to the mature form. However, when they are synthesized in vitro in the absence of their "target membrane", the precursors are not processed and may be detected by immuno precipitation. We have used this technique to identify the precursors of the major polypeptides of the light harvesting chlorophyll a/b protein complex (LHC 2) of pea chloroplast membranes. These polypeptides (Mr = 26,000 and 24,000) were isolated from SDS polyacrylamide gels and used to immunize rabbits. When mRNA isolated from pea leaves was translated in a wheat germ cell free system (Roberts & Paterson, 1973) and the translation products reacted with the anti LHC 2 antibodies, two polypeptides of Mr = 32,000 and 30,000 were specifically precipitated tracks 1 and 2, see Protocols for details of the method. The identity of these precursors was confirmed by peptide mapping and the specificity of the immuno precipitation also permitted the procedure to be used quantitatively to estimate the levels of LHC 2 mRNA present in pea leaves under different conditions of illumination by measuring the recovery of the radio labelled translation product in the immuno precipitate (Cuming and Bennet, 1981). Subsequently, these estimates of the messenger level were confirmed by direct measurement, using a cloned DNA probe (Jenkins, Hartley & Bennett, 1983). The identification of this DNA probe was considerably facilitated by the availability of the LHC 2 specific antibody.

In order to use a cloned DNA sequence to analyse the expression of a particular gene, it is first necessary to demonstrate that the cloned sequence encodes the appropriate gene product. If the amino acid sequence of the polypeptide is known, then the identity of a cloned DNA sequence may be confirmed by DNA sequence analysis. However, this situation seldom obtains and it is more usual for the cloned sequence to be used to select messenger RNA, by hybridization, which subsequently can be analysed by translation, in vitro. The product of this "Hybrid selected translation" can then be identified by immunoprecipitation.

Identification of a cloned sequence encoding the largest polypeptide of LHC 2 is illustrated in (tracks 3 and 4). This identification was facilitated by two additional factors. Firstly, the LHC 2 mRNA was subject to developmental regulation, being tenfold, more abundant in green leaves than in etiolated tissue. Secondly, the messenger was relatively abundant in green leaves, comprising approximate 1 .2% of the total messenger fraction . The first factor permitted a preliminary screening by colony hybridization (Grunstein and Hogness, 1975); only those clones which hybridized strongly with mRNA from green leaves and weakly with mRNA from etiolated tissue being chosen for further screening. The second factor permitted efficient hybridization between the cloned DNA and LHC 2 mRNA. These factors are both important because the technique of hybrid selected translation is both technically demanding and laborious. It is necessary to isolate individual plasmids in pure form, line arise and denature them, and bind them to a suitable support medium (usually nitrocellulose, or diazotized paper). This achieved, the individual filter bound plasmids are incubated with a quantity of mRNA to permit hybridization of complementary sequences, washed thoroughly to remove non specifically bound RNA and treated to elute the hybridized mRNA. This is recovered by ethanol precipitation and translated in vitro for subsequent assay by immuno precipitation.

18.10. THE USE OF ANTIBODIES TO PURIFY MESSENGER RNA

In many cases these favorable conditions will not obtain. Frequently it is desired object to isolate a cloned DNA sequence encoding a polypeptide whose mRNA is neither subject to marked developmental regulation, nor is of great abundance within the cellular messenger population. In such a case an alternative means of obtaining a specific probe must be essayed. One such approach we have utilized successfully is the isolation of a pure messenger species by the immuno purification of polysomes.

Immuno purification of polysomes is a technique designed to purify mRNAs for proteins to which antisera have been raised. Although demonstrated as early as 1961 (Cowie et al., 1961), only in the past few years has this method been developed to be applicable for purifying messengers of different relative abundances isolated from a wide variety of species and tissues. The significance of this technique is that the purified mRNA may then be used as a template for cloning, or to provide a radioactive probe to identify the clone of interest in a "library" of copy (cDNA) or genomic DNA clones. In this way the antiserum is used indirectly to identify the genes coding for the antigenic protein.

The technique relies on the antibodies reacting with antigens on the nascent polypeptides of polysomes. The antibody bound polysomes can then be separated free of the unreacting polysomes and the specific mRNA isolated from them. Originally, the separation of the antibody bound polysomes relied on either immuno precipitation or on binding of the antibody polysomes complex to an antigen column. However, this proved to be successful only for purifying mRNAs encoding abundant proteins, such as ovalbumin and immunoglobulin. A recent improvement of this technique uses Protein A sepharose to isolate the antibody polysome complex by affinity chromatography. Unbound polysomes are washed from the column and the specifically bound fraction is eluted following dissociation with EDTA. The messenger RNA from the dissociated polysomes, being polyadenylated at the 3' terminus is recovered by chromatography over oligod(T) cellulose. The messenger eluted from this column is recovered by ethanol precipitation, and its purity is established by translation, in vitro.

By using these modifications specific messengers encoding seed lectin and Kunitz trypsin inhibitor and the enzyme glutamine synthetase have been isolated from plant tissue. In the last example, glutamine synthetase (GS) mRNA was purified to near homogeneity from polysomes of phaseolus vulgaris root nodules. When translated, this mRNA directed the synthesis of a single polypeptide which co migrated with purified GS when analyzed by SDS/PAGE. This purified GSmRNA was then radioactively labeled and used to screen a "library" of 1,100 cDNA clones prepared from *P. vulgaris* root nodule mRNA. A single clone in this "library" was identified as containing sequences encoding GS. The methods used in this study are described in the protocols. This technique may also be used with Monoclonal antibody and it has recently enabled the purification of messenger sequences of relatively low (0.1%) abundance.

Despite these recent successes, several caveats should be observed. Firstly, the production of antiserum is often a limiting step, particularly for proteins of low abundance or for those which are otherwise difficult to purify. This problem may be overcome by the use of Monoclonal antibody which can be produced in bulk in cell cultures. Secondly, even when antiserum has been obtained, it may not cross†react with the nascent polypeptide. Thirdly, the isolation of polysomes translating the mRNA of interest may also pose problems. Time is well spent in screening for plant material where the desired mRNA is of the highest possible relative abundance. With these limitations in mind, we present the protocol by which the GS mRNA was immuno purified at the end of the chapter.

18.11. DIRECT SCREENING OF cDNA CLONES WITH ANTIBODIES

Antibodies may also be used for the direct screening of a cDNA "library" when the vector†host combination chosen is capable of expressing the cloned DNA sequence as a polypeptide. Such techniques have been successfully carried out in "model" system but have not yet achieved widespread acceptance. These methods depend on in situ lysis of recombinant transformants to release the polypeptide which may then be immobilized for detection with a labeled antibody. Various methods for the immobilization of the antigen have been described, including binding to nitro cellulose filters, covalent attachment to cyanogen bromide activated paper and reaction with specific antibodies bound to polyvinyl discs.

In order for such methods to succeed, several prerequisites must first be satisfied. The DNA sequence must be introduced to a vector in such a way as to ensure its expression. For cloning in bacterial systems, it is therefore necessary to insert the sequence "downstream" of a suitable bacterial promoter. Because "foreign" proteins made in bacterial cells are frequently, rapidly degraded, protocols based on the immuno detection of recombinants have therefore usually in included in the construction of fusion proteins the DNA to be introduced to the vector is cloned at a site in the structural region of a bacterial gene, so that the N terminal portion of the polypeptide product is bacterial and the C terminal portions is "foreign". For an expressed polypeptide to be successfully recognized by an antibody, it is necessary that the introduced DNA sequence is cloned in the correct orientation and in the correct reading frame relative to the bacterial gene. This means that only one†sixth of the recombinants will be likely to produce a recognizable protein. It is therefore advisable, if adopting this cloning strategy, to utilize a system which will yield a large number of recombinants. It is probable that the most successful vectors for this purpose will be a kind to the phage lambda. It is also important to recognize that the detectability of the fusion product will further depend on the nature of the cDNA sequence that has been introduced. Since the antibody antigen association is based on the recognition by the antibody of the conformation of the antigen, fusion in which the antigenic polypeptide constitutes only a small proportion of the hybrid protein are less likely to be detected. Although direct screening of recombinants is still relatively restricted in application, it is to be hoped that further "tuning" of the cloning vehicles will result in such protocols becoming more generally used.

18.12. IMMUNO BLOTTING FOR THE SCREENING OF MONOCLONAL ANTIBODIES

1. Resolve the protein mixture by SDS (–) PAGE (Laemmli, 1970) using a slab gel.
2. Cut a piece of nitrocellulose membrane (45 μ m pore size)to match the shape of the gel and pre soak it in transfer 192 mM glycine; 25 mM Tris HCl; 20% methanol, pH 8.3.
3. Lay the membrane over the gel, being careful to avoid air bubbles and sandwich the gel and membrane between two layers of a soft porous material, e.g., nylon scouring pads or foam plastic.

4. Immerse the whole assembly in transfer buffer, so as to pass an electric current transversely through the gel. Several commercially made apparatuses are obtainable for this purpose.
5. After transfer, cut the filter as required and pre incubate for 30 minutes in phosphate buffered saline (PBS 10 mM Na phosphate; pH 7.2, 150 mM NaCl) containing 8%(w/v) bovine serum albumin (BSA) with shaking at room temperature for 30 minutes.
6. Incubate filters with antibody in PBS and under the same conditions as step 5 for at lest 2 hours (it is often convenient to carry out this step overnight). Keep the volume of the incubation as low as possible, whilst maintaining the filter immersed.
7. Wash the filter 3 4 times with PBS (no BSA) to remove un bound antibody.
8. Incubate the filter in PBS plus BSA containing 10 cpm 125 I labeled protein A for two hours, keeping the volume as low as is commensurate with the immersion of the filter by freely moving liquid.
9. Remove unbound protein A by serial washing in PBS containing 1% Triton X 100.
10. Dry the filter and expose it to X ray film to locate the bound radioactivity.

18.13. PURIFICATION OF ANTIBODIES FOR IMMUNO PRECIPITATION

Immuno precipitation of translation products can be carried out with relatively impure serum preparations. However, optimization of the conditions is better achieved if purified IgG is used for the immuno purification of polysomes, it is essential to use a pure ribonuclease free Ig G fraction. We have used affinity purification or ion exchange chromatography.

1. Crude antiserum (5 cm) is passed over a column of 3 Protein A Sepharose (Pharmacia, 1.7 cm).

2. The column is washed with PBS.

3. Elute and precipitate the IgG by the addition of ammonium sulfate to 40%.

4. Recover the IgG by centrifugation, dissolve the precipitate in PBS and dialyse against PBS at 0°C.

5. Prepare a column 12 cm X 1 cm of DEAE cellulose (DE 52) overlaid with an equal quantity of CM cellulose (CM 52)

6. Equilibrate and sterilize in PBS buffer.

7. Apply the IgG in PBS and elute in the void volume. This fraction is electrophoretically pure, ribonuclease free

Ig G. It can be stored frozen at 20°C, and thawed without loss of activity.

18.14. IMMUNO PRECIPITATION OF POLYPEPTIDES FROM CELL FREE TRANSLATIONS

Translation of mRNA is carried out in small volumes of wheat germs S 30 extract or rabbit reticulocyte lysate. The following protocols apply to a 20 ml reaction mixture.

1. Remove 2 X 1 ml aliquots to determine incorporation of radioactivity into TCA insoluble material.
2. Add 1 ml RNase A (50 mg cm). 1 ml RNase T1 (50 units) and digest at 37°C for 30 minutes.
3. Add 2 ml 10% SDS and boil for 2 minutes.

4. Dilute the mixture to 200 ml by the addition of PBS containing 1% BSA.
5. Add 10 mg Ig G and incubate at 37°C for 2 hours.
6. Weigh out 10 mg Protein A Sepharose (Pharmacia) and swell 3with 0.8 cm PBS containing 1% BSA and 1% NP 40.
7. Add the mixture from Step 6 to the immunoprecipitation 3 reaction mixture in a 1cm Eppendorf micro centrifuge tube and mix at 0°C for 1 hour.
8. Transfer the slurry to a column made from a 1 cm disposable syringe plugged with glass wool and allow to drain.
9. Wash the bed thoroughly with 5 10 cm PBS containing 1%NP 40 (no BSA).
10. Elute the bound immuno precipitate into a 1.5 cm Eppendorf 3 tube with 3 successive washes of 1 M acetic acid (0.4 Cm each).
11. Add 70 ml 100% TCA and incubate at 0°C for 30 minutes.
12. Recover the immunoprecipitate by spinning in a micro centrifuge for 2 minute.
13. Remove supernatant, wash the pellet twice with acetone, and vaccum desiccate.
14. Dissolve the pelet in 25 ml 0.1 M Na_2CO_3 containing 0.1M dithiothreitol.
15. Mix with an equal volume of 4% SDS in 20% glycerol containing bromophenol blue as a tracking dye.
16. Heat at 80°C for 5 minutes.
17. Remove aliquots of the dissolved immunoprecipitate to determine the incorporated radioactivity by liquid scintillation counting and analyse the remainder by SDS PAGE.

18.15. IMMUNO PRECIPITATION OF POLYSOMES

Care must be taken during the isolation of polysomes in minimising the risk of degradation by the ubiquitous ribonucleases. This can best be achieved by the inclusion of ribonuclease inhibitors in the extraction buffers. Heparin, 2' + 3' AMP, detergent and a buffer at high pH and low temperature all help to reduce RNase activity. Trichodermin, a reported inhibitor of translation termination may help to increase polysome yields. All procedures are carried out at 0 – 4°C using sterilized solutions and equipment.

1. To isolate polysomes, freeze tissue in liquid nitrogen, grind to a powder using a coffee grinder and mix well in a mortar with polysome extraction buffer 200 mM Tris HCl, pH 9.0, 120 mM KCl, 50 mM $MgCl_2$, 0.8 mg dithiothreitol, 0.8 mg 2' 3', AMP, 0.1 mg Na Heparin 1 mg trichodermin, 6.8% (w/w) sucrose.
2. Centrifuge in 100 cm tubes at 150 x g for 10 minute and decant the supernatant through 2 layers of sterile muslin.
3. Add Triton X 100 (a stock solution of 10% in polysome buffer) to a concentration of 1% and stir for 10 minute.
4. Transfer to 40 cm tubes and spin at 34,000 Xg for 25min.
5. Aspirate the supernatant and layer over 4 ml cushions of 54% (w/w) sucrose in polysome buffer in 30 cm ultracentrifuge tubes.
6. Spin at 98,000 X g for 4 hours and decant the supernatant.
7. Rince the polysome pellets with suspension buffer 50 mM 3 Tris HCl, pH 7.5, 300 mM NaCl, 5 mM MgC12, 0.1% NP 40, 0.1 mg cm 3Na Heparin,1μg cm trichodermin.

8. Resuspend the polysomes in this buffer to a low 3 concentration corresponding to A_{260} of 5 10 units cm.
9. Remove unresuspended polysomes by centrifugation at 17,000 X g for 10 minute.
10. Incubate the polysomes with the antibodies by gently mixing ribonuclease free Ig G with polysomes (approx, 25 mg per A 260 unit) and incubating overnight at 0°C.
11. Isolate the antibody bound polysomes by passing the antibody polysome solution twice through a column of Protein A Sepharose equilibrated in polysome suspension buffer at a slow 3flow rate of 20 cm /hour.
12. Wash the column with 30 cm buffer.
13. Dissociate and elute the bound polysomes with 15 ml 25 mM Tris HCl, pH 7.5, 20 mM Na2 EDTA.
14. Isolate poly adenylated RNA by making the Sepharose elute 12 mM Tris HCl, pH 7.5, 0.5 M NaCl, 10 mM Na2 EDTA, 0.2% SDS using concentrated solutions of the components.
15. Heat to 60°C for 5 minutes.
16. Cool the solution to room temperature and pass it over a 30.4 cm column of oligo d(T) cellulose equilibrated with step 14 with 3rd step buffer at a flow rate of 15 cm /hour.
17. Wash the column with this buffer until the A_{260} of the eluate returns to its baseline value.
18. Elute the mRNA with 0.8 cm sterile distilled water.
19. Add 5 mg calf thymus tRNA (carrier), 80 ml 4M ammonium acetate, 2 ml ethanol and precipitate at 20°C overnight.
20. Recover the RNA by centrifuging for 15 minute in a micro centrifuge.
21. Wash the pellet 3 times with 75% ethanol.
22. Vaccum desiccate the RNA preparation.
23. Dissolve in 10 ml sterile distilled water.
24. Check the purity of the preparation by translation, *invitro*.

18.16. DERIVATION OF HYBRIDS SECRETING SPECIFIC ANTIBODIES

18.16.1. Myelomas

By physicochemical methods it is possible to separate the Ig fraction of a serum to its component classes and, with some difficulties, subclasses, but the separation of individual Ig, differing only for the variable region, is generally impossible. Tumors of Ig producing cells occur spontaneously in many mammals. They quite frequently arise in man and in particular strains of mice and rats. The associated pathological disorder was originally classified as "multiple myeloma" in man, because multiple tumors develop in the bone marrow (or myelogenous tissue). The disease is generally diagnosed by the appearance in serum and urine of particular proteins, called "myeloma proteins" easily detectable by electrophoresis. Although these proteins were described by Bence Jones in the nineteenth century, only in the 1950's was their relationship with antibodies recognized, and the cell type responsible for their secretion identified as the plasma cell (hence the introduction of the name of plasmacytoma).An essential characteristic of multiple myelomas is the appearance of tumors in several sites of the affected animal, but it has been demonstrated that all of these actually derive from the uncontrolled expansion of one clone, or occasionally a few clones, of plasma cells. The tumor, therefore, is generally monoclonal in origin and the term monoclonal is now used when referring to the myeloma protein as well as to the cells. Furthermore, monoclonality can be ensured in any case in which the malignant cells can be grown in vitro and where they can be subcloned. Nearly all we know about Ig structure and genetics comes from the study of either naturally occurring or chemically induced myelomas.

Over the years a number of immunologists have made considerable efforts to find myeloma proteins with a defined specificity (i.e, antibody activity). Attempts were made to induce specific tumors with carcinogenic chemicals in immunized animals. Alternatively, already existing myelomas were tested for their ability to bind to a wide variety of substances (natural or synthetic) known to be good antigens. Another method, the induction of B cell tumors by certain viruses, was more successful but the tumor lines thus derived are often difficult to handle, because of their poor growth characteristics in vitro. Despite great efforts progress along these lines was very slow. Indeed, the introduction of the highly efficient and robust cell fusion method for specific antibody derivation swept the other approaches aside in the realm of experimental immunology. We should mention, however, that in the case of the production of human monoclonal antibody, the use of in vitro Epstein Barr virus transformation of B lymphocytes is currently, and may remain a popular technique.

18.16.2. Somatic Cell Fusion

Hybrids produced by somatic cell fusion initially possess two sets of chromosomes, one set derived from each of the parent cells. During cell division some of the chromosomes may be unable to duplicate properly and are lost. After a number of cell cycles, however a more stable reduced karyotype is achieved. In inter specific hybrids, this reduction process is not random but involves preferential loss of one of the parental species chromosomes. Best documented combinations are human mouse, human Chinese hamster and mouse rat hybrids from which, respectively, the human, the chinese hamster and the rat chromosomes are preferentially lost.

The phenomenon of differential chromosomal loss has been extensively exploited for mapping of phenotypic characters to different chromosomes and to study gene regulation and dominance. The potential of somatic cell fusion to immortalise in the permanently growing cell line a phenotype otherwise only transiently expressed in normal cells, was not fully realized before the experiment of Kohler and Milstein in 1975. When a tumor cell is fused with a normal cell, hybrids can be derived with growth characteristics of the tumor and expressing some of the phenotypic characters of the differentiated normal cell.

The production of a protein by a cell depends not only upon the presence of the structural gene for that protein but also upon appropriate regulation. Thus, in a hybrid cell, the expression of a phenotype derived from one parent may or may not be apparent depending upon the nature or state of the other parent. For example, hybrids derived from myelomas and fibroblasts never express Ig while, on the contrary hybrids between myelomas of two different species like mouse and rat secrete Ig's of both parents (co dominant expression). The availability of cell lines suitable for fusion has played an essential role in the development of the methodology for the derivation of hybrids secreting monoclonal antibody.

18.16.3. HAT Selection

When using a permanently dividing cell line as a parent for hybrid production one needs to select the newly formed hybrids from the parent. The usual way to achieve this is to use parents carrying markers, such as an enzyme deficiency or a drug resistance. The most commonly used selection system is called "HAT" selection from the use of culture media supplemented with hypoxanthine, aminopterin and thymidine. The method was first described by Szybalska and Szybalsky (1962) and later by Littlefield (1964). Originally it was used as a classic complementation system, in which one cell lacks the enzyme hypoxanthine guanine phosphoribosyl transferase (HGPRT, E.C. 2.4.2.8) and is therefore unable to make use of hypoxanthine for synthesis of purines, and the other cell lacks a different enzyme (thymidine kinase, TK, E.C.2.7.1.21) and cannot utilize thymidine. Both types of cells are killed by aminopterin, a potent inhibitor of de novo synthesis of DNA precursors. Hybrids, on the other hand, can easily survive by complementation of the enzymatic activities from the two parents that enable them to use both hypoxanthine and thymidine.

For monoclonal antibody derivation the HAT system is used in a similar way, but here one of the parents, the myeloma, is lacking either HGPRT or TK and is unable to grow in HAT, while the other parent, the lymphocyte, is unable to grow in vitro for more than a few cell cycles. Hybrids are able to grow in these conditions because they inherit from the myeloma parent the ability to divide in vitro and from the lymphocyte the correct enzymatic complement to grow in HAT.

18.16.4. Selectivity of somatic cell fusion

To derive hybrid myelomas a large number of spleen cells from an immunized animals are fused with cells of a suitable myeloma line by the use of polyethylene glycol (PEG). The population of cells obtained by mechanical disruption of a spleen contains a mixture of different cell types. Of the lymphocytes about 40% are T cells of different sorts and the remaining 60% include all the differentiation stage of B cells from pre B cells to mature plasma cells and memory cells. At any one time less than one cell in a thousand in an immunized spleen produces Ab to the antigen used for the immunization, and less than 5% secretes Ig of any sort. In contrast the population of hybrids derived by fusing spleen cells with a myeloma cell line is much more homogenous in that none of the hybrids appear to have T cell characteristics, and 60-90% of them secrete Ig. If, instead of a myeloma, a T cell lymphoma is used for fusion with spleen cells the resulting hybrids bear T cell markers and none secrete Ig. It appears therefore, that the phenotype of the tumor parent is a major selective influence on the outcome of cell fusion experiments.

18.17. PLANNING OF THE FUSION EXPERIMENT

The goal of a cell fusion experiment is the immortalization of a cell secreting a specific antibody. The overall design of the experiment is schematically represented. At each step of the procedure (immunization, fusion, screening etc.,) multiple choice of methods and approaches are possible and careful planning is important for success. The immunization of an animal with an antigen increases the number of cells in the spleen expressing a specific antibody. The state of immunization of the animal is therefore an important factor in determining the success of the fusion and so in any experiment it is sensible to test the serum of the animals for the presence of the specific antibody and to choose those animals making the strongest response.

Unless an effective immunization protocol is already known for the antigen of interest, it is advisable to inject a number of animals, and preferably animals of different species (mice, rats) or strains. The use of inbred strains of animal is advisable to avoid problems in the production of ascites where histocompatibility between tumor and host is necessary. The more significant scientific reason for choosing to work with rats or mice is the relative responsiveness of these animals to the antigen. In many cases both species respond well and the choice is made on other grounds. The choice of species is more or less restricted to mouse and rat for which in vitro growing plasma cytomas of high fusion potential are available. Cross species fusions can be done, and are reasonably successful between mouse and rat and, between both Syrian and Armenian hamsters and mouse. But with increasing phylogenetic distance the problems of karyotypic instability and low fusion efficiency seems to increase e.g. with rabbit and with man. Where practical, however, intraspecies fusions are to be performed. A list of mouse and rat plasma cytomas in general use is given. As mentioned there, some of these tumor cell lines still produce their own Ig chain and, therefore, the hybrids made from them secrete not only the specific spleen cell derived antibody but also molecules built up by association of Ig chains from both of the hybrids parents. From cases such as this hybrid molecules are being produced, it is possible to derive variants which no longer express one of the chains. The pattern of Ig chain loss is not random but usually follows the rule that the number of different heavy chains produced is equal to or less than the number of different light chains. For general use the non-secreting plasmacytomas are preferable.

Finally, in this consideration of the planning of the fusion experiment, it is worth stressing an issue of practical importance. Before entering into the experiment itself one is at an advantage if one has devised an assay system which can test for antibody activity in hundreds of hybrid culture supernatants per day. The assay, has, therefore, to be fast, reliable and applicable to small samples. A variety of immunological assays are available for the detection of antigen antibody reactions and each type of assay will detect some types of antibody better than others. For example a conventional radioimmunoassay using a labeled ligand will work better with high affinity antibodies, while an agglutination test will depend more on multivalent binding and so will detect preferentially IgM antibodies. It is advisable, therefore, to design the screening system for the fusion to reflect the anticipated use of the specific antibody, if this is compatible with assay speed.

18.17.1. Diary of a successful experiment

In this section we describe the derivation of a set of monoclonal antibody as it was actually done in our laboratory. This project was the result of a collaboration between the Monoclonal antibody center and a group of scientists at Long Ashton Research Station (LARS); the scientific results are published elsewhere. Here we intend to illustrate the time scale involved in such a project and to give some examples of the difficulties encountered and some possible approaches to their solution. A wide variety of methods were used and they have all been described elsewhere here we will only discuss the principles involved in the method used in order to justify our choices.

18.17.2. Antigen preparation

The Ag in question, higher plant nitrate reductase, is a complex molecule formed of two subunits of calcium M.W. 100,000 and a molybdenum containing co factor in the form of the pterin complex. Purification of the entire complex has been possible in only a few species and one of the uses of monoclonal antibody will be the improvement of purification methods. The LARS team had developed good methods of purification before the project started. In general monoclonal antibody derivation purification of the Ag is not an essential prerequisite. In fact probably the most important aspect of the monoclonal antibody technology is the ability to derive Ab to Ag which had been impossible to purify (a classic example being interferon; or that have not even been described before, as a very large variety of literature about cell surface antigens illustrates). Nevertheless, wherever possible, purification of the Ag at least to some extent will greatly help. It is worth pointing out that chemical purity and immunological purity do not necessarily coincide. An Ag chemically contaminated with substances that are not antigenic in a particular animal, is in fact immunologically pure in that particular context. The purity of the Ag is only important in two considerations.

(*a*) If the impurity is more antigenic than the Ag itself, or able in any way to reduce the specific immune response.

(*b*) The method used to test for the response must be capable

of distinguishing between the presence of Ab to the Ag and Ab to the impurity.

18.17.3. Planning the immunization and experiments

Day minus 83 : Four female F344 rats were injected subcutaneously with 200 ml (200 mg) of Ag dissolved in buffer and emulsified in complete Freund's adjuvant (CFA). The choice of animals has already been discussed. Here we did not even try mice as we expected an immediate good response; the Ag is a large plant protein and therefore is likely to be a strong immunogen. Apart from their availability at that time in our animal house, F344 rats were chosen because this inbred strain carries a particular allotype of the Ig k light chain, called "b", that is different from that carried by the myeloma Y3 that we intended to use. This is a mutant myeloma that only secretes light chain (K) and therefore Ab secreting hybrids derived from it, are expected to express not only the specific heavy (H) and light (L) immunoglobulin chains of spleenocyte origin, but also the light chain (K) of the myeloma. The allotypic difference in light chains between F344 rats and LOU rats (from which Y3 was originally derived) offers a tool to differentiate the two light chains secreted by the hybrids.

18.17.4. Day Minus 56

The animals were boosted by a subcutaneous injection of 200 μl (200 μg) of Ag dissolved in buffer and emulsified in incomplete Freund's adjuvant (IFA).

18.17.5. Day Minus 44

Each animal was tail bled and the serum was tested for the presence of specific Ab. This first bleed is used not only to monitor the immune state of the animal, but also, and perhaps more importantly, to set up and optimise the assay for specific Ab. In this example we decided to adopt a double screening procedure, all our antisera and hybrid supernatants would be assayed for their ability

(*a*) to bind to purified Ag in an ELISA assay

(*b*) to inhibit specific enzymatic activity of the antigen.

The ELISA assay consists of coating plastic microtitre plates with the purified Ag. Antisera or hybrid supernatants are then incubated in each well, the excess unbound Ab washed away, and the presence of bound Ab is revealed by the binding of a second, enzyme labeled antiserum specific for rat Ig. Properly set up, this assay is extremely fast and reliable 400-500 samples can be easily tested in 8 hours. The problem is that it requires Ag pure enough to coat the plate. The process of binding to plastic is a more or less non specific step. Most proteins will stick to the plastic, generally in direct proportion to their concentration in the coating solution. If the same impure preparation of Ag is used for immunization and for coating of the ELISA plate, the assay will not distinguish between Ab to the Ag and Ab to the impurity if these are present at a sufficiently high level. In our case the Ag was 90% chemically pure and therefore we expected very few false positives.

The second assay, inhibition of enzymatic activity, is strictly related to the biological properties of our Ag, and is particularly interesting assay because, although more difficult and time consuming than the ELISA (in 8 hours only a few dozen assays could be performed) it immediately gave information about the possible binding sites of the Ab along the Ag molecule and the reassuring feeling, when positive, that we were getting a response against the right specificities. In addition, it could be performed with crude plant extracts containing the Ag. The sera from the first bleeds of our animals were all positive in the enzyme inhibition assay the worst at a dilution of about 1 in 200 and the best at 1 in 1200.

Day minus 31: Second boost of the animals exactly as on Day minus 56.

Day minus 21: second bleed and test of sera. All the sera appear to be positive, but they were all now positive only to a dilution of 1 in 500. We never found a good explanation for this and at the time our ELISA assay was not working reliably because of a problem in antigen purification. Nevertheless, we decided to proceed with rat no. 1 as the immunization seemed acceptable. It is worth noting that in many cases one does not really have experience of what to expect in terms of antiserum titers in rats or mice since conventional antisera are usually prepared in rabbits.

Day minus 5 : to increase fusion efficiency it is essential that the myeloma cells are kept in exponential growth for a few days before fusion. Myeloma cultures are generally fed every other day in our laboratory and not during weekends, but a few days before fusion special care should be taken to get them in top shape. If myeloma cell lines are not kept growing continuously in the laboratory, then cultures should be started at least 10 days before fusion.

Day minus 3 : The selected animal, and a second reserve one, are boosted with an intravenous injection of 100 μl (100 μg) of Ag dissolved in physiological saline.

18.18. CELL FUSION AND SELECTION OF HYBRIDS

18.18.1. Day 0

The animal selected and injected on Day minus 3 is killed, the spleen is removed and cells are used for fusion according to our standard protocol. Half the spleen cells are fused with the rat myeloma Y3 Ag1.2.3 (or Y3 for short) and half with the myeloma YB2/0 (or YO for short). This is generally done because fusion experiments are not very reproducible and it is wise to work with different cell lines to avoid heavy loss in case of contamination. After PEG treatment the cells are seeded into 3 14,24 well plates (2 cm well) in dilutions such that about 4 × 6. 110 spleen cells well are seeded into the wells of the first 6 1 × 6 1 plate, 2 × 10 well in the second, 10 well in the third and 5 15 × 10 well in the last. Each well is numbered individually so that we have wells from 1/1 to 1/96 and from 2/1 to 2/96.

18.18.2. Day 1

All the plates are carefully checked for contamination and number of cells growing. Since we feed with normal medium on Day 0 the parent myeloma cells are dividing normally. Half the supernatant is removed from each well and replaced with selective medium containing HAT.

18.18.3. Day 2

Parental myeloma cells show signs of dying. Plates are fed as at day 1 with HAT medium.

Day 4 : Feed with HAT

Day 7 : All parental myelomas dead. Few if any, cells alive, feed with HAT

Day11 : Foci of cells growing in some of the wells in high concentration plates (no.1 may be hybrids. Feed all with HAT.

Day14 : Hybrids obviously growing in atleast 25% of the wells. Feed all with HAT.

Day 18 : Several wells (33 out of 192) now have confluent cell growth. One of the plates had several wells contaminated with fungus and, as none of the remaining wells showed signs of growing hybrids (this was a plate seeded at high dilution) the plate was discarded to avoid spreading the infection. Supernatants are harvested for testing and half the cells are removed and transferred to "duplicate plates". Original plates are fed with HAT medium and the duplicates with medium containing only HT. From now on plates are fed twice weekly and if new wells become confluent in the original plates, duplicates are prepared. Duplicate wells keep the same number as the original well because they presumably contain the same cell population (polyclonal at this stage) and we identify original wells as "cell lines". Different wells or tubes or flasks that are prepared from the same "line" are given the line name, e.g. 1/26. followed by some other specification in brackets, like plate number of tube number, e.g., 1/26 (P2); 1/26 (P6) and 1/26 (T1).

18.19. SCREENING AND SELECTION OF MONOCLONAL ANTIBODY SECRETING LINES

Day 19/20: Of the supernatants tested more than half are found to be positive by ELISA and of these only half (9/33) show any ability to inhibit the enzyme activity of the Ag (nitrate reductase). At this stage we have many more hybrids from the Y3 fusion (25) than from the YO fusion (8) but the distribution of positives was normal (17/25 and 3/8). Later more hybrids appear in both fusions and we end up with more or less the same number of interesting lines from each fusion.

Day 21 : All confluent wells are fed by removing half of the cells and adding fresh medium. The cells removed are generally discarded unless duplicates need to be prepared, either because the line is an "interesting" one (and therefore deserving of special treatment) or because duplicates do not exist (not yet made or lost because of contamination).

Day 25 : Supernatants from all confluent wells are harvested for assay. A dozen wells are lost because of bacterial contamination, but duplicates already existed. New duplicates are prepared where necessary. All plates are fed.

Day 26 : ELISA and enzyme inhibition assay. More than half the lines tested are positive by ELISA, and about a quarter are also positive for enzyme inhibition.

Day 27 : Five lines are chosen as the most interesting positives comparing the results from the two screens already done (at Day 20 and Day 26) and cells from actively growing wells already scored twice as positive are cloned in soft agar. For a dozen of the positive lines separate cultures are prepared by 3transferring cells from confluent wells into flat saded 10 cm tissue culture tubes.

Day 28 : All plates are fed as usual.

Day 29 : Several of the tubes prepared on

Day 27 : Are confluent, supernatants are separated by centrifugation and the cells are frozen in vials for liquid nitrogen storage. From now on, frozen samples of positive cultures are prepared by this method whenever cells are available as sufficient numbers.

Day 32 : All plates are fed. Supernatant from all confluent cultures are collected for testing. Wells without actively growing cells are drained and washed with alcohol half of the original wells. Of the five lines cloned on Day 27 four are contaminated by bacteria and the agar plates have to be discarded.

Day 33 : On the basis of new tests a second attempt at agar cloning is made for six lines (three already attempted at Day 27 and three new ones).

Day 35 : All plates are fed.

Day 39 : All plates are fed and supernatants collected for testing.

Day 41 : For each of the seven lines cloned (1 day 27 and 6 day 33) 48 individual clones are picked from the agar plate using 3Pasteur pipettes. Each clone is transferred into a 2 cm well of a 24 well plate and fed with medium containing 20% FCS but neither HAT nor HT. A total of 14 new plates are therefore prepared. Each pair of plates containing clones of the same line is labeled with the line name and each well is numbered (from 1 to 48). Individual clones will be identified, therefore, by their "line name" and clone number e.g. from line 1/26 prepare clones 1/26.1 to 1/26.48.

Day 42 : All the original and duplicate plates are fed (not the ones containing clones).

Day 46 : All plates are fed (not the ones containing clones).

Day 49 : As day 46.

Day 53 : Several of the clones picked on Day 41 have become confluent. Supernatants from all confluent cultures are collected for screening and all plates are fed. A dozen wells in the original plates are lost from yeast contamination.

Day 56 : Some original culture of uncloned lines that are still positive are transferred from wells to 25 cm tissue culture flasks for further expansion; all the original and duplicate plates are then stopped. For each of the clone lines, three positive clones are chosen to be kept; for these, duplicate cultures are prepared by transferring half of the cells from the wells to 25 cm tissue culture flasks. All plates containing clones are fed.

Day 57: The flasks prepared on Day 56 are examined for contamination and cell growth. If necessary, more fresh medium 3containing 10% FCS is added (to a maximum of 20 cm).

Day 60: All the plates are fed, supernatants are collected for testing only from wells containing clones chosen as positive. Flasks are examined for cell growth and, in the case of clones, whenever possible duplicate flasks are prepared. For each cloned line (six of them) we have now three selected clones and for each clone we have a growing 2 cm well and two 25 cm flasks. We keep feeding the wells with medium supplemented with 20% FCS, but the flasks are fed first with 10% FCS and later, if the cells grow well, with 5% and then 2.5% FCS. To keep duplicates and fed them with a medium of different composition and therefore from different batches of bottles, is the only way to avoid losses due to contamination.

Day 61: Tests of supernatants show that we have succeeded in deriving six different Monoclonal antibody.

Supernatants from sister clones can now be pooled and more intensive characterization work gets under way. Of course, some preliminary experiments on the specificity of the Ab have already been done, but up to now it has not been wise to invest too much effort in any of the lines as they have been neither monoclonal nor certain of survival.

18.19.1. Conclusion of the experiment

From day 62 onwards, the experiment proceeded as follows. As soon as possible we prepared at least six frozen samples for each clone. Cells from each clone were grown for 18 hours in a medium containing a radio-labeled amino acid and spent supernatants were analyzed by SDS polyacrylamide electrophoresis. Two out of three clones for each line were then discontinued and only one "master" clone per line was kept growing. Only later, when more information was available about the specificity of the different monoclonal antibodies, cells from some of the cloned lines were injected into the appropriate rats for tumor generation and ascites production. When sufficient stocks of supernatant and ascities had been prepared, growth of all the lines was discontinued, frozen cell stocks being available whenever (or if ever) stocks of Ab run out.

A few more hybridomas different from the first six whose derivation has been described here could probably be prepared if necessary by thawing all the samples frozen between Day 29 and Day 60 and cloning them, but before attempting more cloning we like to study the supernatants stored to identify different and useful monoclonal antibody. Thawing and cloning lines frozen at early stages of the experiment is not always successful as only a small percentage of cells are actually recovered from liquid nitrogen storage and, therefore, more intensive cloning is sometimes required.

Thus nine weeks after the cell fusion had been performed, a set of six monoclonal antibody selected for their differential abilities to inhibit the various enzymatic activities of nitrate reductase, where cloned and reasonably secure. This, of course, had been a particularly smooth running experiment in which none of the common snags of this kind of work (e.g., contamination of cultures and assay breakdowns) had interfered.

The final events in the story are as follows. First, once it is clear that these cell lines were to be included in a published paper, they were given "Institutional" names in this case being renamed AFRC MAC 75 to 79. Second, as is now our routine procedure, the cell lines were given an extra layer of security by being deposited in the National Collection of Animal Cell Cultures at Proton Down, Wiltshire, U.K. This is an important safeguard against accidental breakdown of our liquid nitrogen storage system.

The last 10 years has seen the application of monoclonal antibody to all branches of animal physiology and biochemistry and has seen enormous benefits from the new approach. It is now apparent that the potential value of monoclonal antibody to plant science is being recognized a fact exemplified by the contributions to this volume. Although there may be particular problems to be faced, such as the extreme and, therefore, dominating immunogenicity of certain plant glycosidic structures, we can look forward to a flourishing relationship between those unexpected partners, immunology and botany.

QUESTIONS

1. Explain the basic structure of antibody.
2. Discuss the immunoglobulins encoded by multiple gene segments.
3. Write short notes on:
   (*i*) Fab;
   (*ii*) Fc;
   (*iii*) Myeloma proteins;
   (*iv*) Isotype;
   (*v*) Allotype;
   (*vi*) Idiotype;
   (*vii*) Complementarity determining regions.

CHAPTER 19

DESIGNING VACCINES FOR ACTIVE IMMUNIZATION

Several factor must be kept in mind in developing a successful vaccine.First and foremost the development of an immune response doesnot necessarily mean that a state of immunity has been achieved often the because of the immune system that use activated in critical and therefore vaccine designs must recognize the important difference bet activation of the humoral and cell mediated branches.

A seasonal factor is the development of immunologic memory. For eg, a vaccine that induces a protective primate response may fail to induce memory cell formation leaving the host unprotected after the primary response to the vaccine subsides.

The sole of memory cells in immunity depends, in past on the incubation period of though pathogen. In the case of influenza virus, which has a very sheer incubation period (<days), diseases symptoms cue already under way by the time memory cells are activated effective protection agents influenzas therefore depends on maintaining high levels of neutralizing and by repeated re immunizations.

For pathogens with a longer incubation period demonstrable neutralizing and at the time of infectious not necessary. The ***polioviruli*** for eg requires more than 3 days to begin to infect the central nervous system. An incubation period of this length provides the necessary time for memory B cells to respond with production high levels for serum Ab. Thus the vaccine for periods obsignals ti induce high levels of immunologic memory. Following immunization with the stalk vaccine, serum Ab levels pick with in 2 weeks and than decline, but the memory response continues to climb, reaching maximal levels at 6 months and persisting for fears.

If an immunized individual is later expressed the ***poliovirus***. These memory cells will response by differentiating into protect the individual from infection.

19.1. WHOLE ORGANISM VACCINES

Many of the common vaccines currently in use consists of inactivated (killed) or live but attenuated (avirulent) bacterial cells as viral patients the primary herea esistic's of these two types of vaccines are compared in table.

19.2. CLASSIFICATION OF COMMON VACCINES FOR HUMAN

Disease and pathogen.	Type of vaccine
Whole organisms	
Bacterial cells.	
Cholera	Inactivated
Pertains	Inactivated

Plague	Inactivated
Tuberculosis	Attenuated BCG

19.2.1. Viral particles

Influenza	Inactivated
Measles	Attenuated
Humps	Attenuated
Polio (cabin)	Attenuated
Polio (stalk)	Inactivated
Rubella	Inactivated
Varicella zoster (chicken pox)	Attenuated
Yellow fewer	Attenuated

19.2.2. Attenuated viral and bacterial vaccines

In some cases microorganisms can be attenuated so that they lose their ability to cause significant disease (pathogenecity) but retain their capacity for transient growth C in on invocated host.

Attenuation can after be achieved by growing a pathogenic bacterium or virus for prolonged period as under abnormal culture conditions. This produce selects mutants that one better suitable to growth in the abnormal culture conditions and are therefore less explore eg growth in the original host.

For e.g., an attenuated strain of mycobacterium bodies called Bacillus Chalmette Guerin (BCG) was developed by growing *M. bovis* on a medium containing tied concentration of bile after 13 years this strain had adapted to growth with tied bile and had become sufficiently attenuated that it was suitable as a vaccine for tuberculosis the satin polio vaccine and the measles vaccine consists of viral strains that have been successfully attenuated and now serve as successful vaccines.

The *poliovirus* used in the satin vaccine was attenuated by growth, in a monkey kidney epithelial cells.

The measles vaccine contains a strain of *rubella virus* that was growth duck embryo cells and later in human cells lines.

Attenuated vaccines have and some advantages.

Because of their capacity for transient growth, such vaccines provide prolonged immunesystem to the individual epitopes on the attained organism, resulting in used immunogenicity and memory cell production.

As a consequnces these vaccines often require only a single immunization eliminating the need for repeated boosters.

The ability of money attenuated vaccines to replicate with in host cells makes them particularly suitable for including a cell mediated response.

The satin polio vaccine, consisting 3 attenuated straining of poliovirus is administered orally to children on a sugar cube in sugar liquid.

The attenuated viruses colonize the intestine and induce protective immunity to all three stained of ipelociues. The ability of the attenuated sabin vaccine to colonize the intestines enables to induce protection of secretion Ig A. Which serves as an important defence against acquired ***polio virus***. The vaccine also induced Ig M and Ig G classes of Ab.

Unlike most other attenuated vaccine with required signal immunizing dose, the sabin polio vaccine required boosters because, the 3 stains of attenuated *polio virus* in the vaccine with each other replication in the intestine.

With the second immunization the immunity generated by the previous immunization will limit the growth of the previously predominant strain enabling one of the 2 remaining strains to predominate and induce immunity. Final with the third immunization immunity to all 3 strains is achieved.

19.3. DISADVANTAGES

The major disadvantages of attenuated vaccines is the possibility of their reversion to avirulent form.

The rate of reversion of the sabin polio vaccine leading to subsequent paralytic disease is alert one case in $ million doses of vaccine.

Another concerned with attended vaccine in the present of other viruses as contaminants In 1960 it was discovered that the ontogen virus SV 40 had contaminated some monkey kidney cultures used in production of sabin vaccine as a result more stringent vaccine testing was required to eliminate this contaminant.

Attenuated vaccines also may be with complication similar to those seen in the nature disease. A small percentage of recipients of the measles vaccine for eg develop pest vaccine encephalitis. Although these complications are unbearable the risk of such complication in far less than that observed in a naturally acquired measles infection.

In some cases however post vaccine complications may be a potential vaccine unacceptable in this situation e.g., illustrated by trials conducted several years ago with an experimental measures vaccine. This vaccine an attenuated strain of the virus, called the Edmanston Zagnod strain is immunogenicity in infants as young as 4-6 months old. As noted earliest maternal Ab's of the standard measles vaccine is ineffective when it is given. The high rate of vaccine complication with the Edmonton Zagreb strain may result from vaccine mediated immunosuppression since the measles virus is known to cause transient immunogenecity of the Demons ton Zagreb strains ever in the presence of maternal Ab may be associated with potent immune suppressive effects, placing vaccinated infants at lick for infections with other endemic pathogens that unvaccinated infants can resist.

19.4. INACTIVATED VIRAL AND BACTERIAL VACCINES

Another common approaches in vaccine production is to inactivate the pathogen but heated by chemical means so that the pathogen is no longer capable of replication in the host It is critically to maintain the structure of epitopes on surface Ag's during inactivation. Heat inactivation is generally unsatisfactory because it cause extensive protein denaturing. Thus any epitopes that depends on highest orders of portein structure are likely to be alternated.

Chemical inactivation which formulate are variable alkylaking agents has had sucess. The sulk polio vaccine and the pertuses (whooping cough) vaccine are produced by formal dehyde inactivation.

Unlike immunization with attenuated vaccines which generally requires only one dose to induce long lasting immunity repeated boosters of killed vaccines are often needed to maintain the immune status of the host. In addition killed vaccines induce a predominantly humoral Ab response they are less effective than attenuated vaccines in inducing cell mediated immunity and in eliciting a secretory Ig A response.

Even though they contain killed pathogenic inactivated whole organism vaccines are still absolute with certain resles.

Serious complication with the first stalk vaccines was inadequate killing of the virus in z vaccine lots, which cause paralytic polio in a high percentage of recipients. The peruses vaccine high lights another problem that can arise with a complex whole organism vaccine. Encephalitis type axons occurs in a small % of infants receiving this vaccine. The components of the vaccine responsible for these examination have not been determined.

19.5. TOXOID VACCINES

Some bacterial pathogens including those causing diphtheria and tetanus produce Extoxins. These Extoxins produce means of the disease symptoms resulting from infection diphtherias and tetanus vaccines, for e.g. can be made by purifying the bacterial extoxins and then inactivating the toxin with formal dehyde to form a toxoid.

Vaccination with the toxoid induced antitoxoid Ab's which are also capable of binding to the toxin and hallnaloizing its toxic effects. In production of toxoid vaccines the conditions must be closely controlled to achieve defoliation without excessive mediatification of the epitopes structure.

19.6. RECOMBINANT VECTOR VACCINES

It is possible to introduce genes encoding major Abs of especially virulent pathogens into attenuated viruses or bacteria. The attenuated organism serves as a vector replication within the host and expressing the gene product of the pathogen.

A number of organisms product of the pathogen. Vector including vaccine virsus the *conarypoix viruses* attempted *polioviruses adenoviruses* attenuated strains of *salmonella* and the BCD strain of *Mycobacterium bovis*.

Vaccine virus the attenuated vaccine used to eradicate *smallpox* has been widely employed as a vector vaccine.

This large, complex virus with a genome of absent 200 genes can be engineered to carry several dozen foreign genes with out impairing its capacity to infect host cells and replicate the procedure for producing a vicuña vector carrying on foreign gene from a pathogen.

The genetically engineered vaccine expresses high levels of the infested gene product, which can then serve as a potent immunize in an inoculated host. E.poletic has inrested genes from hapatitis B virus herpes simple and *influenza* into vaccine virus. Vaccine in the laboratory have shown that the engineered vaccine induced Antibodies and cell mediated immunity to all B engineered gene products like the small pox vaccine genetically engineered vaccine vector vaccines can be administered simply by dermal screeching causing a limited located infection in host cells. If the foreign gene produce expressed by the vaccine in a viral envelope protein, inserted into the host induced development of cell mediated immunity as well as antibody-mediated immunity.

Other attenuated vector vaccines may prove to the safer than the vaccine. The cannery pox virus has recently been tried as a vector vaccine. Like its relative vaccine the canes pox virus is a large virus that can easily be engineered to carry multiple genes unlike vaccine the conarypoix virus does not appear to be vibrated even in individuals with several immune suppression.

Another possible vector is an associated straining *salmonella typhimirium* which has been engineered genes from the bacterium that causes cholera.

The Advantage of this vector vaccine in that *salmonella* infects cells of the mucus lining of the gut and therefore will induce sector Ig A production. Effective immunity agents, a number diseases including cholera and gonorrhea depends on sed production of secretory at mucous membrane surface one of the *poliovirus* strains used in the sabin vaccine is another candidates for a safe and effective vector vaccine. In their causes the poliovirus vector is genetically engineering so that a portion of the gene encoding the outer capsid protein of heliodor is replaced by DNA encoding the epitopes of choice resulting poliovirus chimera will expresses the deserts epitope in a highly accessible presentation pasturing from the poliovirus nuclocaperid.

19.7. DNA VACCINES

In a recently developed vaccination strategies plastid encoding a viral Ag is injected directly into the nucleus of the recipient. The DNA is taken up by muscle cells and the included protein Ag expressed leading to both a humoral Ab response and cell made altered response to the viral Ag. What is most surprised about this finding is that the injected DNA is taken up and expressed by the muscle cells with much greater efficiently then seen with transfection experiments in tissue culture. The DNA apperture to integrate into the chromosomal DNA or to be maintained for long periods in a episomal form, it is unclear whatever the viral Ag is expressed only by the muscle cells expressed only low level of Class I MHC molecules and fall to expresses co stimulated molecules suggested that dendritic cells in the area also may be involved.

DNA vaccines offer several advantages over many of the existing vaccine. Their ability to induce both humoral and cell mediated immunity control required a live attenuated vaccine is a major benefit. In addition because DNA vaccine allow prolonged expression of two Ag. Significant immunologic memory generated an improved method for administering these vaccines involves coating gold bead is the plasmic DNA and then delivering coates particles through the skin into the underlying muscle with a Gene gun.

A number of companies have been testing DNA Vaccines in animal models. So for the result are encouraging have shown that these vaccines are able to induce protective immunity agents a number of pathogens, include the influence virus.

Enthusion for the potential of DNA vaccines must however, be tempered by several unknowns. It is not known exactly which cells take up the DNA. Not much is known what the consequences of long term Ag expression will be chronic persistent expression of the Ag by the muscle cells.

For e.g., it may eventually lead to a state of immunologic non responsiveness to the Ag on to a site of autonollgic non responsiveness to the Ag or to state of autoimmunity. Finally the safety of introducing foreign DNA molecule at high copy number remains unknown.

19.8. SYNTHETIC PEPTIDE VACCINES

Among the synthetic peptide vaccines currently being evaluated are those for hapatitis B virus, the molecule parasite diptheria toxins influenza and HIV.

Construction of synthetic peptides for use as vaccines to induce either humoral cell mediated immunity require an understanding of the nature of T cells and B cells epitopes.

Although the sequence of many important antigens from pathogen is known so far have been subjected to X ray crystallographic analysis, so their three dimensional structure is unknown. However B cells epitopes must constitute accessible surface regions vaccine designs commonly analyze the primary structure of an antigen to identify strongly hydrophilic sequences, with most likely corresponding to these potential B cells epitopes one than prepared.

Deadly vaccines for individual hormonal immunity should include peptide comparing immunodominant B cells epitopes.

Such epitopes can be identified by determining the dominant Ab in the sera of individuals who are recovering from a disease and than testing various synthetic peptides for their ability to react that Ab with a high affinity. In one such experiment two linger synthetic peptides representing potential B cells epitopes of Hb's Ag were tested for their binding affinity to poolral antisera from individuals who had recovered from hepatitis B. It was found that a synthetic peptide consists of cyclical repeats of amino acids 139 to 147 had a tenfold higher affinity than cyclical repeats of amino acids 124 to 137 the cyclical peptide for a synthetic hepatitis B vaccine.

An effective memory response for both humoral and cells mediated immunity requires generation of a population of memory T cells.

A successful vaccine must therefore include immunodominant T cells epitopes. with synthetic peptides vaccines but is difficult to identify these epitopes with to the sole un predictable as for, if the MHC in influenzing immunodominance for the T cells reptem.

T cells recognize processed peptides that in the majority of cases, appear to represent internal peptides these peptides must have a site (the agrotope) that enables then to interact with MHC molecules as well as a site (the epitope) that enables them to interact with the T cell receptor. Since MHC molecules differ in their ability to present peptides to T cells MHC polymorphism with a recipes influences the level of T cells responsiveness by different individuals to different peptides.

19.9. MULTIVATION SULE UNIT VACCINES

One of the limitations with synthetic peptide vaccines and recombinant protein vaccines is that these vaccines tend to be poorly immunogenic in addition they tend to induce a humoral Ab response but are less likely to induce a cell mediated response. What is needed is some method of structuring a synthetic peptide vaccine to contain both immunodominant B cells and T cells epitope. Further more of CTL response is decided the vaccine must be divided intracellularly so that the peptides can be processed and presented together with class I MHC molecule. A number of innovative techniques are currently being employed to develop multivatin vaccines that can present multiple copies of a given peptide or an mixture of peptides to the immune system.

One approaches is to prepare solid matters Ab Ac (SMAA) complexes by attaching monoclonal antibodies to particulate solid matrices and than saturating the Ab with the desired Ag. The resulting complexes are then used as vaccines. By attaching different monoclonal antibodies to the solid matters it is massible to blend is mixture of peptides of proteins, compensating immuno derivative epitopes for both T cells and bulls, to the solid matrix. These multivalent complexes have been shown to induce humoral and cell mediated responses their particulate nature contributes to their I sed immunogenicity by facilitating phagocytosis by phagocytic cells.

Another means obtained a multivalent vaccine is to utilize detergent to incentrant protein, antigens mutating complexes. Micelles are formed by mixing proteins in detergent and then removing the detergent.

Disposomes containing proteins Antigen are prepared by missing the paths with a suspensis of pheshelied under conditions that form vesicles bounded by a bilayered. The protein sone incompared into the bilayers with the hydrophilic residues exposed.

Immuno stimulating complexed (Iscom S) are lipid carriers prepared by mixing protein peptide Ag's with determinant and glycose called caulin.

Membrane proteins from different pathogens including *influenza virus, measles, hepatitis B* virus and HN have been invoimpartiaed into micelles, liposomes and Is coms and are currently being assessed as potential vaccine.

19.10. PURIFIED MUSCULAR VACCINES

Some of the either associated with attenuate or killed wide organism vaccines can be avoided with vaccines that consists of specific purified macromolecules derived from pathogens. Their general forms of such vaccines are in current use, inactivated exotoxins capsules polysaccharide vaccines.

The virulence of some pathogenic bacteria depends primarily on the antiphagocytic properties of their hydrophages polysaccharide capsid coating of the capsules with antibody complement grately with the ability of macrophages and numopthus to phagocytose such pathogen. These finds antigens provides the retinode for vaccines consisting of principle capsular polycocodes.

The current vaccine for *Streptococcus pneumoniae*, which causes *Pneumococcal Pneumonia*, consists 20 antigenically different capsular polysaccharides. It is marketed as pneumovax 23 by Merck and as Pnuemoneae 23 by lederle laleraforids. The vaccine induces formation of opsonizing Antibodies and is administered to high risk groups such as infants, patients, other immune suppressed individuals.

The vaccine for *Nisseria meningititis*, a common cause of bacterial meningitis, also consists of purified capsule polysaccharides.

One limitation of polysaccharide Vaccines is those invalidity to activate T4 cells. They activate cells in a thymus independent type 2(T1 2) memes resulting in Ig M production but little class switching no affinity maturation, and little, if any, development of memory cells. Several investigations have elevated the induction of Ig E secreting plasma cells in humans receiving Subcutaneous immunization and the pneumococcal polysaccharide vaccine. In this case, since T4 cells are not individual in the response, the vaccine may activate Ig A specific memory B cells previously generated by naturally occurring bacteria Ag's at mucosal surfaces. Because these bacteria have both polysaccharide and protein epitopes, they would activate T4 cells which in turn could mediate class switching and memory cell formation.

One way to involve thymus cells directly in the response to a polysaccharide Ag is to conjugate the Ag to some sort of protein carriers. For eg, the vaccine for *Haemophilus influnzae type b* (Hi b), the major cause of bacterial meningitis in children under which years of age, consists of type 'b' capsules polysaccharide covalently linked too proteins cauders, tetanus toxoid. The polysaccharide protein conjugate is considerably more immunogenic than the polysaccharide alone, and because it activates T cells, it enables class switching from Ig M to Ig G.

Although this type of vaccination induce memory B cells, it cannot induce memory T cells specific for the pathogen. In the case of the Hi b vaccine, it appears that the memory B cells can be activated to some degree in the absence of a population of memory TH cells, thus accounting for the efficacy of this vaccine.

19.11. TRANSGENIC PLANTS FOR PRODUCTION OF VACCINE OF ANTIBODIES

Transgenic plants are emerging as an important tool for the expression of many recombinant proteins of industrial and pharmaceutical value. Specific vaccines have been produced in plants as a result of transient or stable expression of foreign genes.

It has been shown that genes encoding antigens of bacterial *E.coli* heat labile enterotoxin B subunit and cholera toxin B subunit and vaccines using viral vectors (TMV and PMV)] and Viral (HB's Ag) pathogens. The recombinant plant derived proteins are expressed in a form in which they retain faithful preservation of protein folding characteristics and thereby immunogenic properties as well. Transgenic potato tubers expressing a bacterial antigen (edible vaccine) stimulate humoral and mucosal immune responses when they are provided as food. These results are of clear proof for the use of plants as a vehicle to produce vaccines (biofarming).

Plants are also capable of synthesizing and assembling virtually every antibody molecule, ranging from smallest antigen binding domains and fragments to full length and even multimeric antibodies. A number of plant hosts can be used to scale up cheap and plentiful supply of antibodies. Passive immunotherapy is one of the uses of antibodies.

Plants are useful expression system for recombinant and engineered antibody molecules. Bulk production of antibody is possible through transgenic plants. The outlook for plant antibodies for immunotherapy appears to be extremely promising.

Concerns about allergy and immune tolerance of orally applied antigens must be addressed by collaborative research efforts with medical specialists in these fields.

QUESTIONS

1. Explain the classification of common vaccine for humans.
2. Write short notes on :
   (*a*) Attenuated viral and bacterial vaccines;
   (*b*) Inactivated viral and bacterial vaccines;
   (*c*) Toxoid vaccines;
   (*d*) Recombinant vector vaccines;
   (*e*) DNA vaccines;
   (*f*) Synthetic peptide vaccines;
   (*g*) Multivation sule unit vaccine;
   (*h*) Purified muscular vaccines;
3. Explain how transgenic plants are used for production of vaccines.

CHAPTER 20

CANCER, CANCER GENES AND GROWTH FACTORS

Cancer cells are characterized by three properties: 1. Diminished or unrestrained control of growth, invasion of local tissues; and 3 spread, or metastasis to other parts of the body. Cells of benign tumors also show diminished control of growth but do not invade local tissue or spread to other parts of the body.

The uncontrolled growth of cancer cells and their ability to invade and metastasize.

Certain genes controlling growth and interactions with other normal cells are apparently abnormal in structure or regulation in cancer cells. Information on how cell growth both normal and pathologic is controlled or is limited and knowledge of specific genes involved in growth regulation is even more meagre. Little is known as yet about the biochemical basis of metastasis, so that coverage of this topic be brief. Some types of cancer (e.g. Leukemias) can be regarded as example for abnormal differentiation. Again, astonishingly little is known about the molecular basis of differentiation. However, many workers in this area think that further research on oncogenes, tumor suppressor genes, growth factors and their receptors, DNA repair systems and regulation of the cell cycle will provide insight into the nature of the disturbed control of growth, of differentiation (where applicable) and of cell cell interaction exhibited by cancer cells. There has been a recent surge of interest in elucidating the molecular basis of genetic susceptibility to cancer. One example is isolation of the BRCAI gene which increase susceptibility to breast and ovarian cancer.

20.1. BIOCHEMICAL IMPORTANCE

Cancer is the second most common cause of death in the USA after cardiovascular disease. Humans of all ages develop cancer and a wide variety of organs are affected. The incidence of many cancers increases with age so that as people live longer more will develop the disease. Apart from individual sufferings the economic burden to society is immense.

20.2. PHYSICAL, CHEMICAL AND BIOLOGICAL AGENTS

Agents causing cancer fall into three broad groups: radiant energy, chemical compounds and Viruses.

Radiant energy: Ultraviolet rays, X-rays, Gamma-rays and mutagenic and carcinogenic agents, damage DNA in several ways.

Ultraviolet radiation may cause pyrimidine 10 form. Apurinic or apyridimic sites may form by elimination of corresponding bases. Single and double strand break or cross linking of strands may occur. Damage to DNA is presumed to be the basic mechanism of carcinogenecity with radiant energy, but the details are unclear. Apart from direct effects on DNA, X rays and Gamma rays cause free radicals to form in tissues. The resultant OH_0 Superoxide and other radicals can interact with DNA and other macromolecules leading to molecular damage and thereby probably contributing to carcinogenic effects of radiant energy.

20.2.1. Chemicals

A wide variety of chemical compounds are carcinogenic most of the compounds have been tested by administration to rodents in other animals. However, substances are associated with the development of cancer in humans. It is estimated that approximately 80% of human cancers are caused by environmental factors, principally chemicals. Exposure to such compound can occur because of a persons occupation e.g. Benzene asbestos; diet e.g. Aflatoxin B_1 which is produced by the mold Aspergillus flavus and sometimes found as a contaminant of peanuts and other food stuffs; or life style e.g. Cigarette smoking, or in other ways e.g. Certain therapeutic drugs.

20.2.2. TABLE

S.No.	Class	Compound
1.	Polycyclic aromati chydrocarbons	Benzopyrene, dimethyl benzathracene.
2.	Aromatic amines	2 Acetyl aminofluorene,N methyl -4-aminozobenzene(MAB)
3.	Nitrosamines	Dimethyl nitrosamine, diethyl nitrosamine
4.	Various drugs	Alkylating agents(eg.Cyclophoshamide) diethylstil bestrol.
5.	Naturally occuring compounds	Dactinomycin, aflatoxinB_1
6.	Inorganiccompounds	Arsenic,asbestos, beryllium, cadmium, chromium.

20.2.3.Structure

Both organic and inorganic molecules may be carcinogenic. The diversity of these compounds indicates that they do not possess one common structural feature that confess carcinogenecity.

20.2.4.Action

The organic carcinogens are the most thoroughly studied. Some such as mechloroethamine and Beta Propino lactone have been found to interact directly with target molecule (direct carcinogens) but others require prior metabolism to become carcinogenic (Procarcinogens). The process whereby one or more enzyme catalyzed reaction convert procarcinogens to active carcinogens is called metabolic activation. Any intermediate compound formed are proximate carcinogens may be one or more and the final compound that reacts with cellular compounds (e.g. DNA). A possible sequences can be displayed as follows. Procarcinogen, Proximate carcinogen and Ultimate carcinogen

The procarcinogen itself is not a chemically reactive species, whereas the ultimate carcinogen is often highly reactive. At least two reaction are required to convert the procarcinogen 2 Acetylamino fluorine (2AAF) to the ultimate carcinogen, the sulfate ester of N hydroxy AAF. An important generalization is that ultimate carcinogens are usually electrophils (i.e. molecules deficient in electrons) which readily attack nucleophilic (electron rich) groups in DNA, RNA and proteins.

20.3. METABOLISM OF CHEMICAL CARCINOGENS

The metabolism of procarcinogens and other xenobiotics involves monooxygenases and transferases. The enzymes responsible for metabolic activation of procarcinogens are principally species of cytochrome P_{450} located in the endoplasmic reticulum. These are the same enzymes that are involved in the metabolism of other xenobiotics such as drugs and environmental pollutants (e.g. PCBS). The activities of the enzymes metabolizing chemical carcinogens are affected by a number of factors, such as species, genetic consideration, age or sex. The variations in activities of these enzymes help to explain the often appreciable differences in the carcinogenicity of chemicals among different species and individuals of the same species.

20.4. COVALENT BINDING

When chemical carcinogen are administered to animals or placed in cultured cells it can be shown (e.g. by using radioactive carcinogens) that they or their derivatives generally bind covalently to cellular macromolecules, including DNA, RNA and proteins. The chemical natures of the products formed by the interaction of certain ultimate carcinogens with their target molecules have been determined. Most interest has focused on products formed with DNA. Carcinogens have been found to interact with the purine, pyrimidine and phosphodiester groups of DNA. The most common site of attack is guanine and the addition of various carcinogens to the N_2, N_3, N_7, O_6 and O_8 atoms of this base has been observed.

20.5. DAMAGE TO DNA

The covalent interaction of direct carcinogens or ultimate carcinogens with DNA can result in several types of damage which can be repaired. Inspite the existence of repair systems, certain modification of DNA by chemical carcinogens persist for relatively long periods of time. It is possible that these persistent unrepaired lesions are of special importance in generating mutations critical to carcinogenesis.

20.6. MUTAGENS

Most chemical carcinogens are mutagens. It has been demonstrated using ames assay and other tests. As a molecular level, transitions transversions and other types of mutation have been shown to occur following exposure of certain bacteria to ultimate carcinogens. It is assumed that some types of cancer are one to mutation in somatic cells that affect key regulating process in these cells. Direct evidence of this has been obtained.

Since testing the carcinogenecity of chemicals in animals is slow and expensive, assays for screening the potential carcinogenecity of chemical compound are developed. Many are based on detection of the mutagenecity of chemical carcinogens. Such assays are more rapid and less expensive than detecting tumors in animals. The Ames assay has proved useful in screening for potential carcinogens. This assay uses a specially constructed strain of *Salmonella triphimurium* that has a mutation (His) in a gene that codes for one of the enzymes involved in the synthesis of histidine. Thus, these particular Salmonellae cannot synthesize histidine, which must be present in the medium for growth to occur. When a mutation caused by a carcinogen occurs at the site of the His mutation, the latter mutation can restore its reading sequence, converting it to His. The progeny from bacteria containing such a reverse mutation can now synthesize histidine and thus grow in a medium lacking it. Such *Salmonella* can be detected as readily observable and quantifiable colonies growing on agar plates.

One problem with the use of bacteria in mutagenecity test is that they do not contain the spectrum of mono oxygenases found in higher animals. Thus if a compound requires activation to become a mutagenic or carcinogenic sps. this may not occur when bacteria are used ames circumvented this problem by incubating the agents to be tested in a post mitochondrial supernatent of rat liver (the S 9 fraction which is the supernatent fraction after centrifuging a rat homogenate at 9000g for a suitable period of time). The S 9 fraction contains fragments of endoplasmic reticulum and the most of the various monooxygenases and other enzymes required to activate potential mutagens and carcinogens.

The Ames assay identifies approximately 90% of known carcinogens. It is becoming routine to test newly synthesized chemicals by this assay, particularly if they are to be introduced commercially or widely used in industry. Compounds giving a positive reaction should undergo further testing inducing assessment of carcinogenecity of animals.

20.7. INITIATION AND PROMOTION

In certain organs such as skin and liver, carcinogenesis has at least two stages: in skin. Typically identical areas of the skin of a group of mice are painted once with benzo(a)pyrene. If no other subsequent treatment is used no skin tumors develop.

20.8. SOME DNA AND RNA VIRUSES ARE CARCINOGENIC

However, if the application of benzo(a)pyrene is followed by several application of croton oil, many tumors subsequently develop. Appreciation of croton oil alone (i.e., no pretreatment with benzo(a)pyrene) do not result in skin tumors. Many other variants of this basic protocol have been carried out, permitting the following conclusion.

1. The stage of carcinogenesis caused by application of benzo(a)pyrene is called, "initiation", this stage appears to be rapid and irreversible. It is presumed to involve a irreversible modification of DNA, perhaps resulting in one more or more mutation. Benzo(a)pyrene is thus called an "initiating agent".
2. The much slower, second stage (i.e. months or years) stage of carcinogenesis, resulting from application of croton oil is called "promotion"; croton oil is thus a promoting agent, or promoter. Promoters are incapable of causing initiation.
3. Most carcinogens are capable of acting as both initiating and promoting agents.

DNA is the critical macromolecule in carcinogenesis

1. "Cancer cells beget cancer cells" – i.e., the essential changes responsible for cancer are transmitted from mother to daughter cells. This is consistent with the behavior of DNA.
2. Both irradiation and chemical carcinogens damage DNA and are capable of mutation in DNA.
3. Many tumor cells exhibit abnormal chromosomes.
4. Transfection experiments indicate that purified DNA (oncogenes) from cancer cells can transfer normal cells into (potential) cancer cells.
5. Genes that increase susceptibility to cancer have been isolated. However, epigenetic factors (e.g. Methylation of DNA) may also play a role in carcinogenesis.

Some DNA and RNA Viruses are Carcinogenic

Oncogenic virus contain either DNA or RNA as their genome only a few important features of the major members of these two classes are:

Polyomovirus and SV40 viruses have played an important role in the development of current ideas about viral oncogenesis, though they are not implicated in the causation of human tumors. They are both small (containing genome of 5 Kb) and their circular genomes code for only about five or six proteins. Under certain circumstances infection of appropriate cells with these viruses can result in malignant transformation. Specific viral proteins are known to be involved. In the case of SV40 these proteins (often called antigens, because they were detected by immunologic methods) are known as T (large "T") and t (small "t") and in the case of *polymavirus* they are known as T, mid T and t (T refers to the 'fact' that the first of these proteins was detected in a tumor). How these protein cause malignant transformation is still under investigation, the T antigens are known to bind tightly to DNA and cause alterations in gene expression. These proteins show cooperative effects, suggesting that alteration of more than one reaction or process is required for transformation. Hepatitis B virus may be associated with some cases of liver cancer in man.

When cultured cells are infected with certain oncogenic viruses, they may undergo malignant transformation. The most important and biochemical changes occurring upon transformation. Cell shape, motility, adhesiveness to culture dish growth and a number of biochemical process. They reflect the primary processes that cause and the secondary changes that result from – conversion from normal to malignant state. The ability to equate transformation approximately with acquisition of malignant properties has been of tremendous importance in cancer research.

Oncogenes play a crucial role in carcinogenesis

Oncogenes are gene capable of causing cancer were first recognized as unique genes of tumor-causing virus responsible for the process of transformation (viral 1 oncogens).

1. Oncogenes of *Rous sarcoma virus*

The genome of this ***retrovirus*** contains four genes named gag, pol, enr and src.

The gag gene codes for group specific antigens of the virus, pol for the reverse transcriptase that characterizes ***retroviruses*** and enr for certain glycoproteins of the viral envelope. A protein-tyrosine kinase was shown to be the product of src (i.e., the sarcoma causing gene) that is responsible for transformation.

2. Oncogenes of other retroviruses

In addition to oncogens of Rous sarcoma virus, approximately 20 oncogenes of other retroviruses have been recognized. About half of the products of these viral oncogenes are protein kinases, mostly of the tyrosine type.

A. Normal chicken chromosome, showing an inactive myc gene

B. An avian leukemia virus has integrated in the chromosome its proviral from, adjacent to the myc gene. However, in this instance, the site of integration is just downstream of the myc gene and it cannot acts as a promotor. Instead, a certain proviral sequence acts as an enhance element, leading to activation of the upstream myc gene and its transcription.

Oncogenes from tumor cells: Experiments using DNA extracted from tumor have also provided evidence for the existence of oncogenes. The method used for detecting such cellular oncogenes is called gene transfer or DNA transfection. It depends on the fact that certain genes present in tumors can cause transformation of "normal" cultured cells. DNA is isolated from tumor cells and added to recepient cells, often a line of mouse fibroblast known as NIH/3T3 cells. DNA isolated from tumor cells is precipitated with calcium phosphate to facilitate endocytosis and added to NIH/3T3 cells in tissue culture. These cells get transformed within 1-2 week; the cells change their morphology from flat to rounded cells.

The process whereby transcription of a gene is increased from zero or a relatively low level is activation. Certain *retrovirus* lack oncogenes (e.g. Avian leukemia viruses) but may cause cancer a longer period of time-months rather days than those which do contain oncogenes. As for other retrovirus when these particular virus infect cells, a DNA copy (cDNA) of their RNA genome is synthesized by reverse transcriptase and the cDNA is integrated into the host genome. The integrated double stranded cDNA is called "provirus". The cDNA copies of retroviruses are planked at both ends by sequences named long terminal repeats as are certain transposons – "jumping genes" – found in bacteria and plants.

Some changes shown by cultured cells which suggest that malignant transformation has occurred (e.g. After infection of an oncogenic virus)

1. Alterations of morphology: Transformed cells often have a much rounder shape than control cells
2. Increased cell density (loss of contact inhibition of growth): Transformed cells often form multi layers, while control cells usually form a monolayer.
3. Loss of anchorage dependence: Transformed cells can grow without attachment to the surface of the culture of the culture dish and will often grow in agar.
4. Loss of contact inhibition of movement: Transformed cells grow over one another, while normal cells stop moving when they come into contact with each other.
5. A variety of biochemical changes, including an increased rate of glycolysis, alternations of the cell surface (e.g. Changes in the composition of glycoproteins and secretion of certain proteases.

20.9. CHROMOSOMAL TRANSLOCATION

This is a type of chemical change in cancer cells. A piece of one chromosome is split off and then joined to another chromosome. If the second chromosome donates material to the first, the translation is said to "reciprocal".

Schematic representation of the reciprocal translocation involved in Burkitt's lymphoma. The chromosomes involved are 8 and 14. A segment from the end of the q arm of chromosome 8 breaks off and moves to chromosome 14. The reverse process moves a small segment from the q arm of chromosome 14 to chromosome 8. The myc gene is contained in the small piece of chromosome 8 that was transferred to chromosome 14; it is thus place next to genes transcribing the heavy chains of immunoglobulin molecules and itself becomes activated.

20.10. GENE AMPLIFICATION

Amplification of certain genes is found in a number of tumors. One method of bringing about gene amplification in tumors is by administration of the anticancer drug methotrexate, an inhibitor of the enzyme dihydrofolate reductase. Tumor cells can became resistant to the action of this drug. The basis of this phenomenon is that the gene for dihydrofolate reductase becomes amplified, resulting in an increase of the activity of the enzyme upto 400 fold.

20.11. ACTIVATOR OF ONCOGENES

An increase in amount of the product of an oncogene due to increased transcription, thus the increased amount of an oncogene may be sufficient to push a cell toward becoming malignant. However activation of oncogenes is not the sole pathway to malignancy. A combination of their activation in colorectal and probably most cancers. In some cases, their activation may be only a second or occurrence associative with transformation rather than a causal event.

20.12. POLYPEPTIDE GROWTH FACTORS

The study of growth factors is rapidly becoming of major interest. Until recently only very small quantities of most growth factors were available for study. However the genes for number of growth factors have been cloned confirming their separate identities and making usable amounts available through recombinant DNA technology. The growth factors known to date affect many different types of cells, E.g. Cells from the blood, nervous system mesenchymal tissue and epithelial tissue. They exert a mitogenic response on their target cells, though special condition may be necessary to demonstrate this such as depriving cells in culture of return so that they have become quirescent prior to exposure to a growth factor. Growth inhibitory factors also exits (e.g. Transforming growth factors (TGFb) exerts inhibitory effects on the growth of certain cells). Thus chronic exposure to increased amounts of a growth factor or to decreased amount of a growth inhibitory factor could alter the balance of cellular growth.

The effects of growth factors may be endocrine in the sense that like hormones, they may be synthesized else where in the body and pass in the circulate to their target cells. They may be paracrine i.e. synthesized in certain cells and secreted there to affect neighboring cells, though the synthesizing cells are not themselves affected because they lack suitable receptors. And certain growth factors can affect the cells that synthesize them, a mode of action characterized as autocrine.

Relatively little is known about how growth factors operate at molecular level. Like polypeptide hormones, they transmit a message across the plasma membrane to the interior of the cell. In the cause of growth factors the message will ultimately affect one or more processes involved in mitosis. Most growth factors high affinity protein receptors are on the plasma membrane of target cells. The gene for the receptors for many growth factors have been cloned and models of the structure of the receptors constructed.

The products of several oncogenes are either growth factors or parts of the receptors for growth factors. Genes other than oncogenes have been found to play a major role in the causation of at least some types of cancers. These are tumor suppressor genes, sometime called recessive oncogens or anti-oncogens. They operate quite differently from oncogens in that their inactivation (as opposed to activation) removes certain constraints on control of growth.

Schematic representation showing how the translocation involved in Burkitt's lymphoma may activate the myc proto oncogene.

(*a*) A small segment of chromosome 14 prior to the translocation. The segment shown contains the genes encoding regions of heavy chains of immunoglobulin.

(*b*) Following the translocation, the previously inactive myc gene is placed under the influence of enhancer sequences in the genes encoding the heavy chains and thus its activated, resulting in transcription. Only one strand of DNA is depicted, other details are omitted.

This is a malignant tumor of retinal neuroblast which are precursor cells of photoreceptor cells in the retina. In some cases this tumor is inherited.

Another extremely important tumor suppressor gene is P53 encoding a protein (designated P53) of molecular mass 53K DA.

Because of many studies in the past few years, more is know about the occurrence of mutations in this gene in human cancers than about their occurrence in any other gene. A few points.

20.13. MALIGNANCY OF TUMOR CELLS

Once a cell become a tumor cell, the composition and behavior of its progeny do not remain static. Instead, there is a tendency for malignancy to increase. This manifested by increasingly abnormal karyotypes, increasing rates of growth and an increasing tendency to invade and metastasize the important phenomenon of progression appears to reflect a fundamental instability of the genome of tumor cells. It seems likely that mutation in DNA repair genes are involved in this phenomenon creating a mutator phenotype. The activation of additional oncogenes may be involved. Cells with faster rates of growth have a selective advantage.

It is important to distinguish the biochemical profiles of cells that have just been transformed from those of fast-growing, highly malignant tumor cells. The former may show few differences from normal cells apart from key changes leading to cancer (e.g. activation of one or more oncogenes) and those changes generally are associated with transformation. The biochemical profile of highly malignant cells may be very different from that of normal cells.

Biochemical changes often found in fast growing tumor cells:

- Increased activity of ribonucleotide reductase.
- Increased synthesis of RNA and DNA
- Decreased catabolism of pyrimidines
- Increased rates of aerobic and anaerobic glycolysis
- Alterations of isozyme profiles, often to a fetal pattern
- Synthesis of fetal proteins (e.g. Carcino embryonic antigen)
- Loss of differentiated biochemical functions (eg. Diminished synthesis of specialized proteins.
- Inappropriate synthesis of certain growth factors and hormones.

These fast growing cells tend to maximize the anabolic process involved in growth

(eg. DNA and RNA synthesis) cut down on catabolic functions (eg. Catabolism of pyrimidines) and dispense with differentiated functions shown by their normal ancestors. In other words they are concentrating almost exclusively upon growth.

Drugs used in cancer chemotherapy:

Surgery, radio therapy and chemotherapeutic agent are the major modalities used to treat patients with cancer. The development of anticancer drugs is – like numerous other areas of medicine – in many ways an exercise in applied biochemistry. The basic problem is to make available drugs (natural or synthetic) that kill cancer cells effectively but are not excessively toxic to normal cells. In so far as unrestrained cell division is a feature that typifies many malignant tumors, many of these agents are used because they inhibit DNA synthesis. For this reason they are also likely to damage normal tissues whose cells divide continuously (e.g. gut and bone marrow). The growth fraction (percentage of the cells of a tumor that are constantly in cycle) is an important concept in cancer chemotherapy; tumors with a high growth fraction are usually more responsive to cancer chemotherapy than cells that are dormant in the go phase of the cycle.

20.14. METASTASIS THE MOST DANGEROUS PROPERTY OF TUMOR CELLS

Metastasis is the spread of cancer cells from a primary site of origin to other tissue where they grow as secondary tumors and it is the major problem presented by the disease. Metastasis is a complex phenomenon and knowledge of biochemical basis is sparse.

Some changes that have been detected at the surfaces of malignant cells*

* Alterations of permeability
* Alterations in transport properties
* Diminished adhesion
* Increased agglutinability by many lectins
* Alterations of the activities of a number of enzymes (eg. Certain proteases)
* Alteration of surface charge
* Appearance of new antigens
* Loss of certain antigens
* Alterations of the Oligosaccharide chain of glycoproteins
* Changes of glycolipid constituents.

QUESTIONS

1. Explain the biological importance of cancer.
2. What are the physical, chemical and biological agents that cause cancer ?
3. Explain the chemical compounds that cause cancer.
4. Explain the metabolism of chemical carcinogens.
5. Write short notes on.
   (*a*) Covalent binding;
   (*b*) Damage to DNA;
   (*c*) Mutagens;
6. Explain the initiation and promotion of chemical carcinogens in skin.
7. Some DNA and RNA viruses are carcinogenic. Explain.
8. Explain how oncogenes play a crucial role in carcinogenesis.
9. Write short notes on:
   (*a*) Chromosomal translocation;
   (*b*) Gene amplification;
10. Explain the malignancy of tumor cells.
11. Metastasis is the most dangerous property of tumor cells. Explain.

CHAPTER 21

ACQUIRED IMMUNE DEFICIENCY SYNDROME

One disease is described in this chapter the virus-induced immuno deficiency syndrome called acquired immune deficiency of the current status of AIDS through out the world. It is a much greater problem in other regions such as Africa than in the United States. We then go on to describe the structure of the virus that cause AIDS in some detail since this has a direct bearing on the difficulties encountered in controlling the infection. This is followed by a detailed review of the damage that this virus does to the cells of the immune system. This leads to a description of the diseases associated with AIDS. Some other viruses have similar effects on the immune system of animals. So these are briefly described. Finally the complex social issues raised by an infection such as AIDS are mentioned to provide a relatively complete overview of this terrible disease.

The concept that viruses could destroy lymphocytes and so induce a profound immunodeficiency is not new. It had long been recognized that diseases such as measles in humans of distemper in dogs were accompanied by a significant immunosuppression, and as a result infected individuals were more susceptible than normal to other secondary infections. Nevertheless none of these viruses induced an immunodeficiency as severe as that induced by the **human immunodeficiency virus (HIV).**

Beginning in 1979, physicians began to notice a remarkable increase in cause of two unusual conditions among active male homosexuals. These conditions were a rare tumor, known as Kaposi's sarcoma and a severe pneumonia caused by a parasite called *Pnenmocystis carrinni.* Affected individuals were found to be profoundly immunosuppressed as a result of an almost total T cells. Evidence suggested that the disease was a form of immunodeficiency transmitted by an infectious agent, so it was called the acquired immune deficiency syndrome.

21.1. THE CURRENT STATUS OF AIDS

21.1.1. United States

From June 1981 until July 1993, AIDS has been diagnosed in over 289,000 persons in the United States. In addition, it is estimated that between one and two million individuals in the United States are infected with HIV but do not yet show symptoms. About 62% of these infected individuals are homosexual or bisexual men. However, the disease also affects intravenous drug abusers (30%), heterosexual contacts of bisexual men, and infants born to woman with AIDS. The disease is also transmitted by whole-blood or blood products. Before the introduction of effective screening procedures, transmission by blood was a significant problem in hemophiliacs and other who required repeated blood transfusions. This clearly points to AIDS begin caused by an infectious agent.

21.1.2. Worldwide

About 50% of the cause of AIDS reported worldwide has occurred in the United States. However this is very misleading since it is estimated that in the United States about 90% of AIDS cause is reported to the authorities, whereas widespread underreporting is certain in many under developing countries. Taking this underreporting into account, the World Health Organization (WHO) now estimates that approximately seven million causes has occurred worldwide by the end of 1992. This implies that as many as nineteen million people may already be infected with HIV. It also estimated that as many as three million people may already have died from AIDS. Projected estimates of the number of persons infected with HIV by the year 2000 range from 40 million to more than 100 million. Most of this increase will be in Africa and Asia.

Although the first clinical cause of AIDS were reported from California in 1981, there is serological and historical evidence to suggest that it was present in central Africa at least as early as 1959. Thus antibodies to HIV have been detected in blood drawn in Zaire in 1959. In that same year a man in England died from a disease that at the time was diagnosed as leukemia. HIV has been detected in the preserved tissues of this man. Serologic evidence suggests that HIV began to spread in urban areas of Africa during the late 1970s. Isolated cause of AIDS has been diagnosed retro selectively in individuals from Central Africa in the mid-1970s. It has therefore been postulated that the virus originated in Central Africa. Perhaps in monkey populations, and spread from there to Europe and North America. More recently, new AIDS-related viruses have been isolated in Africa.

The prevalence of AIDS has increased significantly in Africa since the mid 1970s. The epidemiology of the disease is very different from that seen in Europe and North America. For example the disease is much more prevalent among heterosexuals is Africa than in Europe or the United States, whereas homosexual transmission or transmission as a result of drug use remains low. The major factor contributing to transmission of the disease is heterosexual concept with many partners. Thus there is nearly an equal sex ratio among AIDS patients. As many as 88% of female prostitutes in some African countries are infected. This level reflects a significant increase from a prevalence of around 0% to 8% in 1980. The differences between the mechanism of transmission in Africa and North America of Europe might be result of the high level of other sub clinical infections in the African population. Thus genital ulcers or infection with *Chlamydia trachomatis* may diseases such as malaria or tuberculosis with activation of the immune system is more common among African adults than Americans. The progression of the disease in Africans is similar to that seen in Caucasians, with about 6% of infected individuals developing disease annually.

In population with a high level of heterosexual transmission and high prevalence of HIV infection in woman, increase perinatal transmission is bound to occur. Thus in Africa, about 30% to 65% of infants born of HIV infected woman are also infected. Postnatal transmission through breast milk has been described in addition to congenital infection. Infant morality in Africa will increase by 30% as a result of AIDS and more than a million uninfected children will be orphaned because their parents will have died from AIDS by the turn of the century. It is estimated that life expectancy at birth in sub-saharan African cities could fall by as much as six years during the next few decades as a result of the impact of AIDS.

A similar situation is now emerging in India and Thailand. In both of these countries, heterosexual prostitution appears to be the main route of transmission and the sex ratio of cases is equal.

21.2. THE AIDS VIRUSES

All responsible opinion holds that AIDS occurs as a result of infection with one of the human immunodeficiency viruses. Very rarely, an immunodeficiency syndrome may develop in humans not infected with HIV. This does not support the idea that there is another significant cause of AIDS not detected by current techniques.

21.2.1. HIV Virus-1.

The primary cause of AIDS is a virus called human immunodeficiency virus-1 (HIV-1) . HIV-1 has been isolated from blood, born marrow, lymphnodes, spleen, plasma, semen, and saliva of infected individuals supporting the concept that AIDS is transmitted by blood, blood production and sexual contact.

HIV-1 is an RNA virus that employs the enzyme reverse transcriptase. It is therefore classified as retrovirus. In retroviruses, the viral RNA is first transcribed into DNA using the reverse transcriptase. This proviral DNA can than use the cell's own protein-synthesizing ability to generate new virons. HIV-1 is related to other membrane of the lentivirus family of retroviruses such as visna virus of sheep, equine infections anemia virus, and felina immunodeficiency virus. On electron microscopy, HIV-1 is seen to be a spherical virus containing a dense core surround by a lipoprotein envelope that is acquired as the virus buds from the surface of an infected cell.

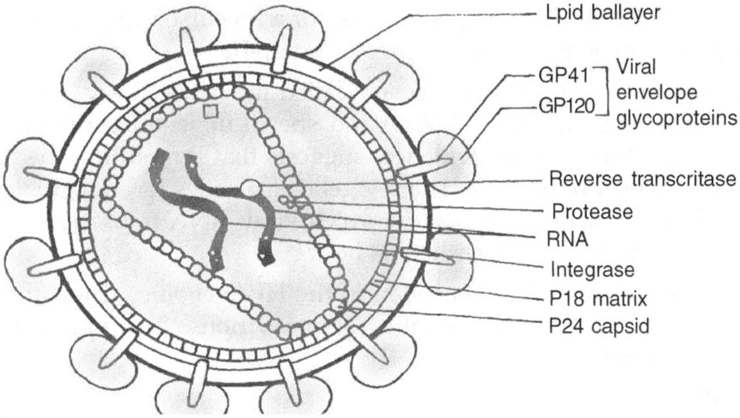

Diagrammatic Sketch of the HIV

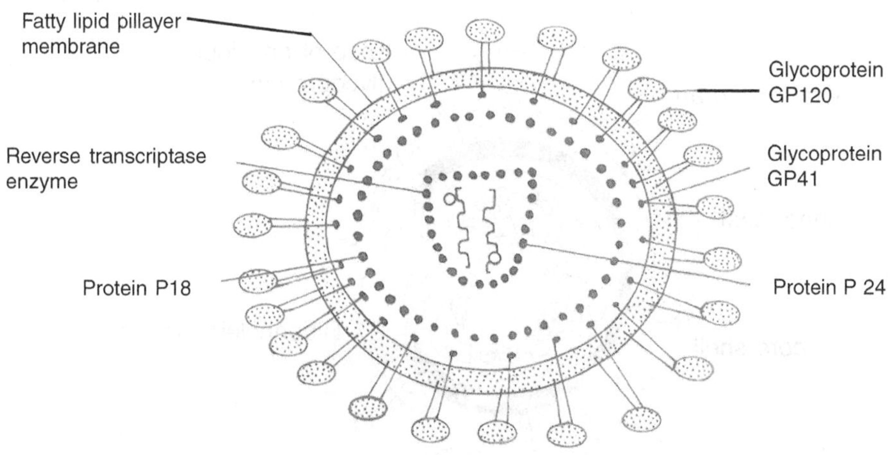

Fig. 21.1

At the center of the virion are two standards of RNA, each containing about 10,000 nucleotides. Attached to these strands are molecules of the reverse transcriptase as well as a protease, a ribonuclease and an integrate. The RNA its associated enzymes are contained within a capsid in the shaped of a hollow truncated cone. The capsid is made of a protein called p24 (p stands for protein and the number is its mass). Surrounding the capsid is a layer of matrix protein called p17. The entire mass is covered by an envelop that consists of a lipid bilayer derived from the outer cell membrane of infected cell. Because of its origin, the envelope may also carry cell surface protein such as MHC molecules. There are many spikes on the viral envelope may also carry cell surface protein such as MHC molecules. There are many spikes on the virus envelop formed by two glycoproteins. One glycoprotein (gp-41) (group stands for glycoprotein) is embedded in the membrane, and four molecules of another glycoprotein (gp 120) form the spike.

The HIV genome is well characterized. The viral RNA has two flanking long terminal repeat (LTR) sequences that act as binding sites for host transcription factors. There are three structure genes called gag, pol, and env. The gag gene codes for proteins that form the core (p24 and p17). The pol gene codes for a precusor protein that is cleaved to yield the four enzymes found is association with the RNA, the reverse transcriptase, the protease, the ribonulease, and the integrase (this is endonuclease that helps integrate viral DNA into host cell DNA). The env gene codes for a precursor glycoprotein that is cleaved to yield the two envelope proteins gp120 and gp41.

There are three regulatory genes called tat, nef, and rev. The tat gene encodes a protein that attaches to the viral RNA and speeds up its transcription. The role of the nef gene is still unclear. Recent evidence suggests that it modifies the cell to make it more suitable for virus production. The rev gene also codes for a regulatory protein. This is responsible for switching the way in which viral RNA transcripts are processed once a cell has been infected for more than 24 hours.

Three other proteins are encoded in the HIV genome. They are called vif, upr and vpu. They play a role in the assembly of new virions. The vpu gene is found only in HIV-1 and not in HIV-2.

21.2.2. HIV-2

HIV-2 is more closely related to the simian (monkey) immunodeficiency virus (SIV) than to HIV-1. Its genome shows about 40% homology to HIV-1 and 70% to SIV. The

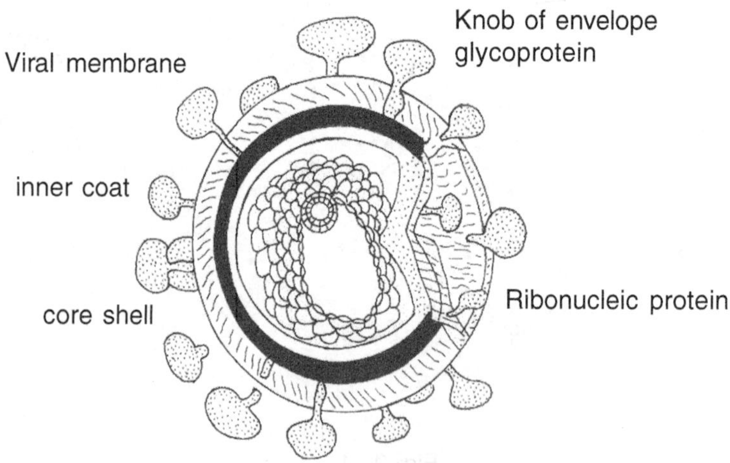

Fig. 21.2

viral proteins of HIV-2 show variations in homology with HIV-1. Thus gag and pol show 50% homology, but the proteins encoded by env, nef and vif show less than 30% homology. HIV-2 remains largely confined to West Africa. The prevalence largely confined to West Africa. The prevalence of antibodies against the virus (seropositivity) ranges from 0.3% to 17%, but in prostitutes it may reach as high as 64% HIV-2 spreads in a similar manner to HIV-1.In Africa, being largely heterosexually transmitted. Despite is similarity to HIV-1, it has reduced virulence as shown by a longer disease-free period.

21.3. GENETIC VARIABILITY OF HIV-1

HIV-1 shows extensive variability in its genetic structure. The most divergent part the genome lies within the env gene, the gene for the major exterior glycoprotein on the envelope. The protein has both conserved and variable sites. The variable has both conserved and variable sites. The variable sites permit the structure of the envelope to very in response to selective pressure from the immune response of an infected individual and permit the survival of viruses that van resist neutralization.

The nucleotide sequences of the enu gene from an HIV isolated obtained from the patient at any given time very by 2% to 3%, and new variants of HIV-1 arise continuously. The reason for this genetic variable is that the virus reverse transcriptases copy virus RNA into proviral DNA. These enzymes are very inaccurate and make on average one mistake in every 2000 nucleotides. There is no mechanism in the virus to correct these mistakes. This remarkable variation has major implications for the development of an HIV vaccine. Thus an HIV isolate put into an experimental animal such as a chimpanzee will give rise to new variants within two weeks. These new variants have antigenically different envelope proteins from the originating strains and cannot neutralized by antibodies directed against the originating strains. Clearly, this rapid production of new antigenic variants will help the virus evade destruction by the immune system. These new HIV variants differ not only in their antigenicity, but also in their pathogenicity. The genetic variation may modify the tissue pathogenicity. The genetic variation may modify the tissue tropism of the virus and account for the great variety of clinical signs of AIDS. The time needed for pathogenic variants to arise may contribute to the long incubation period of the disease.

Studies on the sequences of the gag genes of about 70 virus isolated from across the world have shown that there are seven distinct major genotypes of HIV-1 (A though G). gag codes for internal proteins and is thus relative conserved. Genotype A is found in sub Saharan Africa and Thailand . Genotype B is found in the Americas, Africa Europe and Asia is thus the most widely distributed strain. Genotype which is found in Africa and India. The other genotypes are restricted to sub-Saharan Africa.

21.4. IMMUNOLOGICAL LESIONS IN AIDS

HIV infection affects the cells of the immune system, causing over time severe immune deficiencies. No other infections agents in necessary, although other infections may hasten the development of an immune deficiency. The defect in the immune system is characterized by a loss of lymphocytes (lymphopenia) as a result of a selective deficiency of CD4+ T cells. In normal healthy individuals, the CD4: CD8 ratio is approximately 2.0. That is there are about twice as many CD4+ cells as CD8+ cells. In AIDS patients the ratio drops to around 0.5; that is there are twice as many CD8+ cells as there are CD4+ cells. CD4+ helper T cells are responsible for successful immune responses. As a result of their CD4+ T-cells deficiency AIDS patients have major defects in their cell-mediated immune responses.Their loss of cell-mediated immunity may be demonstrated by decreased in vitro lymphocyte responses to mitogens and antigens; by decreased cytotoxic

T-cell and NK-cell responses, and by a loss of delayed hypersensitivity, a T-cells mediated inflammatory responds to intradermally administered antigen. In addition alterations in B-cells responses occur in AIDS patients. Thus they show elevated spontaneous immunoglobulin secretion, resulting in a hyper gammaglobulinemia.

The T-cells surface glycoprotein CD4 can serve as a high-affinity receptor for HIV-1, HIV-2 binds to the CD 4 molecule through its envelope glycoprotein gp 120 and probably enters the T cells by endocytosis. Alternatively or additionally , the virus envelop may fuse with the T cells membrane. Once inside cells, HIV-1 is uncoated. The viral genomic RNA is transcribed to DNA by the viral reverse transctiptase. This proviral DNA is integrated into the host cell chromosomal DNA in a process that uses the integrase coded for by the pol gene. Large amounts of unintegrated viral DNA also accumulate in infected cells. Once integrated, the virus infection becomes latent until the infected cell in activated by antigen, mitogen, or cytokines. Other viruses such as cytomegalovirus or Epstein-Barr virus may also activity latent HIV infection. Once the infected T cell is activated, the proviral DNA is transcribed to messenger RNA and protein synthesis occurs. The protein in processed, and virions are assembled as they but from surface of the infected cell. Several immune responses may have to occur before all HIV infected cells are activated and $CD4^+$ cells are completely depleted. This could account for the variably long latent period of the disease.

Although it is clear that the major lesions seen in HIV infection are due to the destruction of $CD4^+$ T cells byHIV-1 it has been difficult to account for the extent of the immunodeficiency seen in these patients. Although about 1 in 400 T cells in blood is infected by HIV, only about 1 in 100,000 of these cells actually produces virus. Despite this there is a major loss of $CD4^+$ T cells. It would appear that normal T-cells turnover $CD4^+$ should be able to compensate for this low rate of destruction, yet it does not do so. It is important to point out, however, that blood T cells represent only a small fraction of the total T cell population in the body. The events in the blood are probably not typical of the events occurring elsewhere in the lymphoid system. Indeed, Giuseppe pantaeo and Antony Fauci have shown that HIV preferentially replicates not was the blood but in the germinal centers of lymph nodes. Many of these germinal centers cells contain proviral DNA but not RNA, suggesting that they are latently infected. Thus significant virus replication and T-cells destruction can occur in lymph node cells while the blood T cells appear to be relatively unaffected.

How HIV-1 kills or otherwise turns off T cells is unknown. Infected cells may simple lyse when massive amount of virus bud off their surface. The $CD4^+$ molecule may play a role in this since monocytes and macrophages are also infected with HIV yet are not significantly destroyed. They carry much less $CD4^+$ on their surface than do T cells. If $CD8^+$ T cells are artificially infected with HIV, they too are resistant to its cytopathic effect. The density of $CD4^+$ on the cell surface may therefore influence the destructive effect on HIV.

A second mechanism of T cell loss may be through cell fusion. A high level of env gene products is expressed in cells that are budding viral particles. This protein makes cells fuse and causes syncytium formation with neighboring uninfected $CD4^+$ T cells. Cytolysis with neighboring uninfected $CD4^+$ T Cells. Cytolysis and death of the fused cells occur within 48 hours. The incorporation of uninfected lymphocytes into syncytia in this way may thus contribution to the depletion of $CD4^+$ cells.

A third destructive mechanism may be through immune destruction of helper T cells. $CD4^+$ cells can capture process and present envelop group 120 on their surface so making uninfected T cells a target for HIV-1 specific cytotoxic T-cells. In addition some AIDS patients produce a cytotoxic antibody reactive with activated lymphocytes.

Some investigators have suggested that antigen may trigger apoptosis of HIV infected T cells. This could happen in two ways. Either group 120 binding to $CD4^+$ provides one signal while antigen presented at a later time, triggers an apoptotic response. (Remember that both signals must be provided simultaneously in order to activate T cells). Alternatively, antibody against bound gp120 may trigger apoptosis. Both mechanisms have been demonstrated to act in vitro. They imply that the T cells of HIV-infected individuals already carry one of two apoptosis- inducing signals and that almost any other pathogen may provided the second signal at a time when it would trigger apoptosis rather than an activating response.

Other postulated mechanisms that could account for major T-cell losses may involve the induction of soluble factors with toxic effects on $CD4^+$ cells; loss of a $CD4^+$ stem cell population or loss of a subset of $CD4^+$ cells that are required for the production of the whole T cell population.

21.5.T. CELL ABNORMALITIES

In addition to destroying T cells, HIV impairs the function of surviving $CD4^+$ cells. There are several possible mechanisms for this function defect. First HIV-1 subunits such as the envelope are capable of directly inhibiting T cell function. Second HIV-1 group 120 binds avidly to CD4. In this location, it may interfere with the ability of CD4 to interact with MHC class II molecules and so blocks the ability of T cells to respond to presented antigen. Alternatively group 120 may send a negative desensitizing signal to the T cell, so preventing its response to antigen. T cell that are latently infected with HIV no longer express CD4 on their surface; they are defective in the synthesis and secretion of Il-2 they show decreased proliferation, decreased expression of the Il-2 receptor and depressed production of IFN-r. As a result of these changes , there may be a significant decline in immune function in patients whose CD4 cell numbers are normal.

HIV-1 induces a significant cytotoxic/suppressor T cell response in infected individuals. These $CD8^+$ cells are directed against viral epitopes and so may serve a protective role. The cytotoxic responses diminish with the progression of the disease, but they may contribute to its pathology. Some of the immunesuppression in AIDS patients may be due to stimulation of cytotoxic/supprssor activity as a result of an imbalance between $CD4^+$ and $CD8^+$.

21.6. B CELL ABNORMALITIES

HIV-1-infected individuals have abnormalities in B-cell function. They show polyclonal activation, hyper gamma globulinemia, immune complex deposition, and the development of autoimmunity.At the same time, their B cells show decreased proliferative responses to pokeweed mitogen and to new antigens. Although the loss a helper T-cells function undoubtedly contribution to this defeat there is also evidence of B-cell abnormal. For example, AIDS patients are unable to mount an adequate IgM response. Increased B-cell activity as reflected by elevated serum IgG and IgA levels is a consistent finding in AIDS. Many of the B-cells abnormalities appear to be a result of elevated IL-6 levels. B cells transformed by Epstein Barr virus can be infected with HIV-1 since they express CD4 on their surface. The high incidence of Epstein-Barr virus and cyto megalovirus infection in AIDS patients may therefore contribute to the nonspecific stimulation of B-cell activity.

Among the many suggestions seeking to explain how HIV kills T and B cells is that the viral glycoprotein group 120 may act like a superantigen. Thus in HIV infected individuals, some B cell subpopulations, especially those bearing the immunoglobulin heavy chain domain VH3, are significantly reduced. It has now been shown that group binds to VH3. Indeed, group 120 can activate B cell bearing VH3 in their BCRs. Thus it is conceivable that the binding of group 120 to VH3 will so stimulate this subpopulation of B cells that they will eventually be depleted in the manner of the superantigens that trigger T-cells responses.

21.7. NK-CELL ABNORMALITIES

Activity of natural killer (NK) cells is suppressed in AIDS patients, although NK-cell numbers are normal. These NK cells have reduced cytotoxic ability. This may be resorted to normal by exposure of the NK cells to II-2 or to concanavalin A.This suggests that NK cells are normally activated by T-cells derived II-2 and in its absence show defective function.

21.8. MACROPHAGE ABNORMALITIES

Monocytes and macrophages play a major role in AIDS since they are reservoirs of HIV through all stages of infection. Although CD4 is present in low concentration on blood monocytes, the percentage of monocytes that show it detectably ranges from 5% to 90%. Monocytes expressing CD_4 can be directly infected by HIV, and they may also ingest opsonized HIV. In some tissue such as brain, lymph nodes, and lung, 10% to 50% of macrophages may be infected. On the other hand, the macrophages in the liver (Kupffer cells) have not yet been shown to become infected in AIDS. HIV can replicate in macrophages, although unlike T cells the virions are not readily realized but accumulate in large numbers in cytoplasmic vacuoles. Macrophages may constitute the major reservoir of HIV in the body since once they are infected they can produce large quantities of virus but are not killed by the productive infection. Thus infected macrophages may spread the virus through out the body. Infected ganglial cells and alveolar macrophages may contribute to the development of neurologic and lung lesions in some patients. The numbers of macrophages in AIDS patients are usually normal, even in patients who have very low or undetectable $CD4^+$ T cells.

HIV infected macrophages show functional abnormalities. Their adherence to surfaces is reduced, and they do not respond normally to chemoattractants. Their expression of MHC class II molecules is depressed, suggesting that their antigen presenting ability is impaired. Their phagocytic ability is also impaired and they show depressed clearance of particles from the blood. Persistently infected macrophages may be stimulated to release large amounts of prostaglandins, IL-1 THF-a and IL-6. This cytokine release may account in part for alterations in B-cell and haematopoietic-cell function. In the from of AIDS observed in Africa, patients lose so much weight that the condition is called slim disease. This may be due in part to the excessive and prolonged release of TNF-a.

It is possible that CD-4 bearing dendritic cells lining mucosal surfaces are the first cells infected by sexually transmitted HIV. The follicular dendritic cells in lymph nodes are severly damaged early in HIV infection. The number of Langerhans cells in the skin/Langerhans is reduced in AIDS patients, and this may contribute to the development of Kaposi's sarcoma.

21.9. HIV IN OTHER CELLS

HIV infects and affects many non-T cells, because any $CD4^+$ cell in vulnerable to HIV infection, macrophages, monocytes, Langerhans cells, foilcular dendritic cells, microganglia and endothelial cells in the brain may all be damaged. In addition , HIV causes a granulocytopenia, although neutrophil function is normal.

21.10. AIDS – THE DISEASE

HIV poorly conragious and is spread either when contaminated fluids are injected or when infected lymphocytes are transmitted. The number of lymphocytes in sperm is elevated in individuals with a history of sexually transmitted disease. The virus is known to be transmitted in blood (and in organ grafts), but blood is now screened in developed countries so that the risks of acquiring infection by this route have been significantly reduced. The fact that HIV has been recovered from saliva has given rise to fears that it might be spread by casual contract. It is highly unlikely, however, to be natural route to virus transmission since salivary cells must first be cultured for isolation at tempts to succeed. Normal household (casual) contact is also a highly unlikely mode of transmission. A small percentage of cases occur in heterosexuals with no known mode of transmission. This has given rise to fears in North America and Europe that the infection may increase in incidence in heterosexual and spread rapidly through the population. This now appears very unlikely.

The average time from developing antibodies to HIV to the onset of disease may be as long as ten years. From the experience of those individuals who acquired the infection in the early 1980s, it is possible to estimate that each year approximately 6% of those positive for antibodies to HIV will develop clinical AIDS. A wide range of disease syndromes is now recognized as being associated with HIV infection; they progress from mild asymptomatic infections to persistent generalized lymph node swelling (lymph adenopathy) and to AIDS itself.

Most individuals infected with HIV develop a fever and muscle aches three to six weeks after initial infection. This lasts for several days or weeks and is associated with seroconversion (antibodies to HIV begin to be produced). Large amounts of virus are in the bloodstream at this time, and the disease can be readily transmitted . At the end of this time the patient's immune response eliminates free virus and infected cells. As a result detectable virus disappears. Unfortunately, some infected cells persist, and the virus continues to replicate in low numbers for a variable long period, commonly for several years. During this time the patient is usually fairly health. Eventually, however the virus causes sufficient damage to the immune system that T cells numbers drop below the level necessary to maintain health and infections and tumors begin to develop.

Persistent generalized lymphoadenopathy, a more advanced disease, is characterized by the development of multiple swollen lymph nodes in the absence of other illness, The lymph nodes of these patients show hyperplasia of the follicles in the cortex and infiltration of germinal centers by $CD4^+$ cells. This eventually leads to germinal center depletion and follicular shrinkage (involution). Lympoadenopathy may persist for a long period without the patient showing clinical deterioration, but it may progress inexorably to AIDS itself.

Significant weight loss (HIV wasting syndrome) is also a common feature in AIDS patients. This may be sult in part from a loss appetite, but it can also be due to increased or altered protein metabolism, intestinal malabsorption, chronic production and endocrine abnormalities.

21.11. SECONDARY INFECTIONS

The full HIV-medicated disease syndrome can present in several manifestations. Some individuals simply have prolonged *fever*, diarrohea, weight loss in the absence of concurrent infection. Nevertheless numerous opportunistic infections (Secondary infections) are usually associated with clinical AIDS. These opportunistic pathogens include protozoa such as *P. carinnii.* Toxiplasma gondii, and cryptosopridium muris; viruses such as cytomegalo virus and varicella zoster; bacteria such as *Mycobacterium avium*; and fingi such as *Cryptococcus, Candida,* and *Histopladma.* These organisms which cause few problems in normal persons can cause severe, life threatening infections in HIV-1 infected individuals *P. carinii* is the most common AIDS related infection in the United States and Europe.

21.12. NEUROLOGIC LESIONS

Some AIDS patient may suffer from several neurological problems known as the AIDS dementia complex. These patients may show disturbances in cognition, movement sensation and behavior. Problems can develop even when there is very little brain infection with HIV, and it is not associated with secondary or opportunistic infections. CD4 is found on neurons and ganglial cells in several parts of the brain, so these cells may bind group 120. AIDS dementia complex results from the toxic effects of HIV derived group 120 on neurons. The gp120 causes a rise in neuronal Ca2+ concentrations that eventually triggers cellular apoptosis. AIDS dementia complex has a poor prognosis, and the mean survival time after diagnosis in these cause is 11 to 15 months. About 25% of AIDS patients who show no neurological symptoms when alive have vacuoles in their myelin and macrophage infiltration in their brain on autopsy.

21.13. KAPOSI'S SARCOMA

The occurrence of a rare tumor called Kaposi's sarcoma in young homosexual men in New York California led to the identification of the AIDS epidemic in the United States. Before that time, Koposi's sarcoma was rare skin cancer affecting elderly men of Mediterranean origin. It is now common in homosexuals with AIDS, occurring in about 36% of such patients . Patients with Kaposi's sarcoma may still loss of CD4+ cell function than AIDS patients without the tumor. Some investigators believe that Kaposi's sarcoma is a secondary disease triggered by an infectious agents different from HIV. This is because the tumor is far more prevalent among male homosexual AIDS patients than among AIDS patients from other risk groups. One possible cause is a form of human *papillomavirus.* HIV promotes the development of Kaposi's sarcoma in several ways. Thus it stimulates macrophages to produce large quantities of IL-6 on while the sarcoma is dependent. Two other proteins a T-cell derived cytokine called oncostrain M and the tat gene product of HIV, are also growth factors for Karposi's sarcoma. HIV infects Langerhans cells in the skin and it may therefore promote tumor growth in that region by reducing local defences.

AIDS patient show an increased incidence of some lymphomas that may be associated with Epstein Barr virus infection. The pathogenesis of these lymphomas is complex but may well reflect destruction. An autoimmune thrombocytopenia and bleeding disorder associated in AIDS patients.

21.14. GENETIC FACTORS

Susceptibility to HIV and the course of the disease are partially determined by genetic factors such as the MHC alleles DR5 and B35. In addition HIV susceptibility is associated with certain alleles of a vitamin D-binding protein called Gc. HIV binds more strongly to a highly glycosylated from of Gc protein.

21.15. DIAGNOSIS OF HIV INFECTION

Because HIV may be transmitted by blood and blood products, it is necessary to screen blood for antibodies to this virus. The test used for this purpose is the ELISA. The sensitivity of the test is set at a very high level to ensure that no positive blood is missed. As a result, many false-positive ELISA results are obtained. These positive samples are in turn subjected either to Western blotting or to immunofluorescence assays to arrive at a definitive diagnosis. Between 1 in 6 and 1 in 20 positive ELISA samples are turn positive as detected by western bolting. Nevertheless, a carrier state can exist and HIV may be isolated from healthy, antibody free carriers.

21.16. TREATMENT OF HIV INFECTION

Three antiviral drugs have been approved in the United States for the treatment of AIDS. These are azidothymidine (also called AZT or zidovudine), dideoxycytidine (ddC),and dudeoxyinosine (ddI). All of these can block HIV replication by interfering with reverse transcription. Normally the proviral DNA chain grows by attaching each new nucleotide to a hydroxyl group located at the 3′ end of the chain to from a linkage called a phosphodiester bond. The nucleotide analogs, if present are incorporated into the growing DNA chain. However they lack the 3′ hydroxyl group and as a result no more nucleotide can be added after the chain in blocked. These compounds will therefore terminate proviral DNA chain growth and so inactivate the virus. Obviously these compounds can also block DNA production in uninfected cells. All three drugs are therefore very toxic. A greater problems in the high mutation rate mutation rate of the virus.The virus will eventually produce altered reverse transcriptase that can act even in the presence of these drugs, and their benefits are only temporary.

21.17. AN AIDS VACCINE

At present, no effective vaccine against AIDS is available. There are two main reasons why such a vaccine has not yet been developed. First as described, HIV mutates very rapidly, producing a mixture of antigenically distinct viruses even within one infected individual. Second, HIV is transmitted via mucosal surface both as cell free virions and by HIV infected cells, and different types of responds may be required cells, and different types of response and by HIV-infected cells, and different types of response may be required cells, and different types of response may be required to counteract this. In addition there is a question of usage. Should the vaccine employed to prevent infection if healthy individuals or. Alternatively , to slow progression of the diseases in those already infected with HIV. In experimental primate models , vaccines against HIV have been protective for only a short period and only when the vaccine strain identical to the infecting virus. Likewise it is clear that natural infection with HIV does not confer significant protective immunity.

Traditional approaches to vaccine development have improved immunization with either inactivated virus or recombinant HIV proteins using an adjuvant. These, it is hoped, will stimulate production of antibodies that could block HIV infection of cells. Unfortunately the immunity induced is brief only being effective when antibody levels are very high. An alternative approach, the use of modified life virus in the vaccine, has a major drawback. The mutation rate of the virus is such that the virus could easily become virulent, an unacceptable result. Because of these problems, considerable attention has been paid to newer approaches to the development of an AIDS vaccine. Synthetic HIV protein epitopes have been ineffective since they generally contain too few epitopes. Live recombinant vaccines using HIV recombinant in vaccine, mycobacteria, or enteric bacteria have not stimulated a significant protective response in animals, nor have combined approaches using life recombinant products together with recombinant vaccines been any more successful.

Some potential HIV vaccines are currently being tested for safety in humans. These include vaccines containing recombinant envelope glycoproteins synthetic peptides, and recombinant *pox virus*. Unfortunately none has been protective in animal models.

21.18. AIDS IN ANIMALS

The only primates that are susceptible to HIV-1 infection in the laboratory are the chimpanzee and the gibbob. Although these animals can be infected they do not develop signs of disease. Likewise rhesus macaques can be persistently infected with HIV-2 but without developing disease.

There are several interesting nonprimate models of AIDS. Rabbits are susceptible to a low-level, persistent HIV infection, but like primates, they do not develop disease, *scid* mice reconstituted with human lymphocytes can harbor HIV infections, but the virus destroys their T cells very rapidly in a manner unlike the natural disease. Cats vary rapidly in a manner unlike the natural disease. Cats suffer severe immunosuppression as a result of infection with either the feline *leukemia virus*, an *oncornavirus* or the *feline immunodeficiency virus*, a *lentivirus*.

Several HIV related virus have been isolated from nonhuman primates. These simian immunodeficiency viruses include SIVmac isolated from a rhesus macaque, SIV agm from an African green monkey SIVsm from a sooty mangabey SIV mnd from a mandrill and SIVcpz from a chimpanzee. All these isolated preferentially invade $CD4^+$ cells. Nucleotide homology studies suggest that HIV-2 , SIVmac and SIVstn are very closely related and probably arose from a common ancestor.

When SIVmac infects rhesus monkeys (*Macaca mulatta)* it induces an immunodeficiency syndrome that is similar to human AIDS. It differ from the human disease in that the monkeys are immunosuppressed as a result of depletion of both B and T cells. Consequently it is associated with a profound drop in serum IgG and IgM levels and a severe lymphopenia. Monocyte function in unimpaired but the remaining lymphocytes do not responds to mitogens such as concanavalin A and phytohaemaggulutination. The monkeys are also profound neutropenic, about half the infected animals survive the disease, others die as a result of septicemia or diarrhea with wasting. In many cases, normally innocuous agents such as cytomegalovirus, *Cryptosporidium,* and *Candida albicans* cause infection. Some affected monkeys develop tumors such as fibrosacomas. In monkey colonies, the disease is spread by the bites of carrier animals.

21.19. OTHER HUMAN VIRUSES AFFECTING THE HUMAN SYSTEM

Six retroviruses are known to infect humans, three of these are transforming retroviruses HTL V-I HTI.V-II and HTLV-V (A transforming virus can change the growth characteristics of a cancer cell). These are associated with the development of T cell neoplasms. The other three human retroviruses, HIV-1 HIV-2 and HTLV-IV are not transforming . Both HIV-1 and HIV-2 cause AIDS. No disease has yet been associated with HYL-V-VI.

HTLV-1 is the cause of adult T-cell leukemia a cancer involving $CD4^+$ lymphocytes. Only 1% of individuals exposed to this virus develop the disease, and the incubation period is as long 20 to 30 years. HTLV-1 is also associated with tropical spastic paresis, a neurologic disease characterized by difficulties in walking. HTLV-II has been isolated from several cases of hairy cell leukemia (a B-cell tumor) and HTLV-V has been isolated from a case of mycosis fungicides (a T-cell tumor of the skin). HTLV-IV has been is very similar to HIV-2.

21.20. AIDS PREVENTION AND ITS SOCIAL IMPLICATIONS

Analysis of AIDS patients based on their route of infection indicates that in the United States about 90% of case are in men about 90% of these men, acquired their infection through certain behaviors, notably homosexual behavior and the sharing of needles when using drugs. Clearly avoidance of such behavior will go a long way toward preventing the transmission of HIV. Even in woman approximately half of the cases are associated with intravenous drug abuse, although 36% result from heterosexual contact with infected men. The risks of acquiring AIDS by either homosexual or heterosexual transmission could therefore be minimized by avoidance of sexual contact with individuals from high risk groups of either sex. A more realistic solution is the consistent use of condoms, which prevent the transfer of HIV infected lyhmphocytes in the sexual fluids to uninfected recipients. AIDS infection in intravenous drug abusers is a result in the transfer of infected cells. Clearly there is no easy solution to this problem, although some have advocated providing drug addicts with fresh needles in an attempt to prevent the reuse of contaminated ones. Screening of the blood supply and blood products has greatly reduced the risks of acquiring AIDS from transfusions. Blood for transfusion is screened by ELISA tests and Western blotting.

21.21. PREVENTION OF AIDS-AN INDIAN PERSPECTIVE

AIDS, the Acquired Immuno Deficiency Syndrome, whose damage potential has become known as more lethal than nuclear explosions, is a fatal illness caused by retrovirus known as the Human Immune deficiency Virus (HIV).

It is common that everyone infected with HIV will eventually develop AIDS, because the body's immune (natural defence) system is steadily weaken by HIV. When the AIDS virus enters the body, the immune system again produce suitable anti bodies to fight it. The AIDS virus is extremely difficult to detect but the antibodies which the human body produces to fight the AIDS virus can be detected. When a person found to contain these antibodies, he or she is called HIV positive. It simply means that since AIDS antibodies are present in the person, AIDS virus is also present against which antibodies are fighting. Unable to protect the immune system, most persons die within three four years of the first signs of AIDS appearance. Several cases also found the HIV related diseases soon after their initial infection, nearly all remain well for years. Then, as the virus gradually damages their immune system, they begin to develop illness of increasing severity, characterized by various combinations of symptoms and diseases, such as diarrohea, fever, wasting, fungal infections, tuberculosis, pneumonia, lymphoma, failure to thrive and Kaposl's sarcoma. This causes loss of resistance of human body to every kind of infection.

There are two types of HIV Viruses, currently known as HIV 1 and HIV 2. The virus HIV 1 is dominant in nature worldwide and transmits at faster rate, whereas HIV 2 is less easily transmitted and the time period between the initial infection and illness is longer as compared to HIV 1. According to Prof. Max Essex of the Harvard AIDS Institute, out of the two categories of HIV viruses, the AIDS contracted in western nations is mainly through homosexual contact and intravenous drug use spread by HIV 1 B, whereas in South Asia and Sub Saharan Africa, the AIDS caused by other variety of viruses HIV 1 A, C, and E, which are contracted almost entirely by heterosexual contact. This disease/infection is primarily sexually transmitted diseases (STD), is passed on through unprotected preventive sex. The virus is also transmitted through blood transfusions, use of unsterilized injection equipment or cutting instruments and from an infected women to her foetus or nursing infant. A number of studies have demonstrated that the people with STD infections have a much higher risk of transmitting or acquiring HIV. The HIV risk of infection increases up to 10 times by STD. It is estimated around the world half of the 333 million new STD cases per year occur in young people under age 25 and new infections in young people occur at the rate of 5 per minute. Research findings have indicated STD greatly increases an individual's risk of getting or passing on HIV to his or her partner.

The New England Journal of Medicine has reported that women are twice as likely as men to contract some kinds of STDs. The Medical Council Research program on Uganda reported that the HIV infection rates among 13 to 19 year old girls are 3 times more than that in teenage boys. The characters and numbers of the epidemic affected people varies from region to region, such as, Africa and Asia show more transmission through heterosexual sex while Latin America through homosexual sex and Eastern Europe and newly independent states through drug injection. Besides, biological and social vulnerability also impact the spread of the epidemic. The AIDS virus is 4 times more likely to spread to a woman from an infected man than vice versa. Compared with males, the female reproductive tract is more susceptible to infection with HIV/STDs, a susceptibility that is practically high in the case of young women.

The alarming rise of STD has put India at top position in AIDS scenario in Asian Continent. The World Health Organization (WHO) estimates there could be at least 5 million people carrying the HIV virus in India by the beginning of the next millennium. The World Bank reported that AIDS threatens to reverse the decades of progress in improving the quality of life in the developing countries. In some developing countries like China and India, the HIV has infected 5 per cent of the people who are engaged in high risk behavior. Overall, about 6.4 million people are currently believed to be living with HIV in Asia and Pacific just over 1 in 5 of the world's total. By the end of the year 2000, just a month from now, that proportion is expected to rise to 1 in 4. Around 92,000 children now have HIV in Asia. The prevention measure for spread of STD has the potential to reduce large amount of transmission of HIV/AIDS infection. It is an emerging issue how can the spread of the disease be stopped and what strategies are required for successful planning and management of prevention of the epidemic.

21.22. AN OVERVIEW OF WORLD AIDS

The recent combined estimation of both the UNAIDS Organization and WHO, of the total 30 million people inhabited with HIVAIDS infection, at least one third among them are young people aged 10 24. Everyday, 16,000 newly infected are added to these numbers, which include around 7,000 young people. The finding estimates about 1,600 children under 15 years of age are infected with the virus. And, among infected adults about 40% are women. This means around 2.6 million new infections per year among young people, including 1.7 million in Africa and 7,00,000 in Asia and pacific regions. Overall, young people account for at least 50% of all those who become infected after infancy. There are several countries where the figure exceeds 60% of the total infected persons. The high infection rates among the young people reflect the preponderance of the AIDS threat in the world. The global population is young, and increasing so. The increasing trend of young people during 1960–90 showed that, of the total population grew by 75%, whereas during that period the proportion of the young people increased by 99 per cent. And, among them the developing countries count 85% of world's young people, wherever nine tenths of the epidemic is now concentrated the criticality of the AIDS prevention, gets into the sharp focus.

The most sensitive issue is that majority of the young in these countries do not know whether they are infected. The estimation by the UNAIDS reported that over 27 million people in the world who have no idea they are infected. The rising statistics estimates that the World AIDS/HIV infections will reach 40 million by the beginning of the next millennium. Among the World's young people, some are most exposed to HIV than others. Those living especially in difficult circumstances include young people who are out of school, who live on streets, who share needles with other injecting drug users, engaged in commercial sex, or are sexually and physically abused. The UNAIDS and WHO estimates show that around 2.3 million people died of AIDS in 1997 (Include 46% women and 460 thousand children) i.e., one fifth of the total 11.7 million AIDS deaths since the beginning of the epidemic in the late 1970s.

21.23. AIDS IN INDIA

The UNAIDS estimates the number of HIV infected people in India at present at about 5 million (less than one per cent of total population), this number is greater than in any other country in the world. The number of infected persons has increased steadily from 2.5 million in 1996, spreading mainly through heterosexual sex into general public from high risk groups. Even at the same rate, the rate of infection increases, the number will be around 50 million in next 10 years of time. It also estimates, around 25,000 women and 30,000 new born babies get infected while 20,000 die every year in India. According to the Executive Director of UNAIDS, Dr Peter Piot, though South Africa (1.8 million), Uganda (1.4 million) and Nigeria (1.2 million) have high percentage of HIV infected population, it would be no wonder to see the same position obtaining in India in the coming years. Similarly, India Health Organization, a voluntary agency actively engaged in AIDS control campaign and research, predicted that the HIV infected persons in India would rise between 20 to 50 million by 2000, i.e., the number would double every 18 months. If this were to happen, then by 2000 around 5 per cent of the country's population would be infected and the death toll would be around 6,000 to 12,500 per day.

A recent report of the nationwide survey by the National AIDS Control Organization (NACO) indicates that one out of every ten persons from high risk groups is infected with AIDS causing HIV virus in Maharashtra and Manipur. Maharashtra records a high HIV positive incidence of about 100 per thousand samples screened, while Manipur has about 95 positive cases per thousand samples. Similarly, Punjab and Nagaland reported 44 and 55 cases per thousand samples screened. City wise, the report stated that 25 to 36 per cent of the cases undergoing treatment at the STD clinics have HIV infections, in Mumbai, 3 to 14 per cent in Nagpur, 2 to 5 per cent in Chennai, and 6 to 10 per cent in Madurai. According to experts, the possibility to several cases having gone unreported cannot be ruled out. This is more because of social stigma and lack of awareness among the masses. As per the survey conducted by NACO, till May 1997, out of the total 3.03 million samples screened from high risk groups and clinically suspected cases of AIDS, 56.409 were found to be HIV positive (18.6 per thousand samples). The alarming rise of HIV infected cases from 2.25 to 4.25% in Mumbai, 2 to 4.25% in Pune, 2.25 to 3.75% in Pondicherry, and from 0.45 to 0.81% in Imphal during past five years has been a cause of great concern. In Manipur, where the epidemic took off quickly among male drug injectors, found some drug clinics in the state were registering HIV rates of as high as 73 percent in 1996.

Substantial increase in the number of children infected with HIV/AIDS is another major cause of concern. The medical experts have observed that infection among children progress rapidly. In new born babies or in the symptoms of HIV infection cannot easily detected which hampers restoring to early preventive measures. Sometimes, the infection is also a threat to the survival of a child even after the death of infected mother. The UNICEF has reported that in Mumbai alone 2.5 per cent of all mothers attending anti natal clinics test HIV positive. Another survey by the Maharashtra Government's Health Department revealed that during 1992 96, the number of HIV/AIDS infected pregnant women has gone up four times. The Samraksha, a Bangalore based AIDS control project reported that out of 420 registered cases in its record, 58% found to be married women coming from middle class background. The growth rate of infected women was around 3.3% per annum. The Gujarat AIDS Awareness & Prevention unit reports that the rise of HIV infected cases has gone up by 30% in middle class families. A study by NACO found that out of the 32,00,357 up to October 1997, the HIV positive case was 67,311 and confirmed AIDS case was 5,002. It also included a large number of infected babies. The census figures indicate that if the AIDS menace is not checked, the infant mortality rate due to AIDS could go up to as high as 75%. Heterosexual cases transmission now accounts for over two thirds of all positive cases in India. In India (up to 1992), it was estimated that out of 45.3% heterosexual infected persons, the women constituted about 61.52%. The high rate of infected women in India belongs to this group and most of them belong to adult group. Another survey estimated that out of total 24 million deliveries around 0.2 million cases found stereopositive every year in our country. The stereopositive cases have increased from 2.5 per thousand to 17 per thousand in a decade (i.e., about seven times during 1986-96). According to a study, about 85% of the AIDS cases found were sex related, while 8-18 per cent through blood transfusion and about 4 to 5% from mother to child.

The HIV/AIDS virus was earlier limited to the group with hetero or multi sexual partners and intravenous drug users. Now the disease has spread among people in rural areas also. NACO reported that AIDS is mainly affecting the economically productive and sexually active groups between 14 and 45 years which accounted for over 77% of total AIDS cases in India.

21.24. MOBILISING YOUNG PEOPLE OF AIDS CONTROL

Basically, there are three reasons required for more attention to campaign for prevention of AIDS and particularly with young people. One, the special vulnerability of young people to the epidemic. Of all those infected after infancy, at least half are young people under age 25. Second, young people aged 10 to 24 account for more than 30% of all people in developing world, where the epidemic is concentrated. If HIV prevention in this huge youthful population fails, developing countries will have to face the staggering human and economic costs of vast numbers of adult AIDS cases. Third, the most important, working with young people makes sense because they are a force for change. They are still at the stage of experimentation and can learn more easily than adults to make their behavior safe or to adopt safe practices from the beginning. Besides, the young people can help take the sting and shame out of AIDS where it is still stigmatised, they can bring kindness and practical help to those already infected with HIV or living in a household touched by AIDS. Young people can spur action and move it in the right direction. They are a powerful force for change in their own households, in the lives of their peers, and in the wider community too. In many countries young people are encouraged to take on roles as educators and mobilizers in local community groups, including those in the field of AIDS.

There are special reasons why young people's involvement is essential for action on HIV/AIDS. Young people are a force for change because they are young, resilient, open to change, creative, and often idealistic. Capitalizing on this force for change for young people to work in partnership with adults who encourage their participation and are receptive of their ideas. It can draw young people into productive activities, such as giving emotional support and bringing practical help and care to those affected by the epidemic. Peer support in youth to youth can be aided and encouraged. The United Nations Development Program in its innovative program of exchange tours brings young people from Ukraine to Sweden where they visit clubs that are active in AIDS prevention.

Young people can change the course of the epidemic if appropriate support extended by the adults. With the world's power and resources concentrated in the hands of adult, the onus for AIDS education is on adults as individuals, groups and governments. In many industrialized countries and some developing countries, governments and non government organizations (NGOs) together have risen to this challenge with considerable success. Though a combination of media campaigns, health education on AIDS for those attending school and out of school, the provision of safer sex and other prevention services, and the establishment of supportive policies, countries like Australia, Uganda, etc have helped young people to adopt safer sexual behavior. In Thailand, sustained prevention efforts were aimed at increasing condom use, boosting condom use for women, discouraging men from visiting sex workers, and offering young women better education and other prospects to discourage their entry or sale into commercial sex. As a result of these measures carried out by many government departments in coordination with NGOs and the media. HIV prevalence in 21 year old men in northern Thailand fell from 7.1% in 1992 to 3.5% in mid 1995, and has decreased even further since then. Concerted efforts from various quarters have resulted considerably lower infection rates in young pregnant women attending antenatal clinics in Uganda. The major achievement is the vast decrease in the youngest age group, these aged between 15 to 19. In some clinics, rates have plummeted up from 38% in 1991 to 7.3% in 1996.

Experiments carried all over world show that it is best done not by manipulating the young or using them as tokens, but with their genuine participation. This means listening to the young people, understanding how they think and act, offering assistance and support without dictating the outcome, and engaging them as active participants in activities and decision making (1998 World AIDS Campaign, UNAIDS.

21.25. CAMPAIGN OBJECTIVES

Young people deserve priority attention in the ever growing HIV/AIDS epidemic. The numbers and the characteristics of the epidemic among young people differ from region to region. To curb this epidemic, it is important to formulate cultural sensitive policies and action plans to look beyond AIDS to larger picture of young people's health and development. Force for change, the 1998 World AIDS Campaign with young people, provides an opportunity for the social and private sectors to mobilize young people to promote their highest state of well being.

21.26. INCULCATING AWARENESS

Today, a special challenge in health care is taking an early, comprehensive, concerted AIDS prevention action before a major HIV epidemic gets underway. Substantial measures undertaken by the Government for prevention of rapid transmission of AIDS should be given top priority in health care agenda. A serious action is required to prevent the spread of the disease either by legislating laws or by creating mass awareness. For this, a combination of planned strategy and adequate financial and technical resources is required. The Government should target its efforts more on high risk groups, motivating them to change their sexual behavior and distributing condoms as a primary step to curb the disease, following it up with speedier action plans for early treatment of sexually transmitted diseases. These measures would then be more cost effective, as they would target relatively few people in the core groups (such as sex workers and their clients). Another step is to subside the high charges for clinical services rendered for testing STD as a measure to help the AIDS control program.

"Proper awareness among the masses" is the main measure to be taken for prevention and control of AIDS. Appropriate information put on mass media highlighting the adverse effects of AIDS and its impact on society and how to avoid it, public campaigns on loss to the environment due to the disease, effective advertising, and specialized training for counselors are essential prerequisites for inculcating mass awareness that can prevent spread of AIDS to a larger extent. Besides, programmes for general population on safer sex and prevention and treatment of STD may be organized at public places. Emphasis on private and public finances for supporting AIDS control programmes initiated by the Government and international agencies should be at the top of the policy agenda. It is important for the Government and other agencies engaged in AIDS prevention to provide appropriate information on promoting condom use for commercial sex workers and also to ensure their availability. The primary health centres and family planning units at the grass root level can play an important role in educating the masses about the unsafe sexual activities and promoting campaign for condom use. Counseling about gender issues, STDs and HIV/AIDS, safe motherhood and family planning and other related issues are some other areas where the activities of health care should be planned, programmed, and implemented with explicit honesty.

In a planned strategy, our health care services may combine STD with family planning programmes.This would help in illustrating the problem through pictorial information. In our country, though the AIDS infection is not very rapid in rural areas, it would be appropriate to take preventive measures so early. Comprehensive reproductive health education as a part of family planning programmes particularly targeted for youths with attention on STDs/AIDS should be planned an initiated as a precautionary measures.

It is natural and very common that most of the infected women are asymptomatic to be tested and retested. There are several reasons behind it. The agencies which are dealing with AIDS/HIV testing should create such an atmosphere that the patients do not feel any reluctancy to get checked and rechecked.

India launched its AIDS Control Programme in 1992 with the help of foreign assistance. At present, most of the AIDS control programmes have proved to be inadequate due to financial crunch. Although, an international attention with constant funding would help promoting the AIDS control measures it is the serious efforts of the Government with the help of family and community participation in mobilizing masses that could effectively help in stabilizing the epidemic.

December 1 was celebrated the World over as World AIDS Day.

21.27. CURRENT DEVELOPMENT IN AIDS RESEARCH

About 23 million people worldwide carry AIDS virus. The virus infects 3 million people each year. The virus reproduces at a fantastic speed, producing 10 30 billion new copies of the virus each day. This, of course, enhances mutation rate which is the main obstacle for developing vaccine. India will have the largest number of HIV infected persons in the world , as India enters the 21st century. HIV genes (group 120) inserted into live canary pox virus, harmless to humans, is one of the approaches being studied. Like other retroviruses, HIV contains a single stranded RNA genome, which is replicated through a double stranded DNA intermediate. Infact, viral genes are transcribed, only when they are integrated into the host DNA. The development of naked DNA vaccines is being studied for many infectious diseases like HIV, influenza, tuberculosis and malaria. DNA vaccines are immunogenic. The aim now is o define relatively conserved sequences in these HIV proteins and use them as immunogens. An alternative approach is to use soluble fragments of CD_4 produced by rDNA technology to block gp120. To date live attenuated vaccines have thrown the greatest promise of protector against HIV. This is however of deep concern about safety due to mutation. As an ugandan scientist has said: 'The house is an fire, test what is available even if protection is partial rather than wait and do nothing". HIV 1 was identified in 1983 as the agent that causes AIDS in humans. CD_4 was shown to be the primary cell surface receptor, reflecting the tropism of the virus for CD_4 expressing cells. In vitro HIV 1 replication has been demonstrated in CD_4 T helper cells and specifically differentiated tissue macrophages such as brain microganglia, spinalcord macrophages, alveolar macrophages of the lungs and blood monocytes. Macrophages could be important in transporting the virus to different organs, thus seeding replication in new niches. CD_4 lymphocytes, macrophages and dendritic cells in the secondary lymphoid organs including the lymph nodes, adenoids, tonsils and spleen are heavily infected. Once established, HIV infection persists for life. HIV 1 by changing its phenotype (by shifting tropism and escaping neutralizing antibodies) evades the host immune response. HIV 1 cell tropism are partly determined by the hyper variable loops V1, V2 and V3 in gp120, which also contain epitopes for neutralizing antibodies. The V3 loop is the first region of gp120 to be identified as a target for neutralizing antibodies. V3 is less variable than V1 or V2. Mutations conferring tropism changes and can result in escape from neutralization.

Viral diseases are among the greatest cause of morbidity and mortality in the world. Vaccines have been most helpful in the prevention and eradication of some of these diseases (small pox). The antigenic shift has hampered the development of an effective vaccine. A major problem in developing an AIDS vaccined has been the lack of a suitable host. For instance, contention is that a shortage of captive chimpanzees is impeding the development of vaccines for AIDS. Unlike antibiotics, which can attack bacteria as they replicate extra cellularly, antiviral agents must be capable of entering the host cell and inhibiting viral replication without affecting the host cell. Despite obstacles, considerable advances have been made and exciting developments are expected. The important antiviral drugs for AIDS are 1. Zidovidine (AZT) 2. Dideoxyinosine (ddI) 3. Dideoxy cytidine (ddc) 4. Azidodeoxyridine 5. 2-hydroxyethoxymethy l 6. Phenyl thiothymine (HEPT) 7. tetrahydroimidazo benzodiazepinthionl (TIBO) derivatives. They are specific to HIV 1 and not to HIV 2. It is also known that viruses develop resistance to the drugs in the course of time.

List of experimental AIDS vaccines developed so far

1. Inactivated whole viruses
2. Cloned envelop glycoprotein
3. Recombinant viruses carrying HIV genes
4. Cloned CD_4
5. Synthetic crown sequence peptides
6. Anti idiotype antibody

Another important strategy is the use of gene therapy as a potential treatment. Intracellular immunization strategies involve the expression of foreign RNA or protein as an antidote with potential to fight against AIDS inside the target cell.

Recently an US research team on AIDS (G. Shaw, University of Albama) suggested that AIDS develops when the immune system can no longer replace all the CD_4 cells.

As HIV mutates very rapidly, the key to preventing the development of resistance may lie in destabilizing the protein called zinc fingers. Zinc fingers are responsible for the maintenance of the three dimensional structure of the inner core protein (HIV contains two zinc fingers which do not readily mutate). Several drugs from a family called DIB (disulphide substituted benzamides) can remove zinc from the "fingers" which may ... the structure of the virus.

Cyclophilin A - a common free floating protein in RBC maintains the structural integrity of the capsid protein by binding to it. At the time of reinfection, cyclophilin A zips out capsid proteins and releases the viral genome. Drugs are being investigated to prevent the cyclophilin a binding to the capsid protein.

21.28. BIOTECHNOLOGY POINT OF VIEW

Acquired immuno deficiency syndrome (commonly known as AIDS) is a viral disease. It was first reported accidentally by the center for disease control in Atlanta, Georgia, USA in 1981. This new disease is characterized by severe dysfunction of both T cells and B cells system. The blood of the patient shows that the white blood corpuscles are going haywire. B cells mature into plasma cells and are involved in the production of antibody (Ab), while the T cells are the foot soldiers and are called killer cells. They attack and destroy any germ entering the body before harm is caused. T cells are again subdivided into T helper, suppressor etc. T lymphocytes is generally accepted to be the principal event responsible for the severe immune dysfunction occuring in patients infected with HIV (Fauci A S 1988).

Generally a virus stores its genetic programme in DNA but AIDS virus is a retro virus which means that instead of storing its genetic programme in DNA it uses RNA. Retro virus contains an enzyme called, reverse transcriptase, permitting these little robber threads of genetic material to make their own DNA which then go to and join the chromosomes of the host cell. The virus is labeled as human immuno virus.

21.29. PENETRATION AND DESTRUCTION OF CD4 T CELLS BY HIV

At first the HIV is attached to a molecule called CD that is present on the surface of the T cells. Proteins on the exterior envelope of the virus bind tightly to the CD_4 molecule HIV then enters the cells by fusing to the cell's membrane. After this, HIV hijacks the cell's genetic machinery. Using this genetic machinery and reverse transcriptase, HIV transcribes its own genetic material, DNA from RNA. The genetic material from the virus, now in the form of DNA, inserts itself some where in the genome of the host cell. Once there, it is called a provirus. Once integrated into the host cells genetic material, the provirus may lie dormant for a long period before its DNA instructs the cell to make new viruses. Eventually, some signal prompts the cell to transcribe the viral DNA into RNA. Some of the newly made RNA molecules will serve as genetic material for new generation of HIV.

The T4 lymphocytes thus destroyed

(*i*) cannot induce other immune cells called B lymphocytes to produce disease fighting antibodies;

(*ii*) fail to recognize foreign substances or antigens and fail to proliferate, and finally,

(*iii*) lose their ability to produce interleukin 2 (IL 2), a molecule that enhances the growth of T cells and helps to stimulate their response to foreign antigens.

At the molecular level T cell effector and regulator functions are mediated by secreted lymphokines and monokines, e.g., by interferons, interleukins, migration inhibitory factors, lymphotoxins etc. These soluble substances are the intra cellular messengers that regulate interaction between monocytes, T and B lymphocytes and are involved in homeostatic regulation and expression of immunity and inflammation.

At first, the macrophage is activated by interaction with Ag or by colony stimulating factor (CSF) from Ag activated T cells, and results in secretion of interleukin 1 (IL 1). IL 1 converts an immature Ag responsive precursor inducer T cell to a state where Ag binding causes the release of IL 2 from the inducer T cell. The binding of Ag to a seperate subpopulation of preeffector T cells induces the expression of cell surface receptor for IL 2. Now, interaction of IL 2 with IL 2 receptor causes replication of effector T cells. As a result, clonal proliferation of Ag specific effector cells occurs and eventually they secrete IFN, lymphotoxins (LT), MIF, LIF, and T cell replication factor (TRF) and B cell differentiation factor (BCDF) for B lymphocyte maturation.

The functions of some of these soluble factors and lymphokines are now being investigated in patients having AIDS. The interferons are protein in nature and have antiviral effects including suppression of Ab synthesis, delay in graft rejection, inhibition of delayed hypersensitivity etc. In contrast interferons provide an essential stimulatory signal in the generation of cytotoxic T lymphocytes in the mixed leukocyte reaction and in augmenting natural killer cell (NK) activity. There are three types of IFN, ie., and (according to their nature).

It has been shown that homosexual men with AIDS have high serum levels of IFN and it is labile at pH 2. Viruses are the most potent known inducer of IFN since patients with AIDS are commonly infected with a variety of viral agents, it is possible that the high levels of IFN may then contribute in a secondary way to suppressed immuno responsiveness.

Another lymphokine, IL 2, is also being investigated in patients having AIDS. Human IL 2 maintains long term proliferation of activated human T cell in vitro, enhances T cell dependent mitogenesis and will replace T helper cell activity leading to increased cytotoxic T cell activity and increased natural killer cell activity. The biologic effects of IL 2 leading to proliferation of T cells are dependent on the prior interaction between effector T cell and Ag or mitogen which induces expression of receptor for IL 2. Since there appears to be a loss of T helper/inducer (OKT4) cells in AIDS, one might expect that these patients would have diminished IL 2 synthesis. Several laboratories suggest that there are both diminished IL 2 production along with diminished IL 2 receptor expression by lymphocytes in AIDS affected patients.

We can therefore, conclude that a functional defect probably occurs at the level of the IL 2 producing inducer T cell, a cell which bears the OKT4 Ag. Depletion or dysfunction of this cell leads to diminished mitogen and Ag induced lymphocyte proliferation, diminished generation of cytotoxic lymphocyte effectors and defective Ab synthesis to new antigenic challenge.

21.30. HYPOTHESIS FOR ETIOLOGY OF AIDS

Upto date there is no concrete molecular basis for immune dysfunction, which helps us to define the etiology of AIDS.

Most available evidence is consistent with the hypothesis that it is caused by an infectious agent, possibly a virus with a long latency or incubation period, requiring cell to cell contact for spread. An alternative hypothesis for etiology of AIDS is that plasma from patients with AIDS inhibits the response of normal lymphocytes to stimulation with mitogen or alloantigen, suggesting the presence of circulating immuno suppressive factor.

21.31.Suggestion for future research on etiology and therapy

A proof of each of the etiology hypothesis will require hypothesis will require the demonstration of a new or altered immuno suppressive infectious agents in large number of patients having AIDS, plus confirmation that infection with the agent is associated with the agent is associated with the development of immune dysfunction.

It is clear that definite conclusions regarding the etiology of AIDS and its curative therapy are not yet at hand. Recent announcements of increased support for research on AIDS should provide the motivation in which scientists in basic biological and medical disciplines including immunology, microbiology, biochemistry and molecular biology can collaboratively focus research efforts on the mechanism of pathogenesis of this devastating new disease.

QUESTIONS

1. Discuss the current status of AIDS in India, United states and world wide.
2. Explain about AIDS viruses.
3. Write the immunological lesions in AIDS.
4. What are the secondary infections of AIDS ?
5. Outline the cause and spread of AIDS.
6. Define AIDS.
7. Explain the genetic variability if HIV 1.
8. Explain the T cell abnormalities in AIDS.
9. Write a note on NK cell abnormalities .
10. Outline the macrophage abnormality in AIDS.
11. Justify AIDS has a disease.
12. Write a short notes on:
    (*a*) Neurologic lesions;
    (*b*) Kaposi's sarcoma;
    (*c*) Diagnosis of HIV infection;
    (*d*) Treatment of HIV infection;
    (*e*) Penetration and destruction of CD_4 T cell by HIV;
13. Explain whether AIDS has vaccine .
14. Outline the spread of AIDS in various animals.
15. What are the other human viruses affecting human system ?
16. Suggest the methods for prevention of AIDS and its social implications.
17. What are the measures taken in India to prevent AIDS ?
18. How is the world concentrating to control AIDS ?
19. Enumerate the current developments is AIDS research.
20. What are the biotechnologicial point of views about AIDS ?
21. Comment on injections on future research on etioloy and therapy.
22. Explain the mechanism of AIDS viral DNA in host cell with diagram.
23. Write briefly viral cell $CD4^+$ CD4 shuman cell interaction mechanisms.

CHAPTER 22

IMMUNOLOGICAL AND IMMUNOCHEMICAL TECHNIQUES

22.1. OUCHTERLONY'S DOUBLE IMMUNODIFFUSION TECHNIQUES

The object of Ouchterlony's double diffusion method is to bring together through diffusion, optimal concentrations of antigen and antibody to form visible bands of precipitation. The main advantages of this method lies in the detection of more than one antigen antibody system. Also the method is known for its high degree of sensitivity and a few micrograms of antigen and antibody are sufficient enough to bring about a precipitation reaction.

The principle behind the technique is as follows. Antigen and antibody are both present as solutions in separate wells in agar coated plate or slide. The reactants move towards each other and are precipitated in a medium that originally did not contain either of them. Thus there is a double diffusion and it is two dimensional, wherever optimal concentrations of antigen and antibody are found, separate lines occur. The number of lines indicates the minimum number of antigen antibody systems present. Nevertheless, a single system may give rise to multiple bands under certain conditions: like, in the presence of excessive antigen concentration and in distinctly different molecular weights for the same antigen. The following procedure demonstrates a typical double diffusion reaction. The study is a single antigen system showing the influence of varying ratios of antigen to antibody and the limiting sensitivity of the system.

22.1.1. Materials required

Precoated microslides, phosphate buffered saline containing sodium azide (PBS azide), agar (high endosmosis), gel puncher, pasteur pipettes, petri dishes, boiling water bath, microslides, Human serum albumin (HSA), Bovine serum albumin (BSA), Antihuman albumin, Anti Bovine serum.

22.1.2. Procedure

1. Prepare 1% (w/v) agar solution by heating 1.0 g of agar in 100 ml of PBS azide solution.
2. Allow agar solution to cool to 45°C and carefully pipette 2.5 to 3.0 ml onto required number of precoated microslides placed on an even surface.
3. Place the slide over the template drawn on a white paper and carefully punch the holes using a 3 mm gel puncher. The number and pattern of punches depend on the types of experiment.
4. Carefully remove the agar plugs from the wells by suction
5. Prepare the dilutions of antigens in PBS azide. The dilutions of antigens could be neat, 1/2, 1/4, 1/8, 1/16 & 1/32.

6. Pipette 6 ml of antihuman albumin and antibovine serum into the center wells of different slides.
7. Pipette the dilutions of HSA and BSA in order to concentrations in the surrounding wells.
8. Place the slides in a humid chamber (petridishes containing water soaked cotton) and incubate overnight at room temperature.
9. Examine the slides for immunoprecipitin lines in reflected light.

22.2. OPTIMAL PROPORTION OR ZONE OF EQUIVALENCE OF AN ANTIGEN ANTIBODY REACTION

The object of this test is to determine the optimal concentrations of both antigen and antibody in a precipitin reaction. The test can be regarded as a kind of quantitative test but depends mainly on visual observations. Also very small quantities of materials are used The main purpose is to provide a preliminary evaluation of the antiserum before more time consuming quantitative test is undertaken.

Double diffusion technique used in qualitative analysis of an antigen antiserum reaction can be used effectively in determining the optimal proportion or zone of equivalence. The test could be performed in two ways.

1. The amount of antigen is varied and the amount of antiserum is kept constant (alpha procedure).

2. The amount of antiserum may be varied and the amount of antigen kept constant (beta procedure).

We confine ourselves to the alpha procedure in this experiment.

The procedure for the preparation of agar layered microslides and the punching of antigen and antiserum wells are out lined. The center well is filled with 6 µl of antihuman serum albumin and the surrounding wells with an equal quantity of antigen, viz., HSA. The concentrations of antigen could range from neat to 1/32. Incubate the slides overnight in humid chambers.

With HSA and anti HSA serum mixed in proper proportions three zones of precipitation can be determined.

1. Antigen excess zone
2. Zone of equivalence
3. Antibody excess zone

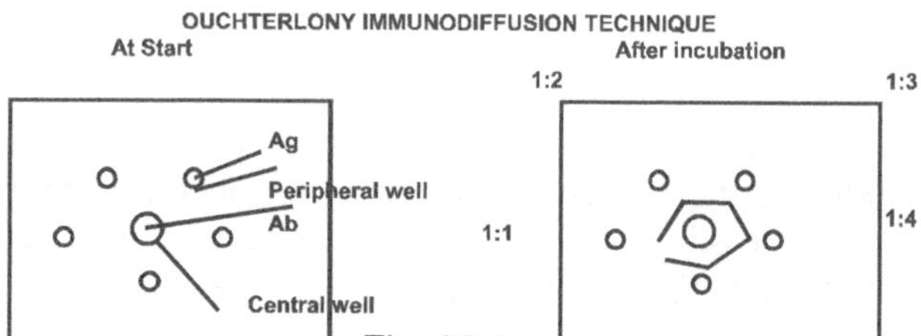

Fig. 22.1.

In antigen excess zone the precipitin line is formed close to the antiserum well and in the antibody excess zone, the precipitin line is formed close to the antigen well. When a precipitin line is formed in the middle of the distance between the two reactants, then it suggests that the reactants are found in optimal proportions. Also in optimal proportion mixture, reaction corresponds reasonably well with the zone of equivalence and lies close to the maximal amount of precipitated antibody in the HSA and anti HSA system. Essentially this is the purpose of the qualitative test to locate the zone of equivalence in a given antigen antibody system.

22.3. OUCHTERLONY'S DOUBLE IMMUNODIFFUSION TECHNIQUE – IDENTITY, PARTIAL IDENTITY AND NON IDENTITY TESTS

The object of this experiment is to detect antigens which are,

1. Identical to one another
2. Not completely identical but share a certain number of common antigenic determinants (partial identity of cross reacting antigens)
3. Distinctly different or non identical antigens.

The ODI technique discussed earlier in chapter 1 will be the basis for the present experiment.

22.3.1. Materials Required

Precoated microslides, phosphate buffered saline containing sodium azide (PBS azide), agar (highendosmosis), gel puncher, pasteur pipettes, petri dishes, boiling water bath, microslides, Human serum albumin (HSA), Bovine serum albumin (BSA), Antihuman albumin, Anti Bovine serum, Rat Serum albumin, horse serum albumin, baboon serum albumin, bovin gamma globulin.

22.3.2. Procedure

Follow step 4 under procedure of 22.12. Instead of 3 mm gel puncher use 4 mm gel puncher and cut 4 mm wells.

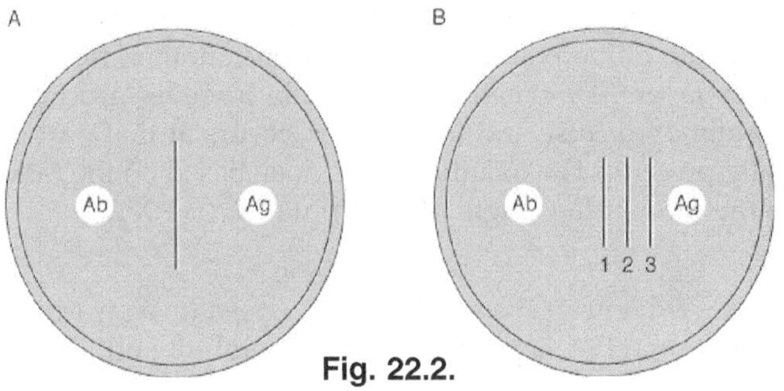

Fig. 22.2.

1. In one slide the center well be filled with 10 ml of anti HSA.The surrounding six wells be filled wixth an equal volume of (1) HSA (2) baboon serum albumin (BASA), (3) horse serum albumin (HOSA), (4) bovine serum albumin (BSA), (5) rat serum albumin (RSA) and (6) HSA in that order.

2. In the second slide the centre well be filled with 10 µl of anti bovine serum. The surrounding six wells be filled with an equal volume of (1) and (2) bovine gamma globulin, (3) and (4) bovine serum, (5) and (6) BSA, in that order.

3. Place the slides in humid chambers and incubate overnight at room temperature.

4. Examine the slides for immunoprecipitin lines in reflected light.

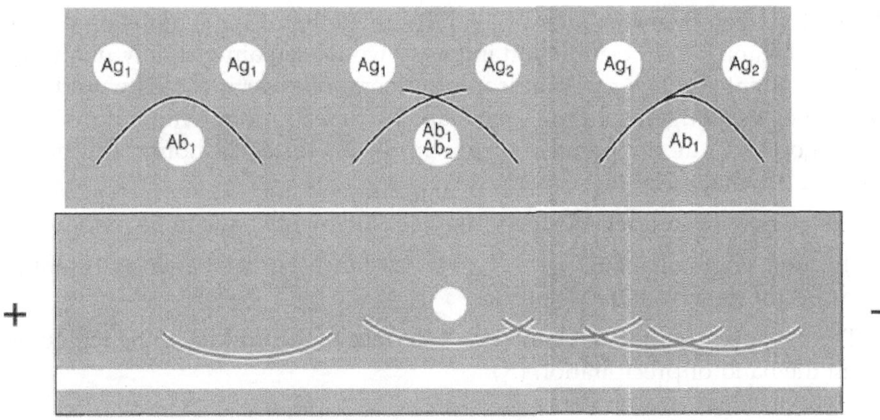

Fig. 22.2.

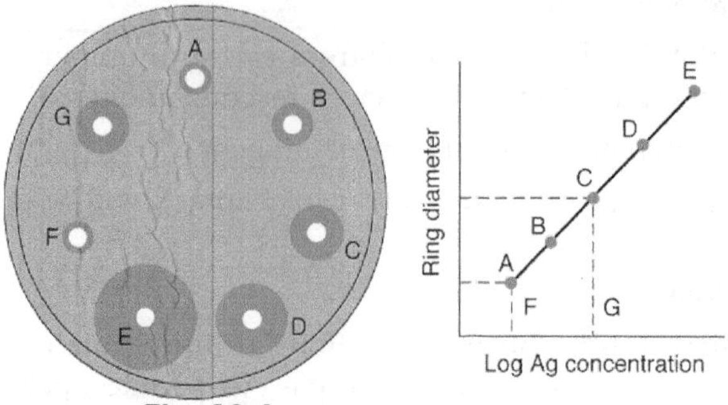

Fig. 22.4.

22.4. Quantitative immunodiffusion techniques

22.4.1. Preer's double immunodiffusion

22.4.1.1. Principle

The technique evolved by John R. Preer involves one dimensional movement of the reactants, viz., the antigen and antibody both of which diffuse through agar plug in a tube. Undiluted antiserum is placed in the bottom of a precoated tube, a layer of agar is placed on the top of the agar layer. Antigen and antibody migrate towards one another, forming bands of (precipitin line in) the agar layer. This method essentially involves a double diffusion, since both antigen and antibody diffuse and the reaction occurs at that part of the tube where neither was originally present. The quantitation is done based on the ratio of distance the antigen diffused to the ratio of the length of agar plug ($X_{Ag} + X_{Ab}$).

22.4.1.2. Materials Required

Precoated glass tubes (3 mm inner diameter 5 cm) closed on one side, agar (high endosmosis grade), PBS – azide solution, HSA and anti HSA, mineral oil and a glass syringe with a 26 gauge needle.

22.4.1.3. Procedure

Deliver 35 ml of anti HSA after testing its zone of equivalence with HSA, into a glass tube to a height of 5 mm using a glass syringe fitted with a 26 gauge needle. Layer the antiserum with double the volume of 0.6% agar prepared in PBS azide solution, to a height of 10 mm. The agar solution should be around 50°C (too high a temperature of agar solution will denature the antibody).After allowing about 10 minute for the agar to solidify add 35 µl of antigen over the layer of agar. It is also that if the lengths of all the 3 layers were kept constant, band position and intensity would remain virtually unaffected. After the addition of antigen layer, add a drop of mineral oil over it as this would prevent the evaporation of the antigen solution. Incubate the tube at room temperature in horizontal position. Look for the appearance of precipitin line after 24 hours. But generally the observations are made 72 hours later.

Once the precipitin line is formed, two measurements have to be made:

1.The total length of agar ($X_{Ag} + X_{Ab}$) the distance Ag Ab between the antigen agar interface and the agar antibody interface.

2. The distance from the antigen agar interface to the middle of the region of greatest density in the band of precipitation (X).

$$P = \frac{X_{Ag}}{X_{Ag} + Ab}$$

By varying the concentration of antigen, it could be seen that the distance at which the precipitatin line is formed also varies.

The log (Ag) is plotted against p and this should be a straight line. The concentration of unknown antigen can be obtained from the graph.

22.4.2. Oudin's single immunodiffusion

22.4.2.1. Principle

The technique evolved by J. Oudin, like Preer's technique, involves one dimensional movement but only of one reagent, viz., the antigen. In tubes of 3 mm inside diameter containing a layer of agar in which antiserum is uniformly dispersed, the antigen added at the top diffuse and forms a zone of precipitin which increases with time. Since only one reagent diffuses, this is called a single immunodiffusion method. A quantitative relationship is developed using three factors namely, time required for complete precipitation of antigen, the height of zone of precipitation and the concentration of antigen.

22.4.2.2. Materials Required

Precoated glass tubes (3 mm inner diameter 5 cm) closed on one side, agar (high endosmosis grade), PBS azide solution, HSA and anti HSA, mineral oil and a glass syringe with a 26 gauge needle.

22.4.2.3. Procedure

Fill up the precoated glass tubes (6 numbers) with a mixture of agarose and antiserum to a height of 3.5 cm (approximately 250 µl). Care should be taken that all the tubes are filled to the same height.

Prepare the antigen in the following concentrations namely, neat, 1/2, 1/4, 1/8, 1/16, 1/32.

Once the agar antiserum mixture solidifies, add 15 to 20 µl of antigen solution and overlay a drop of mineral oil to prevent the evaporation of antigen. Then keep the tubes in horizontal position.

Measure the zone of precipitation periodically at regular intervals for nearly 49 hours in reflected light using a divider and a plastic ruler.

The relationship between the height (h) of precipitin zone and the time (t) taken for precipitation is obtained by the equation, h

$$K=\frac{h}{\sqrt{L}}$$

The K value is plotted against the log (Ag) and this should be a straight line. The concentration of an unknown antigen can be obtained from the standard curve.

22.4.3. Mancini's single radial immunodiffusion

22.4.3.1. Principle

Mancini and his co-workers proposed "when an unknown antigen is allowed to diffuse from a well in a uniformly thin layer of antibody containing agar for sufficient time to allow antigen to combine, the final area reached by the precipitate is directly proportional to the amount of antigen employed and inversely proportional to the concentration of antibody". In other words, there is a linear relationship between the area of the precipitate and the concentration of antigen but this would happen only when all the antigen is allowed to react with the antibody. The method is one of the precise quantitative techniques for the estimation of the unknown quantities of antigen.

22.4.3.2. Materials Required

HSA, anti HSA serum, precoated glass plates (75 × 50 × 1.4 mm), agar (high endosmosis grade), PBS azide solution, 3 mm gel puncher, micropipettes and humid chambers.

22.4.3.3. Procedure

Melt 1.07 g agar in PBS in a boiling water bath and place in a 50'C water bath. Add 2 ml of 1/50 concentration of anti HSA to the agar solution. Mix the two solutions using a stirrer, being careful not to create air bubbles. Leave the mixture in a water bath. Make a sandwich of a precoated agar plate and a non-coated plate with the help of 1 mm spacers (for the sandwiching technique refer to the appendix). Using a pre warmed pipette deliver 5 ml of agar antiserum mixture into the sandwich assembly as shown in figure.2. Even as the agar congeals, prepare 1/2, 1/4, 1/8, 1/16 and 1/32 dilutions of the antigen in PBS azide solution. Once the agar is solidified, dislodge the sandwich assembly by removing the top plate. It is also advisable to smear a trace of siliconised grease on the inner surface of the top plate as this would facilitate the easy dislodging of the sandwich assembly.

On the agar plates cut 3 mm wells using a gel puncheer as shown in figure. Ca should be taken to cut the wells neatly and not to leave any chunks of antibody ag inside the wells. The wells are now filled each with 5 μl of antigen of different concentrations. The plates are left in humid chambers and allowed to incubate at room temperature.

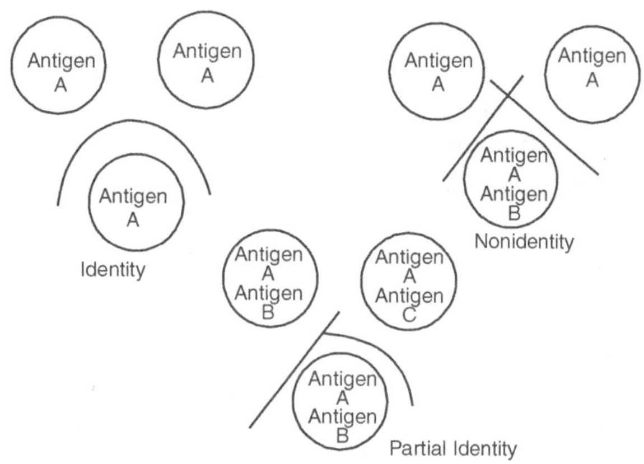

Fig. 22.5.

Two types of measurements are made of the precipitin rings that are formed owing to the antigen antibody reaction.

1. While the experiment is in progress, at periodic intervals, the diameter of the precipitin ring is measured and plotted against the log of antigen concentration.

2. When once the precipitation is complete, that is, when there is no further enlargement of the precipitin ring occurs, the area of the precipitin ring is calculated and plotted against the antigen concentration. This can be done after staining the gels suitably.

The stained slide could be placed in a photographic enlarger and the image pattern be projected on a sheet of white paper. Trace the periphery of each of the projected hollows. All tracings must be in the same sheet of paper to avoid differences on paper weights. The tracings could be cut and weighed and the weights of the tracings cuts could be plotted against the protein concentration of the antigen on a graph paper.

22.4.4. Determination of the diffusion Coefficient of an antigen

By allowing an antigen and an antibody to diffuse from troughs set at right angles and by measuring the angle of precipitin line, it was shown that accurate measurements of diffusion constants of antigen could be obtained. Also studies showed that in an antigen antibody reaction, once the precipitate begins to form, the antigen and the antibody do not obey the laws of diffusion. In such systems no antigen and very little antibody diffuse past the zone of visible precipitation.

22.4.4.1. Materials Required

75 X 50 mm precoated glass plates, agar (high endosmosis grade), PBS azide buffer, trough cutters, humid chamber, HSA and anti HSA serum.

22.4.4.2. Procedure

Prepare a 1% solution of agar in PBS azide buffer and layer it to a height of 4 mm on a 75 X 50 mm precoated glass plate. Cut two thin rectangular troughs in the agar, using template trough cutters, at right angles to one another but separated by 3 mm. Place equal volumes of HSA and anti HSA serum solutions in the troughs but the concentrations of the reactants must be approximately equal as established by zone of equivalence. Place the plate in a humid chamber and incubate overnight at room temperature until the precipitin line is well formed. Once the precipitin line is well formed, using a protractor measure the angle between the precipitin line and the troughs. This could be done either directly or more accurately on an image, enlarged ten times projected onto a screen.

The relationship between the diffusion constant of an antigen and an antibody is given by the equation, 1/2

$$\tan \theta = (D_{Ag}/D_{Ab})^{1/2}$$

where 0 is the angle between the precipitin line and the trough in the experiment. Since the object of the experiment is to determine the diffusion coefficient of an antigen, the equation could be rewritten as 2

$$D_{AG} = (\tan \theta)^2 \times D_{Ab}$$

D_{Ab} for an immunoglobulin G is calculated to be 3.8 X10 cm/sec^{-1}.

22.4.5. Recording and preservation of the gel plates

Although the immunoprecipitin lines produced in agar are often quite visible, intensification and visualisation of weaker lines can be achieved by staining with suitable stains. Such staining also makes photography much easy. The methods to be followed during staining are as follows.

1. The unreacted excess proteins must be thoroughly removed background staining.
2. The plates must be thoroughly dried before staining.

22.4.5.1. Materials Required

Agar gel containing immunoprecipitate, PBS azide solution, Whatmann No:1 filter paper, 0.025% Coomassie Brilliant Blue solution in methanol, water, acetic acid 50:45:5 v/v, destaining solution water, acetic acid, methanol 7:8:5 v/v.

22.4.5.2. Procedure

Wash the gel in atleast 3 changes of excess volume of PBS azide solution for 3 hours. Dry the plate by covering with a sheet of filter paper and leaving at room temperature until the paper flies off the slide.

Immerse the gel plate in stain for 30 minutes or until the stained bands are visible. Destain in 3 or 4 changes of destaining solution until background is clear.

Dry in air at room temperature.

22.4.6. Immunoelectrophoresis (Graber and Williams qualitative analysis)

22.4.6.1. Principle

Agar could be used as a solid support for electrophoresis. The proteins in this medium get separated almost entirely due to charge differences. During electrophoresis in agar, proteins experience two forces:

1. An electrophoretic force, due to applie potential difference. At pH values normally use for agar electrophoresis (above pH 8) mos proteins are negatively charged and th electrophoretic force causes them to migrat towards the anode.

2. An electroendosmosis force which cause the proteins to move towards the cathode.

The overall direction of the movement of protein will be determined by whether th electrophoretic or electro endosmotic force : greater. For most proteins the electrophoreti force is considerably greater and the proteir migrate to the anode. However there ar exceptions to the rule and most immunoglobulins are notable for their cathodal migration.

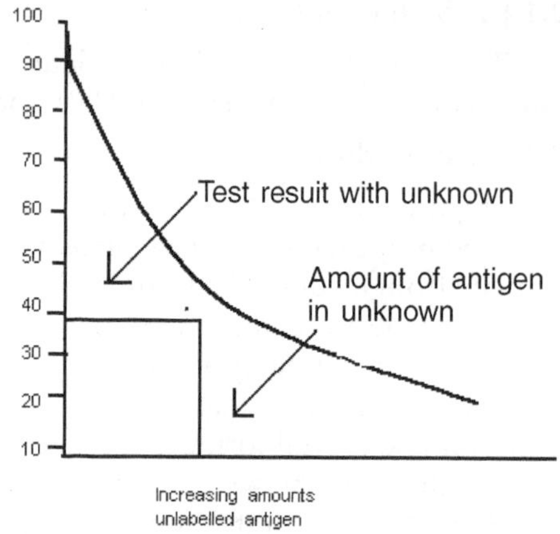

Fig. 22.6.

The ideal nature of agar gels for supporting immunoprecipitin reactions has led to the development of agar derivatives with low electroendosmosis. The outcome of this is the availability of highly purified agar derivatives such as agarose.

Under Graber and Williams technique, proteins in the antigen are first separated by electrophoresis in agarose. Antiserum is then allowed to diffuse from a trough cut in the gel parallel to the direction of electrophoresis. Immunoprecipitin arcs are observed where antigen antibody interactions occur.

IMMUNOELECTROPHORETIC TECHNIQUE

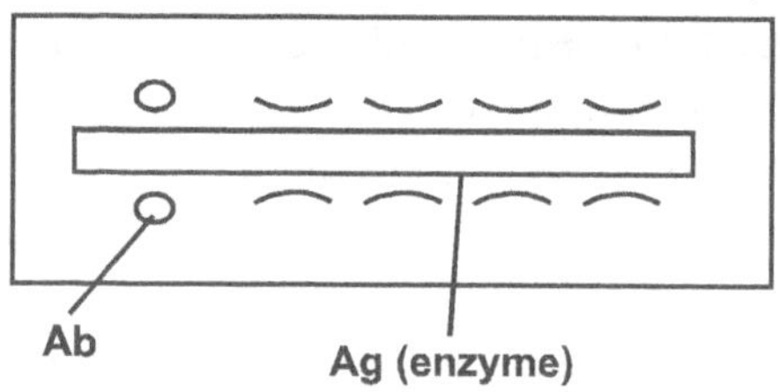

Fig. 22.7.

22.4.6.2.Materials Required

Precoated microslides, agarose (medium endosmosis grade), 3 mm gel puncher, barbitone buffer (pH 8.2), Whatmann No. 1 filter paper, electrophoresis tank and power pack, HSA and anti HSA, pasteur pipette.

22.4.6.3. Procedure

Prepare a layer of 1% w/v agarose on precoated microslides. The 1% agarose solution is prepared by boiling 1.0 g of agarose in 100 ml of barbitone buffer. To obtain uniform thickness, one can make a sandwich of two microslides, one of which is precoated with 1 mm spacers. With a pipette, hot agarose solution can be introduced into the sandwich assembly. Soon after agarose solidifies (around 20 minute), the assembly can be dislodged and the agarose layered slide used for experiment.

Punch out wells and trough in the gel and remove the agar plugs from the wells only using a pasteur pipette. Place the slides in the electrophoresis tank to set up electrical connections between the ends of the slides and electrode chambers. The wells should be nearest the cathode. Load 3 ml of human serum and 1.5 ml of HSA (1 mg/ml) into the wells on either side of the trough and carry out electrophoresis at 6 v/cm. At the end of the run, take out the slides and carefully remove the agar from the trough with antiserum and incubate overnight in a humid chamber. The strong immunoprecipitin arcs will be visible but the weaker components will require staining for visualization.

22.4.7. Counter immuno electrophoresis

This technique is more like double diffusion except an electrophoretic force is used to drive antigen and antibody together instead of simple diffusion. But the method enjoys certain advantages over double diffusion in that

1. a result can be obtained in about 1 hour

2. less antigen and antibody are needed as the electrophoretic force drives all the antigen and antibody together whereas diffusion occurs in every direction.

Essentially the method depends on the positive charges carried by Ig G and Ig M and their cathodal migration plus high electro endosmosis property of the agar used.

22.4.7.1. Materials Required

Precoated microslides, agar (high endosmosis grade), 3 mm gel puncher, pasteur pipette, Tris glycine buffer, Whatman No.1 filter paper, electrophoresis tank and power pack, antigen and antibody.

Examples of Immunolectrophoresis patter s

On left Actual gel, on right, scheme drawn from photograph

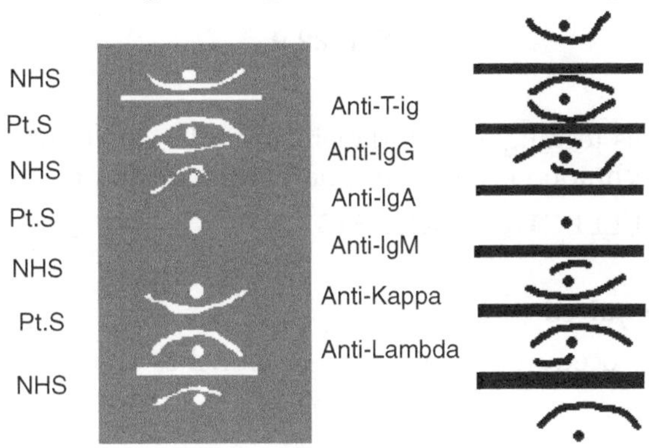

Fig. 22.8.

NHS-Normal human serum, Pt. S-serum from a patient with Ig G(K) myeloma, Anti-T-Ig, antiserum to all (total) immunoglobulins, Anti Ig G, antiserum to human immunoglobulins.

22.4.7.2. Procedure

1. Prepare a 1% agar solution in Tris glycine buffer and layer on a precoated microslide.
2. Bore 2 wells 1 cm apart along the long axis of the slide using the gel puncher. Remove the agar plugs from the wells by suction.
3. Transfer the slide to electrophoresis tank containing Tris glycine buffer as electrode solutions. Soak filter paper wicks in the buffer and use a triple thickness to connect the ends of the slide with electrode compartments.
4. Fill the well close to the cathode with antigen and the other with the antiserum. Carry out electrophoresis at 8 v/cm for 30 minutes.
5. An immunoprecipitate will be visible directly and the same may be intensified using suitable staining procedures.

22.4.8. Rocket immunoelectrophoresis

This technique developed by C.B. Laurell is a quantitative method for the estimation of an antigen and is also termed electroimmunoassay. In this technique the antigen of known concentration is loaded into wells bored into antibody containing agarose and then driven into the gel by electrophoresis. When equivalence is reached immunoprecipitation occurs giving a rocket shaped line. When the precipitation is completed, the length of the rocket or the area under rocket is proportional to the antigen concentration. One can find

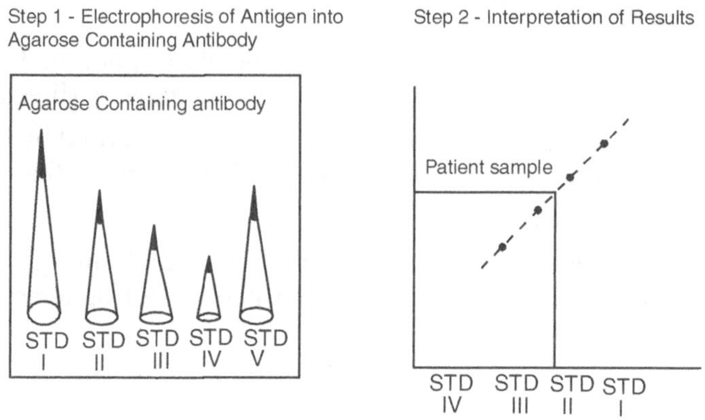

Fig. 22.9.

a parallel between this technique and Mancini's single radial immunodiffusion; in other words, the present technique is an electrophoretic adoption of single immunodiffusion.

ROCKET IMMUNO ELECTROPHORESIS

Std – Standard

22.4.8.1. Materials required

As in previous procedure except 4 mm gel punches and agarose of low endosmosis grade are used.

22.4.8.2. Procedure

1. Melt the agarose in Tris glycine barbitone buffer (see appendix) in a boiling water bath (prepare 1% agarose) and place the solution in a 50°C water bath.
2. Place 1/50 dilution of anti BSA in Tris glycine barbitone buffer in a 50°C water bath.
3. Mix equal volumes of agar and antiserum solutions as thoroughly as possible without introducing air bubbles.
4. Prepare a sandwich assembly of 2 glass plates (75 X 50 mm) of which one is precoated (see appendix for sandwich assembly procedure). Hold the assembly in a slightly slanting position as shown in figure 2 and apply the antiserum agar mixture to the lower corner of the slit. After solidification of the agar (20 minutes) dislodge the sandwich assembly and slid off the top plate carefully. The gel along with the supporting bottom plate is used in the experiment.
5. Place the slide on a template and punch 4 mm holes in a row in the gel. Remove the agar plug from the wells taking care not to leave any chunks inside.
6. Fill the buffer compartments of electrophoretic chamber with chilled Tris -glycine barbitone buffer, making sure that an equal volume is placed in each compartment. The chamber is placed in a refrigerator.
7. Place 0.2 ml of buffer into each of the four small test tubes. To follow visually the electrophoretic migration, a tracking dye such as bromophenol blue may be added to the buffer used for the dilution of BSA. Add 0.1 ml of 0.005% bromophenol blue

per 2 ml of buffer. Starting with 0.1 ml of 0.04% BSA solution (4 mg/10 ml) make four doubling solutions to effect total concentrations of 2.0, 1.0, 0.5 mg BSA per 10 ml.

8. Transfer the slides to the electrophoresis chamber (wells at the cathodal end) and connect the chamber to a power pack.
9. Using micropipettes load 10 ml each of the four BSA dilutions to the wells.
10. A current of 6.5 volts/cm is applied until the tracking dye reaches close to anode end.
11. At the termination of the run, turn off the power supply and disconnect.
12. Inspect the slides for rockets of antigen antibody precipitation.
13. Stain suitably.

22.4.9. Two dimensional crossed immuno electrophoresis

Crossed immuno electrophoresis combines the principles of immuno electrophoresis with the rocket procedure. Proteins in the antigen are first separated by agarose gel electrophoresis. They are then electrophoresed into an antibody containing agarose gel at right angles to the direction of the first electrophoretic separation. The proteins are recognized by the antibody immunoprecipitate as rockets. The technique is essentially used to show

1. The heterogenecity of antigen and antibody
2. Concentration of antigen components in different samples of antigen.

The second idea is possible because the distance from the sample application well would reflect the electrophoretic mobility of the protein.

22.4.9.1. Materials Required

As listed under chapter eight except agar of low or medium endosmosis grade is used.

22.4.9.2. Procedure

1. Prepare a layer of 1% w/v agarose in Tris glycine barbitone buffer on a precoated glass plate (75 X 25 mm).
2. Punch a well 1.5 cm from one end of the glass plate using a 4 mm gel puncher. Remove the agar plug from the well.
3. Transfer the slide to the electrophoresis chamber (well at the cathodal end) and connect the chamber to the power pack.
4. Load 5 μl of sample antigen (whole human serum) and carry out the electrophoresis at 6 v/cm for 1 2 hours. Use bromophenol blue as a marker dye to judge the electrophoresis time.
5. At the end of the run, remove the plate with the help of a razor cut the agarose layer 0.5 cm away from the well towards the length of the plate. That leaves a small strip of agar on the plate containing serum protein separated by electrophoresis. Using this plate prepare a sandwich assembly.
6. Prepare antibody containing agarose gel in barbitone buffer (refer to Mancini's single radial immunodiffusion technique)
7. Pour the agarose antibody mixture into the sandwich assembly. Allow to solidify.
8. Transfer the slides once again to the electrophoresis chamber, set up electrodal contact as before and electrophoresis at 4 v/cm for 6 hours. Rockets must be visible but are best shown by staining.

22.5.1. Precoating of slides

Two types of methods can be used.

1. Place a flask containing a 0.3% aqueous suspension of agar in a boiling water bath. Melt the agar and pipette 3 ml of it into clean slides that have been wiped free of lint. When the agar has hardened, transfer the slides to a 80°C oven and leave them until water has evaporated. A thin film of dried agar remains.

2. Alternatively place a flask containing a 0.5% aqueous suspension of agar into a boiling water bath. Make a sandwich of two microslides or glass plates (as per the requirement of the experiment), dip once into the melted agar, remove and allow to dry. There will be a thin layer of agar on the surface of the slides or plates.

22.5.2. Sandwich assembly

The sand wich assembly of microslides or glass plates is useful in preparing agar layers of uniform thickness. Two microslides or glass plates may be used. There will be a precoated base plate to support the gel upon which is placed a shaped frame made of 1 mm thick acrylic sheets and the three sides of measure 0.5 mm in width. Over this frame is placed a top slide or plate whose lower surface is silicanized. The silicanization is achieved by smearing a very small quantity of silicanized grease. Do not silicanize the base plate. The three pieces of the assembly are tightly held together by clamps.

22.5.3. Phosphate buffered saline Azide solution (PBS azide)

(*a*). 0.15 M $NaH_2 PO_4 . 2H_2 O$ 23.4 g/l

(*b*). 1.15 M Na_2 HPO_4 21.3 g/l

(*c*). Mix 240 ml of'a' and 760 ml of 'b'to obtain a phosphate buffer of pH 7.2.

(*d*). NaCl 18 g/l

(*e*). Mix equal volumes of 'c' and 'd' to obtain a PBS of 0.075 M and disosmotic saline solution.

(*f*).Add 1.0 g of sodium azide to a litre of PBS to get a final azide concentration of 0.1%.

22.5.4.1% Agar

Agar is a high molecular weight complex polysaccharide derived from sea weeds. It will dissolve in aqueous solutions at high temperatures (around 90'C) and forms a gel on cooling. Agarose is more highly purified form of agar. To prepare a 1% agar in solution, suspend 1.0 g of agar in 100 ml of water. Allow the agar to swell for 3 to 5 minutes and then leave the suspension in a boiling water bath. Wait until all the agar is dissolved. Cool the solution to 50 to 55°C before using in an experiment.

22.5.5. Coomassie Brilliant blue staining solution

(*i*) Brilliant blue R 250 — 0.25 g

(*ii*) Methanol —500 ml

(*iii*) Water —450 ml

(*iv*) Glacial acetic acid —50 ml

Mix the solvents (*ii*) to (*iv*) and then add the stain. Stir well and filter before use.

22.5.6. Destaining Solution

Two types of destaining solutions are recommended.

I. (*i*) Water–870 ml

(*ii*) Acetic acid–80 ml

(*iii*) Methanol –50 ml

Mix solvents (*i*) to (*iii*) and use.

II. (*i*) Methanol –500 ml

(*ii*) Water –250 ml

(*iii*) Acetic acid–50 ml

Mix solvents (*i*) to (*iii*) and use. If there is an intense background staining, the second destaining solution is recommended.

22.5.7. Barbital Buffer: (Ionic strength 0.1, pH 8.6)

Barbital sodium	20.6 g
Barbiton	4.0 g
Sodium azide	1.0 g
Distilled water	1 litre

22.5.8. Tris Glycine barbital buffer

22.5.8.1. Buffer A

Barbital Sodium	65.0 g
Barbiton	10.35 g
Distilled water to 5 litre	

22.5.8.2. Buffer B

Glycine	281 g
Tris	226 g
Distilled water	5 litre

Mix equal volumes of buffers A and B.

QUESTIONS

1. Describe the principles of Ouchterlony's double immunodiffusion techniques.
2. Where the ouchterlony double immunodiffusion technique is used.
3. Explain the identity, partial identity and non-identity test.
4. Describe ouchterlony double immunodiffusion technique.
5. Explain the principles of Preer's double immunodiffusion.
6. Explain Mancini's single radio immunodiffusion technique.
7. How will you determine the diffusion co-efficient of an antigen ?
8. Explain the Graber and William's immunoelectrophoretic technique.
9. Enumerate the technique of counter immunoelectrophoresis.
10. Write the procedure behind to perform rocket electrophoresis.
11. What is two dimensional crossed immunoelectrophoresis ?
12. Explain the following procedures;
    (*i*) Precoating of slides;
    (*ii*) Sandwich assembly;
    (*iii*) PBS-azide solution;
    (*iv*) Coomasie brilliant blue straining solution;
    (*v*) Destaining solution;
    (*vi*) Barbital buffer;
    (*vii*) ris-glycine barbital buffer.

CHAPTER 23

STAPHYLOCOCCUS AUREUS PROTEIN – A PURIFICATION

23.1. INTRODUCTION

Although protein A interacts with Ig G's of many species, there are significant quantitative differences in its affinity for Ig G's from different species and for different Ig G subclasses from the same species .

While human and rabbit Ig G bind protein A with high affinity, goat, sheep, and rat Ig G show weak respectively. Classes other than Ig G can bind protein A and the reactivity is also dependent on the animal species and the immunoglobulin subclass .

Formalin fixed, heat killed *S. aureus* bearing protein A can bind free immunoglobulins and immuno complexes, with higher affinity for the latter. The kinetics of binding is fast and the non specific uptake of other material is minimal. As a result of the interaction with the bacteria, sedimentation of bound antigen can be done at low speed centrifugation, with negligible loss during washings. All these characteristics render the use of protein A bearing *S. aureus* more advantageous than other conventional immunoprecipitation methods for the identification of cellular antigens.

Most of the surface antigens from trypanosomatids identified up to now, were isolated through the use of *S. aureus* protein A as immunoadsorbent.

Parasite surface antigens can be labeled either by radio-iodination using lactoperoxidase or iodo gen as catalysts or biosynthetically by incubation of cells in culture media containing radioactive precursors.The latter procedure has an advantage over surface radio-iodination, since in this case the identified antigens are in fact synthesized by the parasite. In the case of radio-iodination, extraneous molecules, strongly adsorbed to the parasite surface may be labeled.

Identification of surface antigens occurs after incubation with specific antisera. In the case of radioiodinated parasites, the incubation may be performed after disruption of the cells, since the labeled antigens represent only surface antigens. In the case of biosynthetically labeled parasites, the incubation with antisera is performed with living parasites, to minimize possible interactions between internal antigens and antibodies. After incubation the excess antibody is washed away, leaving the immunoglobulin bound to the cell surface wherever the antigen is represented. At this step, cells are lysed, and immunocomplexes are recovered, as in the case of radio-iodinated antigens, through the use of *S. aureus.*

In this topic we will describe (a) preparation of *S. aureus* stock solutions for immuno adsorption (b) biosynthetic labeling of parasite antigens; (c) antigen isolation after incubation of intact living cells or of total cell lysates with specific antisera.

23.1. Table 1 Protein 'A' binding specificity

Species	Immunoglobulin	Subclass
Man	Ig G	1,2,4
	g A 2	
	Ig 11	
Rabbit	Ig G	
Mouse	Ig G	1 (weak), 2a, 2b, 3
Sheep	Ig G	1 (weak), 2
Goat	Ig G	1 (weak), 2
Dog	Ig G	a, b, c, d Ig A
	Ig 11	

23.2. EQUIPMENT

Refrigerated centrifuge, Microcentrifuge, Hemocytometer (Neubauer), Rotatory shaker at 28°C, Cell incubator at 37 °C, humidified atmosphere containing 5% CO_2, Gamma counter, Liquid scintillation counter, Vortex.

23.3. PREPARATION OF S. *aureus* STOCK SOLUTION FOR IMMUNO-ADSORPTION

S. *aureus*, strain Cowan I (protein A producer) is cultured in liquid medium as described by Lopes and Vilela. A formalin treated, heat killed suspension of S. *aureus*, is prepared according to Lopes and Villela.The preparation of S. *aureus* stock solution for immunoadsorption will be considered here.

23.3.1. Reagents

10% (w/v) suspension of formalin fixed, heat killed S. aureus, NET buffer: 150 ml NaCl, 5 mM EDTA, 50 mM Tris HCl, pH 7.4, 0.02% (w/v) Nan3, nonidet P 40 (NP 40, Shell chemical co.), 0.5% (v/v) NP 40 in NET buffer, 0.05% v/v NP 40 in NET buffer containing 1 mg/ml bovine serum albumin.

23.3.2. Procedure

Note: This stock solution has to be used within 24 hours.

(*a*) The 10% suspension of S. *aureus* is centrifuged for 20 minutes at 5,000Xg. The choice of volume depends on the number of immunoprecipitations planned. Usually 50 100 ml are used per immunoprecipitation.

(*b*) Suspend the pellet in 0.5% NP 40 in NET buffer. The volume should be the same in (a).

(*c*) Incubate for 15 minute at room temperature with occasional vortexing.

(*d*) Centrifuge for 15 minute at 5,000 Xg.

(*e*) Suspend pellet in 0.05% NP 40 in NET buffer (plus 1 mg/ml bovine serum albumin). Same volume as in (*a*).

(*f*). Repeat (*d*) and (*c*)

(*g*). This stock solution is stored at 0 – 4 ° C.

23.3.3. Technical comments

Formalin fixed, heat killed, protein A bearing S. *aureus* suspensions are now commercially available from a few sources. In our work, we have used pansorbin (Calbiochem) with very good results. Pansorbin suspensions can be kept at 4°C for more than 5 months with no loss of activity. In the preparation of the stock solution for immunoadsorption several detergents such as Tween 20, sodium deoxycholate.

23.4. DOUBLE DIFFUSION IN TWO DIMENSIONS

In this procedure antigen and antibody are allowed to migrate towards each other in a gel and a line of precipitation is formed where the two reactants meet. As this precipitate is soluble in excess antigen, a sharp line is produced at equivalence, its relative position being determined by the concentration of the antigen and antibody in the agar. The local concentration of each reactant depends not only on its absolute concentration in the well, but also on its molecular size and the rate at which it is able to diffuse through the gel.

Multiple lines of precipitation will be present if the antigen and antibody contain several molecular species.

This technique has the particular advantage that several antigens or antisera can be compared around a single well of antibody or antigen.

23.4.1. Material and equipment

2% agar in barbitone buffer, antigen and antibody solutions, Gel punch

23.4.2. Method

1. Melt the agar in a boiling water bath.
2. Pour agar onto pre coated slides, use a labeled surface.
3. Punch pattern required.

The number, size and distance between wells depends on the number and strength of the antisera and antigens.

4. Suck out agar plugs with a Pasteur pipette connected to a water vaccum pump.
5. Fill the wells with antibody or antigen until the meniscus just disappears.
6. Place the slide in a humid chamber and incubate overnight at a constant temperature. (Diffusion and therefore precipitation is quicker at 37 ° C and lines will be seen within 3 hours).

23.5. SUGGESTED ANTIBODY AND ANTIGEN PATTERNS

A straight forward demonstration of identity and non identity can be shown using the antigen mixtures as in figure. In addition, this technique may be used to show the relationships between Ig G molecules and their enzymic digests prepared in chapter 7. Arrange the wells as shown in figure.

Antigen concentration: Initially use 1 mg/ml. Alter concentration, if required.

23.6. ANTISERUM

Use whole anti Ig G, non absorbed, specific anti Ig G sera available commercially are absorbed with light chains to render them class specific.

23.7. INTERPRETATION OF RESULTS

There are three basic patterns of precipitation as shown in the diagrams.

1. The reaction of identity occurs between identical antigenic determinants, the lines of precipitation fuse to give one continuous are:

2. Reaction of non identity. Where two antigens do not contain any common antigenic determinants the two lines are formed independently and cross without any interaction.

3. Reaction of partial identity: This has two components: (a) those antigenic determinants which are common to both antigens give a continuous line of identity, (b)

the extra determinant on one of the antigens gives, in addition, a line of non identity so that a spur is formed. Of course, there may be unique determinants in both antigens, this would give rise to two spurs.

In a and b:Wells 1,4 contain human $I_gG/I_gG_2/I_gG_3$ and I_gG_4 respectively and the central well contains rabbit anti human IgG. (Antisera prepared by immunizing rabbits with Ig G obtained from normal human serum by ion exchange chromatography.

Antiserum (a) recognized sub class differences between IgG1 and IgG4, hence the double spur, but failed to recognize IgG3. Antiserum (b) recognized sub class differences associated with IgG1 alone and so produced a single spur. Both antisera were raised against the same pool of antigen, the variation is due to the rabbits used for immunization.

AD represent antigenic determinants present within a series of protein molecules, 1 6 used as antigens. An antiserum has been raised against two of these determinants, A and B (ad) The precipitation lines have fused completely showing the presence of identical determinants in proteins 1 and 2. Note, however, the detection is purely qualitative and, in addition, does not give any information on the other determinants present. (b).The two precipitation lines cross without any interaction. Proteins 3 and 4 do not, therefore, share any determinants detectable by the antiserum. (c) The two precipitation lines have fused, but in addition a spur has formed towards the well containing protein 5. Hence, proteins 5 and 6 share some common determinants but protein 6 has additional unique determinants detected by the antiserum.

QUESTIONS

1. How is *Staphylococcus aureus* protein purified.
2. Outline protein A binding specificity in various animals including man.
3. How stock solution is prepared for immuno adsorption ?
4. Explain double diffusion in two dimension.

CHAPTER 24

AGGLUTINATION REACTION

When a particulate Ag is mixed with its Ab in the presence of electrolyte at suitable temperature and pH. The particle are clumped (or) agglutinated, agglutination is more sensitive than precipitation for the detection of Ab's. The same principle govern agglutination and precipitation. Agglutination occurs optimally when Ag and Ab react in equivalent proportions. The zone phenomenon may be seen when either an Ab (or) an Ag is in excess. Incomplete (or) monovalent Ab do not cause agglutination. Though they combine with the Ag, they may act as blocking Ag. Ab's inhibiting agglutination by one complete Ab added subsequently.

24. APPLICATION OF AGGLUTINATION REACTION

24.1. SLIDE AGGLUTINATION

When a drop of the appropriate antiserum is added to the smooth, uniform suspension of a particulate Ag in a drop of saline on a slide or a tube, agglutination take place, A + ve result is indicated by clumping together of the particle and the clearing of the drop mixing the Ag 's and the antiserum. With a loop or by gently mixing the slide facilitate the reaction. Depending on the titre of the serum agglutination may occur instantly or with in seconds, clumping occurring after a minute may be due to drying of the fluid and should be disregarding. It is essential to have on the same slide a control consisting of the Ag suspension is satire, but the antiserum to ensure that the Ag is not agglutination. Agglutination is usually visible through naked eye, but may sometime require to confirm under the microscope slide. Agglutination is a routine procedure for the identification of many bacterial isolate from clinical specimen. It is also the method used for blood grouping and cross matching.

SLIDE AGGLUTINATION TEST

(Antiserum plus antigen and normal serum plus antigen)

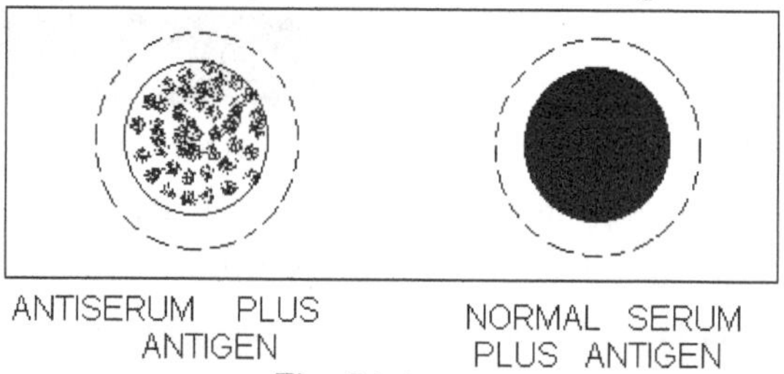

Fig. 24.1.

24.2. TUBE AGGLUTINATION

This is a standard quantitative method for measurement of Ab's when a fixed volume of an particulate Ag suspension is added to an equal volume of serial dilution of an antiserum in test tube, the agglutination titre of the serum can be estimated.Tube agglutination is routinely employed for the serological diagnosis of typhoid, brucellosis and typhoid fever.

TUBE AGGLUTINATION TEST

Transfer Serially 0.5 ml from each tube and then Add 0.5 ml of Antigen to Fach Tube

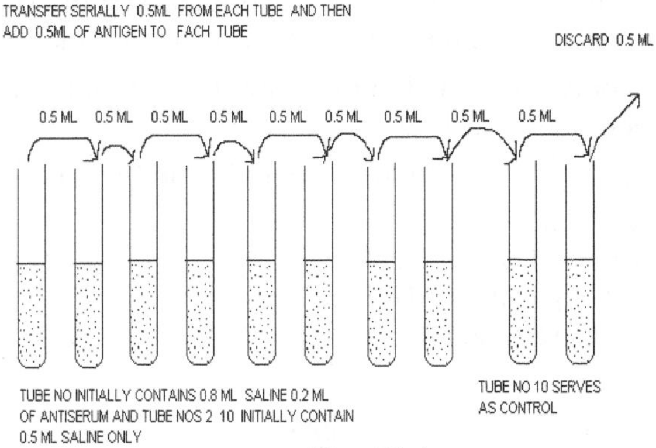

Fig. 24.1

In the widal test used in typhoid, two types of Ag's are used. The 'H' Ag are the flagella and the O are somatic Ag of the typhoid bacillus. The H Ag are a formalized suspension of the bacillus and an combination with the Ab, forms large, loose fluffy resembling wisps of cotton wool, conical dryers tubes are used for 'H' agglutination. The 'O' Ag is prepared by treating the bacterial suspension called alcohol. It forms tight, compact deposits resembling chalk powder Round-bottomed relax tube and used for 'O' Ag. Agglutination bacillus spread out in disk like pattern at the bottom of tube. The tube agglutination test for brucellosis may be complicated by the prozone phenomenon and the presence of blocking Ab's. Several dilution of the serum should be tested to prevent false –ve results due to prozone. Incomplete unblocking Ab's may be detected by doing the test is hypertonic [5%] saline or albumin. saline or more reliably by the anti globulin [coombs]test, Incomplete Abs are also met with in sera containing Rh Ab's. Agglutinate can often be obtained if Rh +ve erythrocyte, treated with proteolytic enzymes (papain) are used as the Ag.

The weil flair reaction for sera diagnosis of typhus fever is a heterophil agglutination test and is based on the sharing of a common Ag between typhi and some strain of Proteus bacilli.

Samples of test agglutination test using red cells as Ag's are the Paul-bunnels test's and the cold agglutination test. The former is based on the appearance in the sera of infections mononuclear patient of agglutination to sheep erythrocyte which are absorbed by (or) red cells but not by guinea pig. Kidney extract, the cold agglutination test is the in 1°C a typical pneumonia the patient sera agglutination human 'O' group of erythrocyte at tic the agglutinate being reversible at 37°C.

24.3. THE WIDAL TEST

This bacterial agglutination test was originally introduced by widal for diagnosis of typhoid fever later on its application was extended to diagnosis of many other disease like undulant fever, dysentery, cholera etc. It is now commonly used for diagnosis of enteric fever's brucellosis. Enteric fever is the typhoid fever caused by *S.typhi* and paratyphoid fever caused by *S.paratyphi* A, B or C as amongst the paratyphoid fever. Paratyphoid A and B are common. In routine test for the search of antibodies of *S.typhi*, and *S.paratyphi* A and B is made that for *S.paratyphi* which is excluded.

24.4. TEST PROCEDURE

The test procedure involves incubation of bacterial suspension with serial dilution patients serum to find out titre of agglutination in the latter. To titre of an antiserum is measure of the lot of Ab units per unit volume of original serum. Thus is test is + ve upto a dilution of 1/64 the titre of serum in 6 units of Ab/unit volume.

The test is commonly performed in two ways by a doubt dilution method and a drop method . Drop methyl of widal test.

Following four Ag suspensions are used routinely in the widal test.

1. *S.typhi* O antigen called as 'To'.
2. *S. typhi* 'H' Ag called TH.
3. *S. paratyphi* 'A' 'H' Ag called 'AH'.
4. *S. paratyphi* 'B' 'H' Ag called 'BH'.

24.4.1. Procedure

Arrange 20 agglutination tubes in a suitable race 4 hrs of 5 tubes each to first tubes of the each row l0drops of 1 in 10 diluted serum are added to the 2,3,and 4th tube of each row 5,2 and 1 drop of diluted serum are added respectively. No serum is added to 5th tube of the each row. Therefore to each tube saline is added to make a total of 10 drops in each tube than 15 drops of 70 Ag is added to each tube of first row. In tubes of the 2,3 and 4th row 15 drops of TH, AH and BH Ag are added respectively in each tube of corresponding row. The drop of serum to saline the Ag added in tubes of each row.

		Drops added			
Tube	1	2	3	4	5
Serum	10	5	2	3	1
Saline	—	5	8	7	9
Ag[bacterial]	—	15	15	15	15
Total		25	25	25	25
Final dilution of Serum		1:25	1:50	1:125	1:250 controlled.

The reaction containing those tubes are then placed in water bath at 37°C. Some prefer temp at 50-55°C. However it has been found that prolonged exposure at this temperature favors the agglutination reaction of O Ag. The water level in water bath should be such that only one third to half of the fluid level in agglutination tube remain submerged. The allows convation occurrence to develop during incubator which facilitate agglutination by learning Ag and Ab in close contact. In tropical countries during summer, when room temperature is alone 37°C, the temperature of water raised by 8-10°C above the room temperature. The tubes of 'O' Ag are kept at 4°C for 24 hrs and those of H Ag at temperature for 30 min following which they are examined for agglutination.

The flakes of agglutination (cotton wool type of agglutination) of H Ag will precipitate in narrow tubes with naked eye. The agglutination of O Ag consists of small granules which are easily observed with hand lens in good illuminate or using a concave mirror. The bottom of the tube is examined with out shaking. In tubes with + ve reaction, granular deposit from irregular ring or deposition is in random manner while in tube with -ve reaction, the organism settle to form a regular smooth

24.5. THE WELFLIX REACTION

This is used for diagnosis of typhoid fever. This was first described by level S Zelix. It is based upon the fact that rickettsial species causing epidermis endemic and scrab typhi or spotted fever and have a carbohydrate happen with somatic constituent of non flagellate strains of proteins. The Abs product in rickettsial disease readily agglutinate these examine.

24.6. HEME AGGLUTINATION TEST:

24.6.1. Blood grouping and cross matching

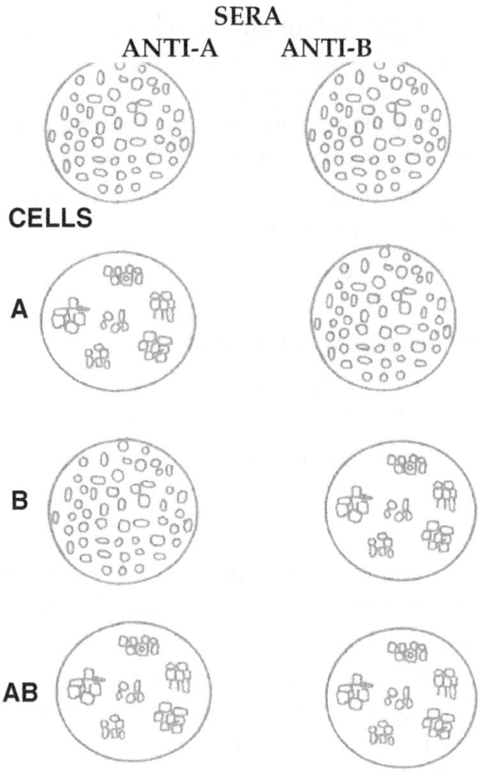

RBC posses certain Ag called haem agglutination or iso Ag (because of this presence in some species). An individual may have A, B both A and B or none of the 2 Ag in his RBC's correspondingly he will have iso agglutination anti B[b], anti A[a], none (or) both anti A and B respectively. Blood groups of individual is determined by finding out which of the iso agglutination will agglutinate his RBC. Two drops of washed RBC'S are mixed separately with antisera a and b. RBC'S agglutinate by 'a' belongs to group A, agglutinated by b belongs to group 'B', agglutinated by a and b are of AB group and those not agglutinate Ab all by either of two are of group 'O'. Later on, it has been shown that there of 28 AB types of group A. A1 and A2 accordingly there will be A1B and A2 B group also besides ABO system. RBC's also possess other Ag eg., Rh, MNO system and some minor blood group system.

The blood group determination is mainly regulated for blood transfusion before deciding transfusion even if donor and recipient belongs to same blood group. Cross matching is done by mixing RBC'S of the donor and the serum of recipient and vice versa, if no agglutination occur transfusion can be given.

24.7. COLD AGGLUTINATION TEST

The patient of 1 A typical pneumonia certain cold agglutination capable of agglutinating RBC in cold, i.e. at low temperature. Demonstrated of such Abs in a suspected case of P.A.P in a diagnostic significance. A series of double dilution of patient serum are made and mixed with equal volume of 0.2% suspension of group 'O' RBC [human].The tubes containing these are placed at 4°C overnight and reading are taken while the tubes care cool the agglutination is reversible at higher temp. A titre of 1/33 (or) 1/64 after final dilution is considered as significant.

Before separating serum, the clotted blood should not be chilled. The clotting and clot retraction should be allowed to occur at 37°C once the serum is separated from RBC'S it can be stored at4°C if test is to be done at a latter time.

24.8. PAUL BUNNEL TEST

This test based on the fact that pcasea suffering from or suffered reactor from infectious mononucleosis posses high level of agglutination for sheep cells.

A series of double dilution of patient serum are inactivated at 55°C for 10 minutes.To these equal volume of 1% suspension of washed sheep RBC's are added and incubated of 37°C for 4 hrs and read. After incubated they may be kept at 4 °C for overnight and reading taken each day after keeping them again at 37°C for 1 hour by cold agglutination. A titre of 1/128 is of diagnostic significance. Repeated test reveals increasing title.

High levels of lowsman's Ab's has been observed following serum therapy and also in human being. Absorption studies with guinea pig kidney Enclose RBC have been shown that normal heterophil. Ab's are observed by guinea pig kidney but not by ox RBC'S and the Ab appearing after serum therapy are absorbed by both, that of infections mononuclear type are absorbed by ox RBC'S and not by guinea pig kidney.

The Ab appearing after return therapy is partially absorbed by ox RBC'S and resembles Ab of normal serum.

After absorptive guinea pig kidney are a titre of 1/28 and above is the +ve or significant.

24.9. ROSE WAALARS TEST

Patients of rheumatoid arthritis, shown serum factor which agglutinate sheep RBC'S sensitized with rabbit anti sheep erythrocyte serum.

The serum factor is an Ab to denatured globulin , there react with Ab's present on sensitized sheep RBC'S and being Agglutinated.

24.9.1. Test

Patient serum is heated at 56°C for 30 minutes. Two sets of doubling detection or it was prepared from ½ to 1/1024.To one these equal volume of washed 2% sheep RBC'S area added and to other equal volume of sensitized sheep RBC'S are added.

The tubes are incubated for 1hr at 37°C and placed in refrigerator for overnight, then kept at room temperature for 30 minutes and read. In titre of 16 or more in 2 sets. The tests is minutes +ve in 70-80% of case of and also in some causes of dissiminated lupus ery-thromatous. and scleroderma and erythroma multiform.

24.10. MODIFIED ROSE WAALAR TEST

In activated patients serum in treated equal volume of the washed and packed red cells (sheep) after 1 hr at 37° C to absorb any sheep red cells Ab.

The mixture is centrifuged at 4000 rpm for 10 min to separate the serum several dilution of serum is treated with sensitized Red cells. A titre 1:12 or more is +ve.

Instead of serum, serum globulin solution can also because, *r* globulin are obtained by treating 0.4 ml of rerun of $NH_3 SO_4$ (1.40 ml) prepared by dissolving (9.3g of anhydrous NH_3SO_4 and 40 g of NaCl in distributed water to make 1000ml.

The precipitate is dissolved in 1.6 ml saline and treated as whole reason a titre of 1.09 is +ve.

24.11. COOMBS TEST

To demonstrate non agglutination test yet specific Ab to red cells when sometime appear in hemolytic anemia (or) Rh incompatibility. These Ab serve in complete blocking or univalent Ab's.

24.11.1. Principle

Preformed for dilution of Anti' –Rh Ab's that do not agglutinate the Rh+ve erythrocyte is saline.

When sera containing the incomplete Anti Rh Ab's were mixed with Rh +ve cells the Ab globulin coats the surface of erythrocytes, though they are not agglutinated when such erythrocytes coated with Ab globulin are treated with Rabbit anti serum. human gamma globulin the cells are agglutinated . This is the principles of antiglobulin test.

Coomb's test may be direct of indirect.

Indirect test, sensitization of erythrocyte with incomplete Ab's takes place in vivo as in hemolytic disease of the new born due to an incompatibility.

The indirect coombs test resistant of RBC'S Ab globulin is performed in invites The employed, direction of Anti Rh Ab'S are used to incomplete or non agglutinate Ab

24.11.2. Direct coombs test

To detect hemolytic autoimmune disease in newborn infants is erythroblastois foetalis.

- The blood plasma of newborn is centrifuged and plasma discarded.
- Packed RBC'S are washed 2-3 times with saline to remove the unwanted protein bound on the cells.
- Washed erythrocyte are treated with coomb's serum (Ab to reble + r globulin) then mixture is incubated at 37°C for short time.
- Visible clumping indicate +ve veb and the agglutination in a measure of sensitization of enzymes cytes by Rh Ab's produced by mother.

24.11.3. Indirect Coombs test

For the sensitization of erythrocytes and performed to measure Rh Rb 's in sensitive mother.

- When RBC is Rh –ve the serum sample is converted and treated with sheep RBC'S and incubated at 37°C for sometimes and headeal with antiglobulin.
- Agg indicate Rh Ab's when test is done in normal healthy patients who does not have Rh –ve Ab's are final that the sheep erythrocyte are not coated with Ab's. When this mixture is treated with antiglobulin no agglutination occur. It is performed in invite condition.

(1) Rh +ve erythrocyte are mixed with incomplete Ab.

(2) The Ab coats the cells.

(3) But being incomplete can't produce egg on the antiglobulin serum which is complete Ab to Ig Agglutination takes place.

24.12. COATED PARTICLE AGGLUTINATION TEST

■ Soluble Ag's are coated on inest particles like latex, coliorden , bentoside, cellulate.

They are used to detects quantitate. Ab's these coated particles can be stored at – 2°C

or by freeze drying to use at a later date of various coated particle test later Agglutination test more popular and heme agglutination in most sensitive. The latter can detect upto 0.003mg of Ab8 .

■ To coat different Ag's are tried on same particle to detect more than one Ab's by single Ag precipitium.

24.13. LATEX AGGLUTINATION TEST

A 1:10 suspension polystyrene uniform size latex particle of 0.8 μ diameter is made and filtered a hatman filter paper.

It is mixed with Ag concentration in appropriate buffer.

Excess antigen is removed by centrifugation coated latex particles are available along with +ve, -ve control sera fear a number of biological test.

The lab for screening test dilution of patients serum is made with buffen. A drop of dilute serum is mixed with Ag coated latex particles.

→The zone mixed which stick in +ve test. Agglutination occurs with in minutes. For quantitation of Ab's serial dilution of serum are made and tested.

→ Highest dilution gives (+) ve results in the litre of Ab in serum.

Eg. of latex agglutination test are.

24.13.1. Latex agglutination test for rheumatoid fever

- Ag in the test is human P globulin coated latex particles cycline buffer is added.
- Serum is detected with 1:20 as directed by the manufacture of Ag.
- (+) agglutination occur in less than 5 minute.
- Flash + ve results occurs in 2% of healthy cases.
- False – ve results occurs in 3.5% of patients.
- The test is used to diagnose Rheumatoid arthritis in S Anti-human and globulin (Rheumatoid fever) is found in patient serum.

24.13.2. Latex agglutination test for detective of with reactive protein

- Latex particles are coated with anti 'C' reactive protein.
- Buffer used in glycine buffer pH = 8.2.
- Serum is diluted 1:20 with buffer or as direct by manufacture of the reagent.
- Agglutination occur in 2 min.
- Serial dilution of serum can be used to quartz tube the with Reactive protein.
- C.R.Protein is an dilute phase protein capable of reaching with somatic polysaccharide of *Pnemococcus*.
- In concentration up to 5 mg/ml in serum of acute inflammation and high level persist during active phase of the disease.

The test is used to diagnose.

- Active Rheumatic fever.
- Raised C-R protein found in culute myocardial.
- Rheumatoid as there h's.
- Bacterial viral infections.
- Malignancies of meta states.

Latex agglutination test for detection of anti streptolysine 'O'

- Latex particles coated with streptolyrine 'O'.
- Serum is diluted to 1:20 by manufactures Agglutination occurs in 2 min.

To find the titre of A serial dilution of serum 1:30 to 1:100% ending to 200 to 1000 ml/ml of serum may be prepared and tested.

– (+) ve test indicate 200 cm of anti-strepto lysine 'O' in serum it is suggestive of a present acute streptococcus infection.

• to Coccus and infection.

• High titre are found in Rheumatic fever. The test is used to diagnose the Rheumatic fever.

24.13.3. Latex agglutination test for Rota virus

Latex particle are coated with Ab to Rota virus. (Rota virus)

- A drop of filtered stoos of the patient is mixed with latex reagent.
- Rota virus of present in stool will produce agglutination of latex particle. Latex agglutination test for identification of pathogenic

24.13.4. Staphylococcus.

- A number of test are used for identification of staphylo coccus, slide coagulation Test tube coagulation test, production of DNA's and mannitol fermentation.
- Slide coagulate test gives false (+) results and (-) results.
- Tube coagulation test takes 4-2 hrs.

Both these test are differ to interpret on suppy of plasma of suitable qualitied.

• DNA ase test require overnight incubation and give false (+) results. Mannitol fermentation gives false (-) ve and false (+) ve results.

One drop of coated latex peptide suspension is mixed with few calories of the organism on a slide.

• After shaking with sticks and rocking the results are read with in 30 sec clumping of particle is of presence of dumping factor protein A .

LATEX AT FOR TOXOPLASMOLYSIS

• This test utilizes the Ag coated latex particle for dilution and quantitation of Abs to Toxo plasmosis particle serum.

• A + ve test does not differentiate active or latent infection it is a rapid screening test.

(*vii*) **Latex AT for soluble bacterial Ag**

Soluble Ag's of bacteria causing meningitis are found in CSF, veins and serum causing acute lower respiratory tract unfection are found in spectrum, serum and wine are also found in filtrate of culture of bacteria.

- This is done by precipitating Ag by alcohol and dissolving the precipitate in buffered saline.
- These test are specific and rapid for diagnosis of bacteria unfection.

24.14. MISCELLANEOUS AGGLUTINATION TEST

Hence Agglutination inhibition test for virus. Many viruses agglutinate RBC'S of vertebrate chicken and human.

24.15. VIRUS AGGLUTINATED MONKEY'S RBC'S

HAIT gives results similar neutralization lies in haemagglutination site on the virus is same to that concerned in early stage of infection. In flaviruses Haem agglutination Ab's are distinct from nuteralizing Ab's.

Haemagglutination of many viruses are inhibited by substance other than Ab's present in serum. They should be removed before performing haemagglunation inhibition test.

Heterophic Ab if present in serum should be removed by treating serum with Ag before testing 10^6 viruses. Particle are needed to produce demonstrate haemagglutination that enfleza with virus agglutinate RBC to hightitre at 4°C.

24.16. HAEM AGGLUTINATION INHIBITION TEST FOR INFLUENZA

Chicken red cells blood is an anticoagulant and it is centrifuged at 3000 rpm for minutes.

The pocket cells are washed 3 times with saline and stored at 4°C as 10% suspension as solution of dextrose gelatin versonal buffer.

For tests 0.5% red cells suspension is made in saline. The virus suspension obtained from infected chick embryo in Alantoin fly is diluted 1:10 served 2 folds dilution are made patient serum.

Equal volume of serum and viral suspension is added after 1-2 hrs agglutination in the tubes are recorded highest dilution of the serum in which fail of Agglutination occurs in the title of agglutination inhibition and Ab.

QUESTIONS

1. Explain the application of agglutination reaction.
2. Describe the widal test.
3. Describe welflix test.
4. Explain the blood groupings and cross matching.
5. Write short notes on:
   (*a*) Cold agglutination test;
   (*b*) Paul bunnel test;
   (*c*) Rose walar's test;
   (*d*) Coomb test;
   (*e*) Coated particle agglutination reaction;
   (*f*) Latex agglutination reaction;
   (*g*) Miscellaneous agglutination reaction.
6. Explain haem agglutination inhibition test for influenza.

CHAPTER 25

PURIFICATION OF Ig BY ION EXCHANGE CHROMATOGRAPHY

25.1. 1. Principle

When human serum is added to an ion exchange column, by using a buffer with a pH greater than 6.5, Ig G is not absorbed and is eluted in the first fractions. The other proteins then removed from the column by a stepwise elution with testing concentration of salt. The isolated fractions are than analyzed by cellulase acetate electrophoresis and immuno electrophoresis.

25.1.2. Materials

1. DEAE Sephadex A50 120. 10/2g
2. Disposable syringes (10ml) 5 Nos
3. Tris buffer (10 mmol/liter pH 80). 3 liters.
4. Buchner filtration apparatus.
5. NaOH (1mol/litre) 1 litre
6. HCl (1mol/litre) 1 litre
7. Glass wool or plastics inters
8. Peristaltic pumps that can pumps
9. Human serum 5ml.
10. UVs spectrophotometers or column monitoring equipment
11. Buffered saline solutions (prepare the following concentrations of saline in the tris HCl 250 ml.
    (*a*) 20 mmol/litre.
    (*b*) 50 mmol/litre.
    (*c*) 100 mmol/litre.
    (*d*) 150 mmol/litre.
    (*e*) 200 mmol/litre.
    (*f*) 300 mmol/litre.
12. Solid polythene glycol 500mg
13. Materials for 5.2
14. Materials for E.P 64 method

25.1.3. Preparation of column

Add 2 g of DEAE sephadex A, 50, 120 to 70ml of water and leave until it swell. Take the suspension under vaccum and leave to equilibrate to room temperature. Place some glass wool in the bottom of 10 ml disposable syringe as a support and prepare a column of the 1M exchange material wash the column with 1 mol liter NaOH than repeatedly with water until the pH falls to about 7 wash with saline then mol/litre HCl and final with the this HCl buffer until the eluted registers pH 8.0, Alternatively the material can be washed on a buchner funnel.

25.1.4. Separation of Ig's

To add 0.5 ml of your own serum on to the DEAE sephadex of collect fine fractions of about mt 2.5 ml each by pumping the this HCl buffer through the column. In the meantime prepare 25ml of a serum of fluids containing sodium chloride at the following concentration in the buffer 20 mmol/litre, 50 mmol/liter, 100 mmol/litre 150 mmol/litre, 200 mmol/ litre, 300 mmol/litre. Pump these solutions through the column in thus stating with the lowest salt concentration and collect fractions until all the proteins has been eluted with that solution than move on to a higher concentration of salt carefully monitor the proteins eluted by measuring the absorbance of each fraction of 280 nm and plot the extinction against fraction number. Concentrate each fraction between 5 and 10 times by dialysing against solid polyethylene glycol for m about 3h and examine their proteins concentrates by cellulose acetate electrophoresis, immunodiffusion, immunoelectrophoresis.

25.2. IMMUNOFLOURESCENT ASSAY

Fluorescent dye are commonly employed as labels in primary binding tests, the most being fluses in solution cyanide (FITC). FITC is a yellow compound that is reading conjugated to Ig with out affecting their reactivity.

When *r* radiated with invisible ultraviolet at blue light at 290 nm and 145 nm, it re-emits visible green light at about 325.

FITC labeled Immunoglobulin may be employed in no of techniques, the most important of which are the direct and indirect fluorescent Abtests.

DIRECT FLOURESCENT ANTIBODY TEST

The test is used to identify the presence of Ag, Ab directed against a specific Ag. Such as bacterium or virus may be labelled with FITC. A tissue or smear containing the organism is fixed to a glass slide and incubated with the labelled antiserum it is then washed to remove unbound Ab.

If examined by dark field illumination under microscope with an ultra violet light source, the antigen with particle that have bound the labeled Ab's are sera of flourescence brightly. This direct test can be used to identify bacteria with their numbers are very low, as when examining the spectrum of person suspected of shedding ***Mycobacterium tuberculosis*** or when examining smear from lean ts for the presence of the close radial organism. It may also be employed to defect virsus type infection is tissues culture in tissues from infected animals, such as rabies virus in the brain infected animals as or Ag of human immuno deficiency virus on the surface of infected cells.

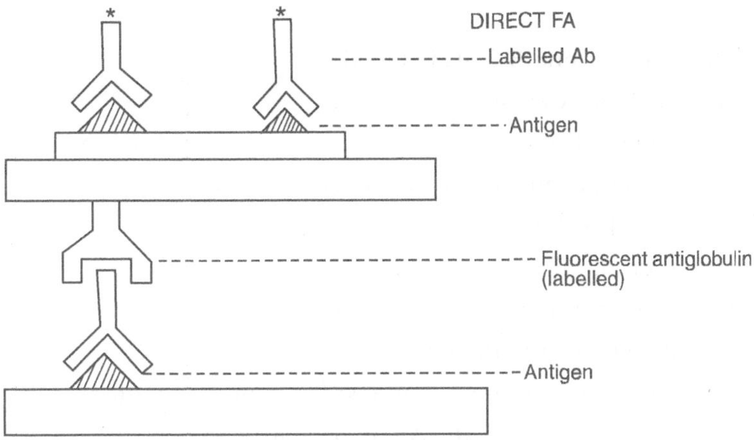

Fig. 25.1.

25.3. Indirect flourescent Ab test

The indirect flourescent Ab test may be used for the detection of antibodies in serum on the demonstration and identification of Ag. Ag is employed in the form of tissues smears, section of cell cultures on a slide or coverslip this is inclunated in a serum suspect of contaminations Ab's to that Ag and the serum is then washed off leaving only specific Ab's bound to Ag.

These bound Ab's may be visualized after incubation of the smears in FITC. Labeled antiglobulin is removed by washing and the slide is examined for flourecense which indicates that Ab was present in the test serum. The quantity of Ab in the test serum may be estimated by examined increasing detection of serum on a number of different Ag preparate number.

The indirect flourescent Ag test has number of advantages one the several labeled Antiglobulin molecule the flourescences will be considerably brighter than in the direct test similarly by using antiglobulin sera specific for each Ig isotype, the isotype of specific Ab present in the serum may be determined all test of their nature must be accompanied by the use of appropriated contract.

25.4. COMPARISON OF TECHNIQUES INVOLVED IN DIRECT AND INDIRECT FLOURESCENT AB'S TESTS

Test	Direct	Indirect
Requirements	Ag	Ag
	Flouresent labelled Ab	Flouresent labelled anti globulin Ab
Method	(i) Put FITC labelled Ab on to suspected	(i) pu suspected Anti serum on to known Ag preparation
	Ag preparation	Known Ag
Preparation.	(*ii*) Wash and Examine	(*ii*) Wash then cover with FITC labelled antiglobulin (iii) Wash and Examine
Detects	fluorescence indicates presence of Ag	Fluoresecence indicates presence of Abs in serum

25.4.1. Application

(*i*) Identification of T and B Cells in blood.

(*ii*) Detection of Auto Ab is serum.

(*iii*) Detection of Ig in tissue.

(*iv*) Determination of complement component in tissues.

(*v*) specific tissue freed Ab.

(*vi*) Determination of microorganism in tissues.

(*vii*) Localization of hormones and enzymes.

(*viii*) Quantification of serum Pudent and Ab.

(*ix*) Identification of chromosomes and specific bounding pattern.

(*x*) Identification of tumor specific Ag on neoplastic tissues.

(*xi*) Identification of transplantation of Ag on various organs.

QUESTIONS

1. Explain the method of purification of Ig by Ion exchange chromatography.
2. What are the principles behind during purification of Ig by Ion exchange ?
3. Write short notes on :

   (*a*) Materials required for preparation of Ig ;

   (*b*) Preparation of Ig;

   (*c*) Preparation of coloumn;

   (*d*) Indirect fluroscent Ab test;
4. What are the application of this method ?

CHAPTER 26

RADIO IMMUNOASSAY AND RELATED TECHNIQUES

26.1. INTRODUCTION

Several versions of the technique named displacement analysis by Yalow and Berson (1960) or saturation analysis by Ekins (1960) have been developed and diversified in recent years. These related techniques may be applied to the measurement of antigen (Ag) or antibody (Ab) and involve the addition of a labeled reagent either as labeled Ag or labeled Ab. The final measurement is of the label following a seperative step. In order to be applicable to the measurement of small quantities of Ag or Ab the properties of the label must be such that it can be detected at very low concentrations. For many purposes radioactive labels are convenient and have been widely used. More recently, alternatives to radio labels have been used including enzymes. Other labels have been detected by the properties of fluorescence, luminescence, electron spin resonance (the label is a radical containing an unpaired electron) or the biological properties of a bacteriophage. The label may also be a relatively large particle such as a latex particle or an erythrocyte whose properties are altered during the combination of Ag and Ab on its surface.

Specific binders other than antibodies may be used, and when these are naturally occurring plasma proteins the assay is often referred to as a competitive protein binding method. An example is the use of thyroxine binding globulin for the determination of thyroxine. Sometimes the binder is a relatively crude preparation of cell receptors for which the antigen has a biological affinity. Such methods are called receptor assays, e.g., the use of a rabbit uterine cytosol preparation for the assay of plasma oestrogens. In both types of assay the affinity of the binder may be less than for an antibody and the specificity variable.

Potentially greater binding affinity and specificity is obtained if the binder is an antibody raised against the desired antigen. Methods using such antibody binders are referred to collectively as immunoassays and are conveniently considered under sub headings depending on the nature of the label, e.g., radioimmunoassay and enzyme immunoassay. In all protein binding assays the material bound is called/ligand. However, in immunoassays the ligand may be more specifically termed as antigen.

26.2. RADIO IMMUNO ASSAY

There are several reasons for the rapid growth in the use of radio-immunoassay (RIA) other than the need to replace the cumbersome and time consuming chemical and bio assays previously in use.

The method offers considerable sensitivity, perhaps best considered as the smallest amount of substances being measured which can be distinguished from Ag with antibody of high affinity labeled antigen of high specific activity RIA has the ability to measure the substances at concentration as low as 1 ppm/l.

The methods offers specificity, the ability to discriminate between the desired substance and related substances. This arises from the highly specific nature of the antigen antibody interaction. Where a highly specific antiserum is available it is often possible to determine the required substance in serum without any preliminary purification procedure, thereby considerably simplifying the whole analysis.

RIA also has the advantage of a level of precision comparable to that of many additional chemical methods.

Accuracy is difficult to assess since RIA can only give a measure of immunochemical activity which need not imply measurement of biological activity and complete molecular structure. In addition, high quality standard material may be difficult to obtain. Nevertheless, reasonable accuracies may be inferred from correlations with other analytical methods such as bio assays or chemical procedures, and also from recovery experiments.

The technique is of general applicability to any substance against which antibodies can be produced, so that the same equipment and operating systems can be used for many different substances, and if necessary the techniques can be automated to handle larger work loads.

In RIA the essential components of the analytical system are the antigen (Ag) to be determined, a fixed amount of labeled antigen (Ag^*) and a fixed limited amount of antibody (Ab). The reaction is then:

$$Ag + Ag^* + Ab < \rightleftharpoons AgAb + Ag^*Ab$$

It will be apparent that provided the labeling procedure still allows the binding of Ag^* by the antibody, then there will be competition between Ag and Ag^* for the limited number of antibody binding sites. Once equilibrium has been reached, Ag and Ag^* will either be present in the 'free' form (F) or combined with Ab in the 'bound' form (B). As the amounts of Ag^* and Ab are constant, an increasing amount of Ag will result in less label being present in the bound form. If a satisfactory method of separating the bound and free forms is available then the radioactivity of either or both fractions can be used to produce a calibration curve. With many antigens, particularly the large peptide hormones, the sensitivity of the assay can be improved by delaying the addition of the labeled antigen. This is known as non equilibrium RIA or sequential saturation analysis. The amount of label can be much greater than that which would be used for the previously described simultaneous reagent addition system. When considering the use of this assay system, the requirement for sensitivity has to be balanced against the introduction of an extra manipulation and the increased use of labeled reagent.

The following section deals specifically with RIA. However, many aspects are relevant to other assay systems which will be described later.

26.2.1. Technical requirements for Radio-immunoassay

There are a number of basic requirements for establishing a satisfactory RIA, and these will be considered under separate sub headings. Antigen is required in the preparation of all the main reagents, namely standard, label and antiserum. Antigen purification can be an exicting process especially in the case of those hormones which are labile or present in minute amounts in tissues and plasma, such materials often expensive, or even unobtainable all but a few specialized centers. Conversely, small antigens such as steroids and drugs are usually readily.

Once the antibody has been produced, the serum must be assessed per quality and quantity. Information must be obtained on the titre or concentration, specificity and the avidity of the antibody. Methods for obtaining this information will be described later under assay optimization, since reagents other than antisera are involved.

26.2.2. Monoclonal Antibodies

These offer a potentially unlimited supply of identical antibodies, the ability to target specificity and the possibility of a purer antibody solution for labeling. The relatively modest avidity constants of monoclonal antibodies could limit the sensitivity of conventionally designed RIA's.

The essential stages in the production of a monoclonal antibody are:

1. A mouse or rat plasma cytoma cell line is selected for its ability to grow in vitro and to fuse with immunoglobulin secreting cells, while possessing a property by means of which it can be selected against. Usually the latter involves choosing cell lines lacking the enzyme hypoxanthine phosphoribosyl transferase. When these cells are grown in a medium containing aminopterin, which blocks the main pathway of nucleic acid synthesis, they are unable to utilize an alternative pathway whereas normal cells are able to do so provided thymidine and hypoxanthine are also available. Thus, by the addition of hypoxanthine, aminopterin and thymidine to conventional media a selective (HAT) medium is produced.

2. Immunization of mice and preliminary testing for antibody content.

3. The hybridization or fusion step, which involves mixing spleen cells from the immunized mouse and plasmacytoma cells in the presence of a high concentration of polyethylene glycol to produce hybrid cells with the features of both parents. The cell suspension is then washed before diluting in selective growth medium and pipetting into microtitre wells.

4. At this stage any remaining non fused cells should be incapable of growth, because mouse spleen cells do not grow in vitro, and plasmacytoma cells lack the necessary enzyme for growth in the selective medium, whereas the hybridoma cells possess both the ability for in vitro growth and the necessary enzymes as a result of the fusion of both cell types.

5. The contents of the well found to be producing antibody are recultured to separate each individual cell type. This process is known as cloning.

TECHNIQUE FOR THE PRODUCTION OF MONOCLONAL ANTIBODIES

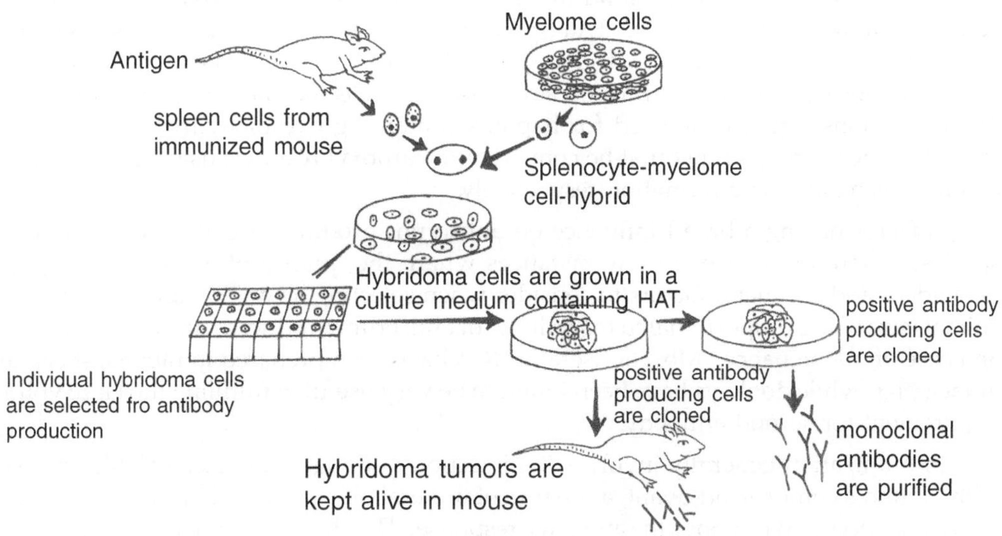

Fig. 26.1.

6. Stable clones producing antibody are then used for large scale production of monoclonal antibody. The large scale production can be by either in vitro culture in vivo growth in mice or rats.

It is often desirable to obtain pure monoclonal antibody preparations especially if they are to be labeled. Unfortunately both methods for large scale production produce a background of either serum proteins from the culture fluid or natural proteins from the mouse ascitic fluid. DEAE chromatography or affinity purification should enable a suitably pure preparation to be obtained.

26.2.3. Labelled Antigen

The ideal Ag* is a homogenous preparation of the antigen which has been labeled, without any loss of affinity for the antibody being used. The choice of radioisotope in practice usually lies between 125I and 3H except in special cases. The factors influencing the final choice include the case of introducing the radiolabel into the antigen, the chemical stability and specific activity of the product, the half life, and the ease of counting. Further details of radioactivity and counting methods. Iodine 125 has half life of 60 days and the labeled antigen has to be prepared every few months as a result whereas 3H has a much long half life of 12.3 years. However 125 I be incorporated into available in pure form, but difficulty may be experienced in obtaining conjugated forms of these molecules.

26.2.4. The Antibody

The properties of the antiserum dictate the specificity of a radio-immunoassay and therefore much attention is paid to its production.

In general, proteins having a molecular weight in excess of 5000 readily stimulate antibody production, ie., they are immunogenic. Small molecules such as thyroid hormones, steroids, drugs and peptide hormones must be coupled to larger carrier molecules in order to become immunogenic. However, the resultant antisera will react with the free form of such molecules which are known as haptens. Examples of carrier molecules include bovine serum albumin, thyroglobulin and keyhole limpet haemocyanin. It is not entirely clear whether the choice of carrier has a significant effect upon the antibody response. It is important to consider the site of attachment of hapten to carrier and also the number of haptens per carrier, bearing in mind the need to leave important groups in the molecule accessible to antibody response. Examples of conjugation reactions include the use of carbon amides or the mixed anhydride reaction, both of which mediate in the formation of peptide bonds between the carboxyl residues of one molecule and the amine of the other. These reactions can also be used for haptens containing oxo or hydroxyl groups, e.g., steroids. These groups must first be converted to carboxyl residues using carboxymethyl oxime or hemisuccinate formation respectively.

A factor having a broad influence on antiserum production is the choice of animal species. There are relatively few instances where the quality of antiserum is species dependent and the major factors to consider when selecting an animal are: the size of the animal (size being proportional to the volume of antiserum produced), the ease of handling and ease of maintenance. Most antisera for RIA have been prepared in rabbits, sheep and guinea pigs, while donkeys have been found to be very useful in fulfilling the large volume requirement for second antibody.

Immunization generally involves the injection of antigen, held in a stable emulsion with Freund's complete adjuvant, a mixture of mineral oil, waxes and inactivated bacteria, which enhances and prolongs the antigenic response. The choice of route is usually between intra muscular, subcutaneous or intradermal immunization, each method having its own strengths and weaknesses. Whichever route is chosen the schedule usually includes a primary injection followed by boosters at monthly intervals. Dose levels chosen range from 10 ug to 1 mg depending on such factors as the expected immunogenicity of the material being injected and the size of the animal. Peak antibody response occurs about 7 to 14 days following a booster injection, and harvesting of antisera should be timed to take advantage of this. Rabbits are bled from a lateral ear vein, guinea pigs by heart puncture and larger animals, such as sheep, by jugular vein cannulation.

Most antisera are stable at 20°C for at least one year or at 4°C after treatment with a bacteriostatic agent such as azide or merthiolate. Repeated freezing and thawing should be avoided, especially if the antiserum has been diluted. Freeze drying or storage below the eutectic point of saline (26°C) should be used for prolonged storage.

26.2.5. Radio-iodination Procedures

These are frequently used for radiolabeling polypeptides in which a tyrosine residue is present. The method of Hunter and Greenwood (1964) involves the exposure of the antigen for a defined short period of time to the oxidising agent, chloramine T, and 125 I in the form of carrier free iodide. Depending on the reaction conditions, one or two 125 I atoms are incorporated at one or more tyrosine residues. Sometimes a histidine residue may be iodinated.

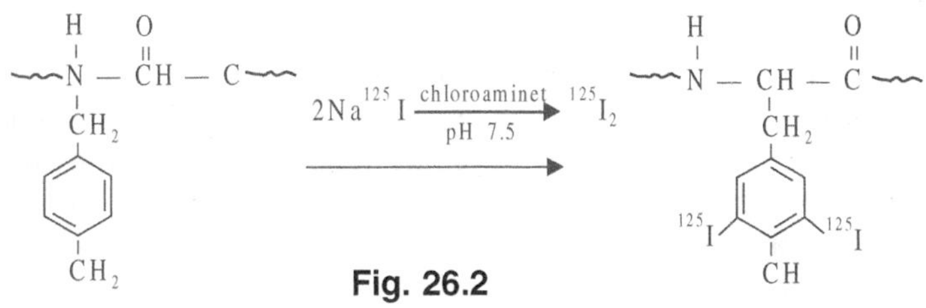

Fig. 26.2

The reaction is terminated by adding sodium metabisulphite. The oxidising properties of chloramine T can also lead to oxidation of sulphydryl, thioether, imidazole and indole groups and possible denaturation of the antigen. Such products may have different affinities for the antiserum than the native antigen. Milder iodination conditions reducing the risk of denaturation include the oxidation of iodide by lactoperoxidase or electrolysis.

Immediately after iodination a purification procedure is used in order to remove excess iodide and also to separate the labelled antigen from any damaged material. Changes in the antigen can result from:

1. Exposure to reagents used in labeling, notably the oxidant.
2. The presence of the iodide group.
3. The handling and exposure to surfaces, of dilute protein solutions.
4. Radiation damage.

It is usual to assess the degree of incorporation of radio-iodine into a compound by determining the specific activity. All methods of calculating specific activity make assumptions with varying degrees of validity. For details of methods see Bolton (1985).

Some antigens contain no residue which can be iodinated or unacceptable damage may occur during the labelling process. Such molecules may be condensed at a carboxyl group with the amino group of tyrosine methyl ester, tyramine or histamine to give a product which can then be labelled with ^{125}I. Alternatively one of these three compounds can be converted to their radio-iodinated forms before the condensation step. An alternative method uses 3(4' hydroxy 3' iodophenyl)propionic acid, N hydroxy succinimide ester in which the iodine atom is ^{125}I. This substance, available from Amersham International, reacts with primary amino groups in the antigen to form a covalently linked radioactive residue.

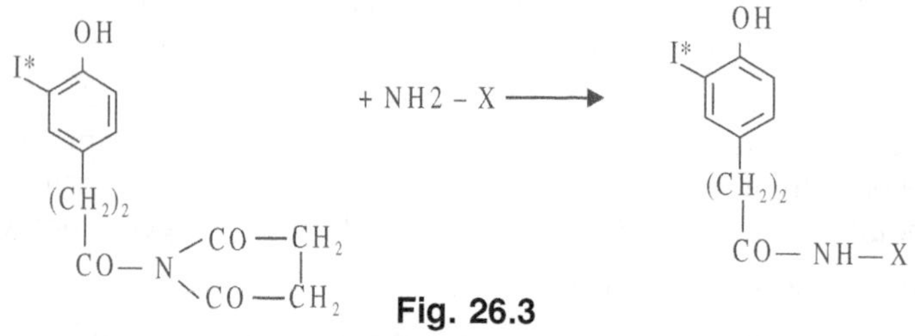

Fig. 26.3

26.2.6. Assessment of the quality of Radiolabels

The immuno reactivity of a labelled preparation can be assessed by determining the percentage of label capable of being bound in the presence of excess antibody. It is usual to aim for binding greater than 90%. However, labels exhibiting lower binding may prove to be acceptable. When testing a new batch of label for use in a well established assay, it is generally sufficient to demonstrate its satisfactory performance in a test assay consisting of a standard curve and quality control samples.

26.2.7. Storage

The stability during storage can be affected by several factors which include temperature, pH, the type and concentration of protein carrier, and the presence of protective agents, eg., Trasylol. Commonly, iodinated preparations are stored in the presence of bovine serum albumin at 20°C, preferably in small aliquots each sufficient for a complete assay. Repeated freezing and thawing is to be avoided. Some antigens are better stored at 4°C and some may preferably be kept at temperatures below 20°C. Freeze†dried preparations of tracers may also be satisfactory and are the preparations of choice when they are to be distributed to other laboratories.

26.2.8. Safety

All radio iodination procedures must be performed in a ventilated or total enclosure and not on the open bench. It cannot be stressed too strongly that careful, organized and tidy laboratory work is required both for successful iodination and for personal safety. For current recommendations concerning the design and classification of the working area, the advice of the radiation protection adviser should be sought.

26.2.9. Separation of free and bound antigen

The separation of bound antigen (B) from free antigen (F) is an essential step in RIAs. It is time consuming, technically demanding and a major source of imprecision. There is a wide range of methods now available and the factors considered in the selection of a method for a particular RIA include efficiency and general practicability.

26.2.10. Efficiency

The method chosen should be capable of completely separating the B and F fractions with a reasonable margin for variation of reaction conditions, while not upsetting the distribution of Ag and Ag* between the B and F forms. The selected separation method should also have minimal sensitivity to the composition of the reaction method should also have minimal sensitivity to the composition of the reaction mixture, i.e., lack of a 'matrix' effect.

26.2.11. Practicability

The method chosen should be quick, simple, safe, cheap, suitable for large batches and widely applicable without the need for extensive reoptimization.

26.2.12. Separation methods

They may be divided as follows;

1. Methods based on differential m Various chromatographic and electrophoretic techniques have been performed but in general such methods lack practicability.

2. Adsorption of free Antigen: Where the properties of the free antigen permit. This may be absorbed onto a solid support such as dextran coated charcoal, talc or ion exchange resins. Although reasonably practical for smaller antigens, e.g., steroids, their 'matrix' effects and interference in the primary antigen antibody reaction militate against their continued use.

3. Fractional Precipitation of bound Fraction: This is achieved chemically if the molecular size of the B and F forms differ considerably. Ammonium sulfate, ethanol, isopropanol and polyethylene glycol have all been used. These methods, though practicable for certain analytes, still require careful optimization.

4. Double Antibody Precipitation of Bound Fraction: This is widely applicable method, whereby an antibody to the specific antibody is used to precipitate B. If the specific antiserum (first antibody) was raised in a rabbit, then an anti rabbit antibody prepared in another animal is suitable (second antibody). This method has been shown to be both efficient and reasonably practicable. The major disadvantages are reagent cost and the prolonged incubation and centrifugation stages. These problems can be overcome to some extent by the addition of polyethylene glycol or by using second antibody attached to a solid phase.

5. Solid Phase First or Second Antibody: Here the antibody is linked to a solid phase so that once equilibrium is reached, separation of B is quick and simple. Early coupling methods such as the simple adsorption of antibodies onto polystyrene tubes had some problems, including efficiency and reproducibility. Some of the practical difficulties have been overcome by materials such as finely divided celluloses and plastic beads, to which antibodies can be covalently coupled by simple and non hazardous methods.

The choice of suitable solid phase materials facilitates the application of separation techniques such as sucrose layering or magnetic separation which lend themselves to a degree of automation not possible with centrifugation. These methods also lead to an increase in assay precision because B can be easily be washed free of F.

26.2.13. Standards

The accuracy of a laboratory's results and the relationship to those obtained by other laboratories depend on the use of suitable common standard materials.

For small antigens such as steroids, drugs or thyroid hormones, pure preparations are readily available from commercial sources. Many polypeptide hormone standards are available from commercial sources. Many polypeptide hormone standards are available from commercial companies and primary standards may be obtained from organizations such as the World Health Organization and The National Institute of Biological Standard and Control, Holly Hill, London. These materials are particularly useful for calibrating secondary standards for routine use.

Standard preparations should be contained in a matrix as closely resembling human serum as possible. For some RIAs a suitable substitute could be peptidase free buffered albumin solution, but for many there is no substitute for human serum. Analyte free human serum can be obtained either by selecting a suitable human population. e.g., males where the assay is for a pregnancy specific hormone, or by removing endogenous antigens, or by suppression of the endogenous material with, eg., dexamethasone for the production of adrenocorticotrophic hormone (ACTH) or cortisol free serum. A series of standards may be diluted and stored deep frozen in small aliquots, each sufficient for a single assay. For long term storage it may be necessary to use temperatures below 20'C or even lyophilization.

26.2.14. Assay optimization

Theoretical analysis (Ekins et al., 1968) would suggest that the equilibrium constant (Ka) of the antibody can be used to optimize RIA by giving an indication of the required concentration of antiserum and labeled antigen (Ag*). In practice most assayists follow a more pragmatic approach, possibly using the theoretical predictions for 'fine tuning'. The optimization of any RIA system can be relatively complex since alteration of any reagents or conditions (time, temperature, buffer type, protein carrier, etc.) is likely to require modification of at least one other component. The following order is usual.

26.2.15. Assay conditions

It is convenient initially to choose conditions which have been shown by experience to support acceptable assays for similar substances, bearing in mind that changes may be made subsequently. A moderately long incubation time could be used to reach equilibrium reasonably confidently, also a low temperature (4°C) plus inhibitors in order to lessen any damage by free radicals, oxidants, enzymes and bacteria. Also assay volumes should be kept to a minimum in order to promote an efficient immunological reaction.

26.2.16. Labeled antigen concentration(Ag*)

Maximum assay sensitivity is approached when both Ag and antibody tend towards zero. Extreme dilution, however, necessitates prolonged incubation and count times and so a relatively arbitrary concentration of Ag is chosen. Many workers use a concentration similar to that of the analyte at the lower end of the assay range. However, this is very much influenced by the specific activity of the label, counting time required and any difference in affinity for the antibody between Ag and Ag.

26.2.17. Separation system

A choicer of separation system must first be made and then it must be optimized. In the case of a second antibody system the concentration (titre) of the first antiserum must be established before the second antibody can be titred. A convenient, preliminary approach would be to use an excess of a liquid phase second antibody bearing in mind the need to add carrier serum when the first antiserum is used at high dilution.

26.2.18. Antibody Concentration (Titre)

Antiserum titration curves are performed by incubating increasing dilutions of antiserum with labeled antigen in the absence of unlabeled antigen. The choice of dilution range may be assisted by knowledge of previous bleeds from the same animal or from guidelines provided with commercially available antisera. In the absence of any prior indication a reasonable preliminary approach would be, to use 10 fold dilutions from 1/10 to 1/100.

The amount of labelled antigen bound to the antibody at each dilution can be measured and expressed as a percentage of the labelled antigen added (%B). The %B values can then be plotted against the antiserum dilution to produce the type of curve shown in the diagram. Conventionally the antiserum dilution giving approximately 50% B has been chosen at this stage, but some of the sensitive assays have been established using antiserum dilutions. However, when antiserum dilutions giving zero analyte binding below 50% are used, it is crucial to have a very precise separation stage in order to achieve the minimum non specific effects.

26.2.19. Standard Curve

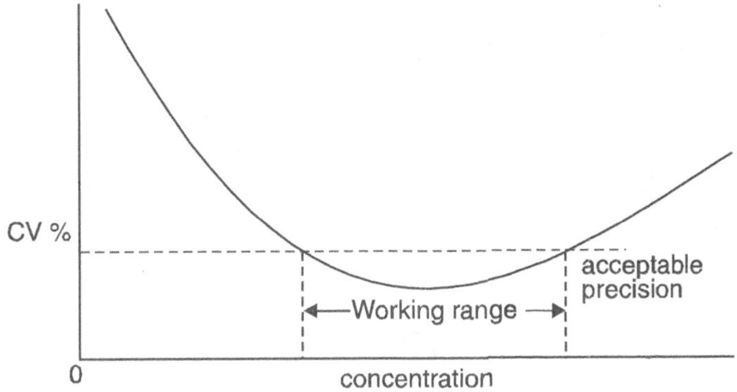

Fig. 26.4

26.2.20. Validation of a Radio Immunoassay

Having optimized the assay conditions the next stage is to assess whether the assay is valid in the analytical situations for which it was designed.

In the case of 'in house' methods, validation can be considered an extension of assay optimization, and is indeed likely to lead to minor changes in assay design. However, in the case of commercial kits, provided the protocol is being strictly observed, no assay optimization is required.

The validation process should involve the assessment of sensitivity, specificity, precision, accuracy, linearity and stability. It is important to relate the performance of a particular assay to the specific requirements and circumstances of the individual laboratory. In the case of commercial kits, only a limited assessment is possible within the individual laboratory and this information should be augmented by the experiences of other users and if possible from more detailed evaluations performed elsewhere, possibly under the auspices of External Quality Assessment Scheme organizers.

26.2.21. Assessment of Sensitivity (Detection Limit)

Sensitivity of defined as the lowest concentration which can be reliably differentiated from zero concentration. A reliable but arbitrary way of assessing sensitivity involves determining the mean and standard deviation of the counts from a minimum of ten 'zero standards'. The sensitivity is taken as the concentration corresponding to the mean counts minus three standard deviations. It should be noted that routine assays do not normally employ sufficient replicates of either patients' samples or zeros for the above to be strictly valid, and that certain pathological sera can adversely affect sensitivity.

A more realistic estimate of the lowest reporting limit of an assay is gained from the precision profile or from baseline studies in External Quality Assessment Schemes.

26.2.22. Assessment of Specificity

The specificity of an assay should be investigated using structurally related molecules, the choice of which is influenced both by knowledge gained from existing assays and from the probability of such substances being present in biological samples. A simple test of cross reactivity consists of assaying a known amount of Ags and Abs.

26.2.23. Standard Curve

It is now necessary to investigate the response of the assay system to unlabelled antigen standard. A series of standard solutions constructed to span adequately the whole of the intended working range (usually a minimum of 6 points for a conventional RIA). The response (%B) is plotted against the logarithm of the standard concentration.

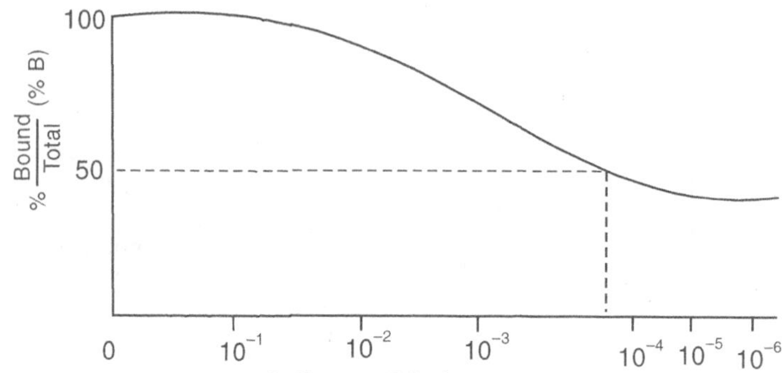

Antiserum Dilution Curve

Fig. 26.5

A study of the preliminary standard curve should enable the assayist to judge the necessity to change reagent concentrations or incubation conditions. It is recommended that any changes at this stage be made one at a time and the results assessed by the use of precision profiles.

26.2.24. Precision Profiles

The aim of precision profiles is to assess the performance of assays throughout the whole concentration range. The experimental preparation of a precision profile requires repetative analysis of analyte concentrations covering the whole of the assay range. Within batch precision profiles are useful but more realistic information is gained from combining the results from several batches.

During method development a within assay profile is sufficient, and a reasonably simple method is to analyse each standard concentration. The results can then be expressed graphically.

The choice of acceptable precision when deciding the assay working range is arbitrary. The aim should be to have a coefficient of variation below 5% for the clinically critical areas of the working range, but perhaps to accept 10% for the less critical. If an improved precision anywhere along the standard range was required then it would be necessary to adjust reaction conditions.

The concept of precision profiles has been discussed by Ekins (1984) and the reader is referred there for a more rigorous explanation.

26.2.25. Assessment of Accuracy (Bias)

Assessment of relative bias is easily made by using samples previously assayed by another well characterized method and calculating the correlation. Quality control samples, particularly those from national schemes can give valuable information in relation to several different methods. Recovery experiments formed by adding known amounts of analyte to either serum samples with confidently known values, or to analyte free serum, can also be useful.

26.2.26. Assessment of Precision

The routine assessment of precision is described in the section on quality control. The addition of a precision profile provides extra valuable information.

26.2.27. Assessment of Linearity (Parallelism)

This is normally performed by serial dilutions on sera containing high analyte concentrations to give values throughout the standard range. The assayed concentrations are then plotted against the dilutions used. The response should be linear, ideally passing through the zero position. A lack of linearity may indicate matrix interference.

26.2.28. Other Aspects

In addition to these technical evaluations, kits should be examined in order to check whether packaging, labelling and instructions are acceptable.

Having successfully completed the above stages, it is necessary to ensure, before routine clinical application, that:

1. A reference range is established. There is a tendency to measure the analyte on personnel drawn from the laboratory staff. These are generally young, healthy adults, and the range obtained may not be relevant for the bulk of the population under clinical investigation. Although it is considerably more difficult, it is preferable to use age and sex matched populations. Co-operation with other laboratories can help here.

2. There are adequate numbers of trained staff to perform the assay.

3. Adequate instrumentation exists.

4. There are adequate supplies of reagents.

26.2.29. Counting and Curve Fitting procedures

The details of radioisotope counting techniques are discussed in chapter 9. Many workers count each tube until a pre set number of counts or pre set counting time has been reached. The pre set number of counts is often 10,000 yielding a coefficient of variation of 1%. This strategy, although safe, can often lead to a drastic waste of counter time, possibly to the extend that a department would purchase an unnecessary extra counter. The requirement is to control error in the final analyte concentration not in the raw counts. Ekins and other workers have developed an approach, the practical impact of which can be better appreciated by considering some figures. Suppose that the experimental error in a given assay tube is 5% (not unusual in RIA!), and the counting error, assuming 10,000 counts, is 1%. However, the two errors are not simply additive, and when one uses the correct statistical formula the total coefficient of variation is only 5.1%. If the tube is only counted for 1000 counts, the counting error is 3.2% but the total error would be 5.9% which represents an increase of only 0.8% in imprecision at a cost of ten times as much counting time (ideally using micro processor controlled counters) without a significant worsening of assay performance. Following radioactivity measurement, many curve fitting procedures correct for non specific binding before relating it to the concentration of analyte. Non specific binding refers to the amount of labeled antigen which is bound in the absence of antiserum. There are many ways of relating corrected counts to the concentration of standard. Manual methods include plotting the bound counts as a fraction of the total counts on the ordinate against antigen concentration or the logarithm of concentration on the abscissa. Automated calculation of results saves time and eliminates many potential errors. It is becoming increasingly untenable to object to automation, particularly in view of the satisfactory programs now available even for hand held calculators. Automated curve fitting procedures may be arbitrarily divided into two categories:

1. **Empirical** : These are so named because their use is not based upon theoretical reaction mechanism, but tests entirely upon success in practice. This group includes linear interpolation, splines and logistic techniques. Many of the better packages are based on the four or five parameter logistic method.

2. **Model based** : These procedures are based upon considerations of the law of mass action as related to ligand binding assays. The number of such models are unlimited. Whichever technique is selected it is worth noting that good data handling does not compensate for poor raw data. Also modifying the assay design to suit the data processing technique could entail a great sacrifice in assay quality. It is a pre requisite of computerised data handling that counts be reliably entered into the computer, much of the potential advantage being lost if manual keyboard entry is used.

26.2.30. Quality Assessment

The quality assessment procedures required in RIA are no different in principle from those widely practiced in clinical biochemistry. However, because of its extreme sensitivity and the nature of the reaction, RIA is more susceptible than most techniques to problems related to quality, particularly when comparing results between laboratories. Typical co efficient of variation figures for RIA would be 5% for within batch and 10% for between batch analysis. The usual strategy for quality assessment is to have an internal system of quality control, on which are based the day to day assay management decisions.

26.2.31. Internal Quality Control

In common with other large batch analyses performed in clinical biochemistry, it is usual to include both within and between batch controls and to perform statistical analyses on these. The non linear reaction kinetics in RIA have led most laboratories to use control samples with at least three levels of analyte concentration in order to cover more comprehensively the assay range. With some analytes the calculation of assay means can provide an extremely sensitive indicator of systematic error. However, for many substances measured by RIA, patients' means are too variable for them to be of practical use. Aspects of quality control individual to RIA include such measures as observing the binding of label in the absence of standard (Bo), non specific binding (NSB), and parameters relating to the curve fit. Statistical packages are available which not only perform results calculation plus the quality control measures mentioned above, but also calculate precision profiles using data from existing assay samples. The latter is not only a sensitive means of quality control but it also enables each patient result to be accompanied by meaningful confidence limits. Any well designed internal quality control scheme will have a simple but effective means of recording data so that the current and past performance can be readily compared.

26.2.32. External Quality Assessment

The aims of EQA schemes and their implementation are described. The value of such schemes to the laboratory lies in providing both long term independent support for the internal procedures and in their ability to provide information which is difficult to obtain by a single laboratory. For the laboratory manager, data on assay bias is of greatest interest since precision can be more reliably assessed by the more timely and frequent internal quality control techniques. Most EQA schemes express bias relative to the mean result for all the participants, but in the absence of evidence that this is a reliable estimate of the 'true' analyte concentration, data on bias need to be interpreted with caution. For certain haptens, gas chromatography linked to mass spectrometry can be used to provide the 'true' target. Tests of base line security and recovery experiments performed through the EQA schemes can provide invaluable information on an individual laboratory's accuracy.

26.2.33. Automation

Given the success of automation in routine clinical biochemistry and the large number of labour intensive RIAs performed, it might be expected that the technique would become a prime candidate for automation. However, with the exception of automatic pipetting devices, this has not been the case, particularly in the UK. The reasons possibly include the type of laboratory originally involved in RIA and also the inherently difficult problems for automation, particularly the separation step. The major advantages of automation, are: improved precision, increased sample throughput, decreased cost per sample and identical incubation time for each tube making possible the use of shorter non equilibrium conditions. In order to remove the problem of centrifugation, separation methods such as filtration, magnetic particles or solid phases have been used. Alternatively, a homogeneous assay system which does not require separation may be chosen. The choice of automation as in routine clinical biochemistry is essentially between continuous flow and discrete analysers. Continuous flow has the advantages of reliability, unattended operation and limited cost with the disadvantages of limited incubation times and carry over. Discrete systems reduce carry over and often enable flexible incubation periods to be used, making them particularly suitable for analytes with a wide range of circulating levels, and for assays requiring high sensitivity. However, they do tend to be more expensive, particularly in maintenance costs. Whichever instrument is chosen, the capital cost will be high and in many cases the user is tied to manufacturers' reagents and assay protocols, which can be a problem. Any prospective purchasers would be well advised to seek independent evaluation and to insist on a trial period within their own laboratories. Given the increasing use of routine immunoassay it is inevitable in the longer term that some degree of automation will occur in many laboratories. The eventual direction could be greatly influenced by developments such as labeled antibody techniques, particularly those using non isotope labels.

26.2.34. Free Hormone Measurement

Certain hormones, e.g., thyroid and steroid hormones exist partially in protein bound form. It is generally held that the concentration of free hormone in serum more closely represents the physiological activity of the hormone than the corresponding total hormone concentration. The measurement of free hormone would be expected to be particularly useful where abnormalities of binding proteins exist often leading to grossly misleading total hormone concentrations. Several of the techniques for free hormone measurement introduce analytical principles which are conceptually different from those underlying the immunoassay of the total hormone concentrations in serum. They may usefully be divided into'direct' and 'indirect' methods.

26.2.35. Indirect Methods

Indirect methods rely on the measurement of two separate and independent variables one of which is the total hormone concentration, the other being an assessment of hormone binding which can be used to sub divide the methods which are as follows:

1. **Tracer Dilution** : In this technique a known trace amount of radio-labeled hormone is added to serum. After subsequent separation of free and bound fractions by techniques such as equilibrium dialysis, the radioactivity in the free fraction in combination with measurement of the total hormone concentration gives an estimate of the free hormone concentration.

2. **Free Hormone Indices** : This technique is analogous to tracer dilution with the substitution of a solid adsorbent separation stage.

3. **Hormone/Binding Protein Ratio** : This uses the relationship of the total hormone to its principal binding protein. the ratio gives a reliable and relatively simple measure of free hormone concentration. Binding proteins can be measured by several immunological techniques including RIA and immunodiffusion.

26.2.36. Direct Methods

These methods rely on the single direct assay of free hormone without the need or total hormone measurement. Examples of direct methods include

1. **Equilibrium Dialysis** : Serum is allowed to equilibrate in a small dialysis cell, the free hormone in the dialysate is then measured by a sensitive RIA.

2. **Direct RIA** : Such methods rely on the proposition that the introduction into serum of a 'vanishingly' small amount of antibody results in a fractional antibody occupancy which is dependent on the free hormone concentration. It is the case that even in the situation where the ratio of total hormone to antibody binding sites is in excess of 1000 there will nevertheless be free binding sites at equilibrium. The original version of this technique consists of two stages where, following the incubation of sample with a trace amount of antibody, the antibody is separated from the incubate and then back titrated with labeled hormone in order to determine the degree of occupancy by free hormone. The above 'two stage' procedure, while theoretically sound, may suffer from poor precision mainly as a result of unavoidable variation in timing, particularly in the washing and second incubation stage. More pertinently it has the practical inconvenience of a two stage procedure with the attendant washing step. The introduction of a further methodological refinement which uses a labelled hormone analogue means that a single step procedure can be used. The hormone derivative, while retaining the ability to bind to antibody, is unable to react with serum binding proteins thus leaving the equilibrium of free to bound hormone undisturbed. the method would be invalid if the labelled analogue were to bind to serum proteins.

3. **Other Methods** : These include the use of micro encapsulated antibody, the differential distribution of labelled antibody between free hormone and hormone analogue, and methods involving consideration of the rate of binding of free hormone by antibody. Hormone levels determined in urine and saliva have also been used to provide an indication of free hormone concentration.

26.2.37. Techniques related to Radio-immunoassay

RIA is now widely used for the quantitation of many analytes in serum, however it suffers from several disadvantages. These include:

1. Prolonged reaction times as a consequence of having to use reagents at high dilution.
2. Problems resulting from using a gamma emitting radio isotope, namely a relatively short shelf life and possible health hazard;
3. The need to add each reagent very precisely;
4. Limited assay range.
5. Lack of direct linear relationship between analyte concentration and signal response.
6. Difficulty of automation.
7. Lengthy counting time.

26.2.38. Labeled Antibody Immunoassay

Labeled antibody assay, often referred to as immuno radiometric assay (IRMA), offers several advantages over RIA. In particular, the introduction of the two site or sandwich assay procedure in which the antigen is reacted sequentially or simultaneously with solid phase antibodies and labeled antibodies, has led to improvements in specificity, sensitivity, and speed over conventional RIA. Improvement in specificity result from the need for a substance to possess antigenic sites for both of the carefully selected antibodies in order to be measured. In RIA, the sensitivity is essentially limited by the equilibrium constant of the antibody, whereas this theoretical constraint does not apply to labeled antibody methods, the ultimate sensitivity limits of which are primarily imposed by classical considerations of signal to noise ratio. Rapid reactions result from the use of reagent excess, which also contributes to increased precision because only the analyte being measured needs to be added precisely. The non competitive nature of the reaction gives a direct linear relationship between analyte concentration and response, which also leads to an extension of the useful assay range. The main disadvantage, namely the requirement for relatively large amounts of purified antiserum, may be expected to be largely overcome by the use of monoclonal antibodies. Another constraint is the limited specific activity of125 I labeled antibodies if immuno-reactivity is to be retained. This could be overcome by the use of non isotopic labels, leading to improvements in immunoassay performance in terms of stability and sensitivity.

26.3. ENZYME IMMUNOASSAY (EIA)

Enzymes can act as labels because their catalytic properties allow the detection of extremely small quantities of labeled immune reactants. They have the advantage of a long shelf life, no radiation hazard, no requirement for specific instrumentation not already available in clinical laboratories, and no lengthy reading times. Enzyme immunoassay can be subdivided into heterogeneous and homogenous assays. The former consist of assays which are analogous to RIAs or IRMAs, whereas in homogenous assays, the reaction between enzyme labeled antigen and antibody alters the activity of the enzyme. These assays do not require a separation step and have no analogy in RIA. Homogenous assays are almost invariably used to quantitate smaller molecules, the key reagent being an enzyme hapten conjugate (E H), the activity of which is inhibited by binding to antibody.

$$\underset{\text{Active}}{\text{E H + Ab}} \quad \bullet \quad \underset{\text{Inactive}}{\text{E H Ab}}$$

The antibody is thought either sterically hinder the active site or to induce conformational changes in the enzyme. Unlabelled hapten (H) competes with enzyme labelled hapten for the limited quantity of antibody and so reduces the inhibition.

$$\text{H} + \text{E}^{-}\text{H} + \text{Ab} \quad \bullet \quad \text{H Ab} + \text{E}^{-}\text{H Ab} + \text{E}^{-}\text{H}$$

This assay has been termed the enzyme multiplied immunoassay technique (EMIT). Problems have occurred when trying to measure high molecular weight antigens by EMIT. It would appear that in most cases the antibody when bound to the enzyme labeled antigen is too far removed from the enzyme's active site cause steric hindrance. The choice of enzyme depends on specific activity, conjugating characteristic absence from biological fluids, availability and a satisfactory method of measurement. Alkaline phosphatase, horseradish peroxidase and beta galactosidase are the most widely used enzymes in heterogeneous EIAs. Glucose 8 phosphate dehydrogenase has been frequently used for EMIT. Unlike radioisotopes, enzyme activity can be affected by its environment. This is likely to be a particular problem with homogeneous assays, where the enzyme activity is measured in the presence of the biological fluid under examination. In addition the critical manipulations involved in enzyme measurement can potentially lower the performance of the assay.

Homogeneous EIA (EMIT) would appear to be well suited to the measurement of haptens such as drugs, whereas heterogeneous EIAs would be useful for the semi quantitative assay of larger molecules. Enzymeimmunoassay in the form of a two site labeled antibody technique, immunoenzymometric assay (IEMA) may prove to be valuable in high sensitivity, quantitative measurement of high molecular weight analytes, provided that the reliability and precision prove satisfactory.

26.4. FLUORO-IMMUNOASSAY (FIA)

In fluorescence or photoluminescence light of the appropriate energy (wavelength) excites the molecule from its ground state to a higher energy state. The return to the ground state is accompanied by the release of energy in the form of light of a longer wavelength. The use of fluorescent substances as labels offers the same advantages as enzymes without the problems inherent in biological methods of detection. Unfortunately, various factors are present in serum which can result in an apparent increase or decrease in fluorescence by phenomena such as light scattering, quenching or by the effects of endogenous fluorophores. Because of this, FIAs using conventional fluorophores and fluorimetry have failed to achieve the sensitivities accorded to RIA. Developments have occurred in several areas which should greatly increase the impact of FIA: 1. Labeled antibody techniques have the same potential advantages as for RIA & EIA plus the ability to remove interfering factors by washing the solid phase immediately prior to reading the fluorescence. 2. Instrumental improvements have led to the use of pulsed, light, time resolved fluorescence, by which the fluorescent signals of interest can be distinguished on the basis of their time decay characteristics from naturally much shorter lived background fluorescence. 3. Certain rare earth chelates, particularly those of europium, have certain advantages. They have a prolonged fluorescence dacay time which helps to separate the desired signals from that of the background and also they possess a wide difference between the excitation and emission spectra (Stokes shift) which reduces the effects of non specific fluorescence. 4. Fluorescence polarization has been used to distinguish between fluorescent antigen bound to antibody and unbound antibody also present in the reaction mixture, thereby simplifying the immunoassay. Molecules of the large antigen antibody complex rotate more slowly than those of the small antigen so that emitted light from the complex, but not unbound antigen, is preferentially polarized in the same plane as that of the exciting beam.

26.5. LUMINESCENCE -IMMUNOASSAY (LIA)

The potential of chemiluminescence as a label is illustrated in table. However, before this potential is realized, certain problems have to be overcome. These include interference from serum constituents, the vigorous conditions required for labeling antigens and for producing the final luminescence, and inefficient detection of the short burst of light.

Approximate Detection Limits for Different Immunoassay tables

12.5_I	10^{-17} mol
Fluorescence	10^{-15} mol
chemiluminescence	10^{-19} mol

As in the case of FIA newer developments have occurred which largely overcome the disadvantages. These include:

1. The use of solid phase labeled antibody techniques, whereby the products of the immunological reaction can be washed free of potential interfering substances.

2. Newer chemiluminescent probes such as acridinium esters are insensitive to serum catalysts such as haem proteins, and because the mechanism involves dissociation of the excited species from the rest of the molecule the nature of the emission is largely independent of micro environmental effects.

3. The newer labels require less vigorous conditions both for coupling and for producing the final luminescence. The latter results in a lower background and hence an improved signal to noise ratio.

4. As a result of using luminescence enhancing compounds the signal is much more prolonged, with the effect of making the reagent addition and final reading steps much less critical, thus leading to more efficient luminescence measurement.

26.5.1. Conclusion

The properties of immunoassays will ensure an increase in demand for the foreseeable future. Radioisotopes will probably remain the label of choice for conventional competitive, limited reagent immunoassay, with exceptions such as EIA for qualitative and quantitative assays where a visual colour change offers a very convenient end point. However, the advantages of labeled antibody techniques, particularly when using monoclonal antibodies, will probably lead to them largely superseding competitive immunoassays. The choice of label for these assays may be dominated by non isotopic labels such as fluorescence or luminescence, this being much influenced by the state of the art in instrumentation and automation.

26.6. RADIO IMMUNO ASSAY

Radio Immuno Assay [RIA] procedures are capable of measuring trace amounts of any substances that can serve as an antigen or happen. Resolution is limited principally by the specific activity of the radio actively labeled standard and by the specificity of the antibody preparation. At present the practical limit is on the order of 10^{11} to 10^{9} g antigen per sample. In the past many biologically important molecules such as drugs and hormones could be assayed only as low resolution using predominantly bio assay procedures. Today an ever increasing number of this enzymatically inactive molecules are being reliably quantitated in samples of biological fluids such as blood urine and spinal fluid. The following tube lists some molecules now being assayed out linely in this manner.

An experimental sample of unlabelled antigen is mixed with a constant known amount of the same antigen of radio actively labeled. To this mixture is added limited amount of rabbit antibody prepared against the antigen being measured. It is critical to add sufficiently small amount of antibody. So that the system is in a condition of antigen excess even when unlabelled antigen is absent. Since the system is antibody limited, the amount of labelled antigen bound to the rabbit antibodies.

29.6.1. Table. Compounds quantitated by radio-immuno assay

- Adreno Corticotropic hormones
- Aldosterone
- Brodykinin
- Calcitonin
- Carcino embryonic antigen
- Cyclic AMP
- Digalis

- Estrosliod
- Gastrin
- Glucagen
- Growth hormone
- Human IgE
- Insulin
- Marphine
- Parathyroid hormone
- Prosaglandin
- Secretine
- Testosterone
- Thyrogloblin
- Vassopressin

26.6.2. Conceptual operation of the radio immuno assay

RIA the soluble Ab.Ag complex is both a product of the first step and a reactant in the second step of the assay is universally proportional to the quantity of non radio active antigen present. This basically an isotope dilution type of experiment the final step of the procedure is separation of antigen antibody complexes from free antigen a prerequisite for measuring the amount of antigen bard. Although any methods are available the one most widely used experimentally is immuno precipitation. This method of separation can be used successfully because

1. Antigenic determinants of an antibody molecule are not located very close to its antigen binding sites and

2. Immunoprecipitation of an antibody molecule does not adversely affect its ability to bind antigen. A good example would be Goat antibody obtained following immunization of the animal with the globulin fraction of rabbit serum. Some caution is needed in selecting the pair of antibody preparation to be used because some do not work well together, as the case for donkey antibodies prepared against Guinea Fig Gamma globulin.

Precipitation a standard radio active antigen is the presence of increasing amount of non radio active competitor. A curve such as this is required for standardization of radio immuno assays.

26.6.3. ^{125}I or ^{131}I Labeling procedures

Two principal methods exists for incorporation of ^{125}I or ^{131}I into purified antigens. Both involve oxidizing iodide "active iodine" presumably a cationic form of iodine (I+), which then reacts with ionized tyrosine residues of the antigen. The first method (12) empby lacto peroxidase and hydrogen peroxide and the second method (14) uses hypochlorous acid generated by the flow decomposition.

Synthesis of B1I iodotyrosine with the lactoperoxide oxidation system of chloromine T is aqueous solution these reactions are carried out in extremely small volumes ranging between 0.05 ad 1.0 ml following incorporation of iodine into the antigen the reaction mixture is transfered to a molecular sieve column to separate un reacted iodine from that band to protein more convenient and gentle methods of protein iodination are beginning to develop and to be made available commercially.

Extreme care is required during these operations of loss of antigen immuno reactivity is to be avoided. Alteration of the antigen resulting from replacement of the tyrosine protein by iodine. 2) chemical damage caused by oxidizing reagents and the reducing agents subsequently added to neutralized them and 3) radiation damage to the protein. Covalent addition of a large atom such as iodine to the surface of a protein can potentially change its conformation. Addition of more than one molecule greatly increases the probability of such adverse conformational changes. Therefore incorporation should be limited to one iodine molecule or less per molecule of protein. The values significantly below this level are common. Chemical damage is most often observed when the chloramine T method is used. It can be avoided to a great extent is chloramine T is added in small aligouts untill incorporation has reached the desired level. Using such a filtration method decreases the concentration of chloramine T present at any one time and eliminates the need to add a reducing agent to neutralize excess oxidizing agent. Alternatively the lactoperoxidase procedure may be used chemical damage is reduced but so is the efficiency of labeling radiation damage can be minimized by avoiding concentrated solutions of radio activity for prolonged periods of time. The iodination reaction requires only a short time to reach completion, following which the reaction mixture is diluted during molecular sieve chromatography. Thereafter it is advisable to keep the solution dilute even if serum albumin must be added to stabilize the antigen. Care must be used, however, to select a source of albumin that doesnot in any way affect the immuno chemical reactions.

All the problems discussed above require that the labelled antigen be checked for possible damage following the iodination procedure. Inspite of all possible precautions the presence of one mole of iodine per mole of protein may still have an effect. Therefore it is most important to determine whether or not iodination alters the antigen with respect to its intended uses. Deleterious iodination effects may be detected as decreased enzyme specific activities or altered physico chemical properties of the antigen. The most sensitive problems however involve antigen binding to other molecule for example does the antigen bind as well to an affinity column or to its cognate antibody as it did prior to iodination of the antigen is a hormone it bind as well to its receptor site and once bound, does it still elicit the same physiological changes as before iodination much effort and careful consideration are required if the evaluation is to be at all rigorous.

Prior to any evaluation of the functional integrity of labeled molecules, they must be separated from unlabelled molecules that may still be present. This is especially important for hormones which can elicit physiological effects of minute concentrations. Separation is usually based on a physiochemical charge in the molecule as a result of iodination depicts one such separation using ion exchange chromatography and table 88 lists a number of other methods that have also been used to successfully isolate various iodinated hormones.

26.6.4. Standardization of Radio Immuno Assay

Several potentially serious conceptional and technical problems may arise during standardization and use of a radio immuno assay. The first of these involves losses of antigen during assay. In most cases the protein concentrations found is an assay mixture are quite low especially it purified preparations are used throughout under these conditions losses resulting from non specific absorption to glass and plastic vessels are substantial and must be eliminated. This may be done by adding serum to the reaction mixture (2% final concentration) Serum should be chosen that (1) does not contain any materials which would react with antibodies found in the assay and (2) does not affect the stability of the test antigens. The latter is exemplified by glycogen which is unactivated by crystalline barine serum albumin but not horse serum.

The second problem is one correctly selecting the appropriate quantities of antiserum. Both antibody preparations must carefully titrated. Antibody against the antigen being tested must always be used at a concentration where there is antigen excess. The second antibody prepared against gamma globulin analagous to the first antibody preparations, must always be used in moderately large excess. If the first antibody preparation is used at too high a concentration, all of the antigen labeled standard and unlabelled experimental sample will bind to antibody, precluding competition on the other hand, if the second preparation is used at too low a concentration, there will be incomplete precipitation of the first antibody, resulting in decreased assay resolution.

Since the filter of antiserum changes from animal to animal and with in the same animal with time, it is advisable to use a single preparation for each set at end points. When this is not possible each new preparation must be completely standardized. In the case of secondary antibody preparation, only its ability to precipitate the primary antibody dies must be quantitated. Primary antibody preparation however must be titrated for their ability serve as both antibodies, antigens. The amount of radioactivity in the final immuno precipitate is determined using a scintillation counter if ^{14} C, 3 H or ^{35} S is used as the labeling isotope. If however ^{125} I is used as a second option is available. Since ^{125} I gives of Gamma rays it is possible to use a gamma ray scintillation counter. This instrument operates in a manner similar to that of a standard scintillation counter but uses a scintillation system.

The sample to be counted may either be suspended in water or left as pellet. When lowered the sample well of the counter, the sample is positioned close to a large AI crystal containing trace amounts of thalium chloride gamma rays from the sample, pass through the vial or the test tube wall in the crystal where they interact with Ag I crystals producing flashes of light which are detected by an adjacent photo-multiplier tube. There are two advantages of using UV counter for this type of assay first the time required to accumulate statistically significant counting safes is usually quite small thus increasing the number of samples that can be processed per unit time. Second gamma ray counters are usually less complex electronically do not require the use of scintillation vial or scintillation fluids are therefore considerably less expensive to purchase or operate.

Schematic diagram describing the physical organization a operation of the UV rays sensing elements of a gamma ray scintillation spectrometer PMT indicates the photo multiplier tubes.

The data obtained from radio-immuno assay experiments may be plotted in a number of different ways. If a quantitative measurement of an unknown antigen is desired the amount of radio active standard precipitated may be plotted as a function of the amount of unlabelled test sample added. In this e.g to competition is observed when buffer is substituted for unlabelled experimental sample extract desired from a wild type organism containing lactate dehydrogenase the antigen being measured completes very well as evidenced by a dramatic decrease of radio activity in the immuno precipitate. Extract derived from a mutant strain of devoid of lactate dehydrogeneous also decreases the immuno precipitate radioactivity although less successfully, indicating that the mutant extract contains a protein that does have determine sites uncommon with purified lactate dehydrogenase activity. However from this data it is difficult to compare the two competative curves quantitatively. An alternative method of data presentation is shown in the figure below.

Here the percentage of total radio activity bound to the immuno-precipitate is plotted on semi logarithmic coordinates as a function of the amount of the competilor added. The advantage of such a plot is that in many cases the contact portion of the curve is a straight line, which aids in comparing the degrees of competition exhibited by standard and unknown samples. Also shown in the figure are the results observed when standard radio active antigen (dioxy casticies ferone) is challenged with various non radioactive steroid analogues. Significant competition is observed with progesone dioxy cestisol a costicosterone. The effect of competition is significant when it is considered that happen against which the antibody was produced is not likely to contain multiple determinants.

Although the latter method of analysis is better suited to quantitative interpretation, the data presented high light perhaps the most serious conceptual problem involved with careful interpretation of radio immuno assays results what exactly is being measured. The merits of immuno-specificity are to be leveled specific it is unfortunately not absolutely specific. Antigenic determinants of different proteins may often do share some similarities; since a single protein has more than one protein type of determinant the chances of finding such a similarity increases even more dramatically. In the case of an enzyme active site only one or a very few substrates bind to the active site well a yet a variety of analogues may usually be found which bind to a limited extent (0.5 1% as well). The immerse sensitivity of radio immuno assay techniques however easily detects such slight binding. One must always ask therefore, whether the observed competition is the result of a small amount of competitor whose identity is the same as the standard antigen or a some what larger quantity of material that has a similar antigenic determinant but is not the desired antigen at all. It must be remembered that inspite of its great flexibility requires a knowledge of identity of the precipitated material information that is often difficult is not impossible to obtain.

QUESTIONS

1. Explain the method of Radio Immuno Assay.
2. Outline the technique of Radio Immuno Assay.
3. Explain the technical requirements of Radio Immuno Assay.
4. Explain monoclonal antibodies.
5. Explain the procedures for Radio Iodination.
6. Write short notes on :
   (*a*) Labelled Antigen;
   (*b*) The antibody;
   (*c*) Assessment of the quality of Radio labels;
   (*d*) Storage;
   (*e*) Safety;
   (*f*) Separation of free and bound antigen;
   (*g*) Seperation methods;
   (*h*) Standards;
   (*i*) Assay;
   (*j*) Optimization;
   (*k*) Assay conditions;
   (*l*) Labelled Antigen concentration;
   (*m*) Antibody concentration;
   (*n*) Validation of Radio immuno assay;
   (*o*) Assessment of sensitivity (detection limit);
   (*p*) Assessment of specificity;
   (*q*) Standard curve ;

(*r*) Precision profile;
(*s*) Assessment of accuracy;
(*t*) Assessment of precision;
(*u*) Assessment of linearity;
(*v*) Counting and curve fitting procedures;
(*w*) Quality assessment of RIA;
(*x*) Internal Quality control of RIA;
(*y*) External Quality control of RIA;
(*z*) Automation of RIA;
aa. Prehormone measurement;

7. What are the techniqus related to Hormone Assay ?
8. Outline the procedure for Radio Immuno Assay.
9. Explain the procedure for Fluro immuno Assay.
10. Explain the procedure of Luminescence Immuno Assay.
11. What are the compounds quantitated by Radio immuno assay ?
12. Explain the standardization of Radio immuno Assay.

CHAPTER 27

ELISA (ENZYME LINKED IMMUNO SORBENT ASSAY)

There are different types of immunoassays (IA) depending on the label (tag/probe) used to mark the immuno reactive components of the assay system. The label may be radioactivity (^{14}C etc) fluorescence (fluoroscein, etc), enzyme (peroxidase, etc); particle (RBCs, latex); luminescence (luminol, etc); and miscellaneous (free radicals, coenzymes, etc). They are differently called as radio-immunoassay (RIA), fluorescence immunoassay (FIA), enzyme immunoassay (EIA) etc.

The immunoassays have many advantages over the other assay procedures because they are relatively simple, sensitive, precise, quick, cheap, applicable to many types of analyte (m.wt > 150) and with scope for high sample throughout. Some of the disadvantages however are cross specificity, requirement of absolute assay conditions and initial cost of assay development; if monoclonals are used and that too in a combination method (for example combining with HPLC), some of these disadvantages may be nullified or minimized.

Immunoassays have proved to be very valuable in many areas like agriculture, food science, medicine, pharmacology, genetics, Epidemology, Veterinary Science, Industries and Forensic Science.

Essentially immunoassay is based on the principle of competitive binding/saturation analysis. Fixed amount of labeled analyte and unknown amount of unlabelled analyte compete with each other to bind to a limited number of sites on the antibody. After incubation (equilibrium reached) the free or unbound analytes are washed off in what is known as a phase separation method. The free and the bound forms of the analyte can be monitored by the distribution of the label in the two phases; references to standard calibration curve gives the assay of the analyte in the test specimen.

For all immuno-assays there are three pre requisites viz, antiserum with desired specificity; immunoactive labeled form of the analyte; and the separation technique.

27.1. EIA (ENZYME IMMUNO ASSAY)

Immuno assays involving enzyme label on an antibody or an antigen and measurement of enzyme activity/inhibition through immunochemical reactions are known as EIA. Advance in enzymology and protein chemistry coupled with requirement for specific and sensitive assay methods have ultimately resulted in the development of EIAs. The important land marks in the history of EIA are:

(*i*) Fluorescein labeling of antibody (Coons, 1942)

(*ii*) Coupling two protein molecules with no disturbance to their biological and chemical reactivity (Singer and Schick, 1960).

(*iii*) Coupling of antibody and peroxidase using glutaraldehyde (enzyme labeling became practical Avrameas, 1969).

(*iv*) Demonstration of use of enzyme instead of isotopes in immunometry (Miles and Hales, 1969).

Enzyme immunoassay has edge over other immunoassays like RIA and FIA in terms of

(*a*) rapidity (few hours vs days)

(*b*) stability of reagents (long shelf life vs short/reasonable shelf life).

(*c*) requirement of less sophisticated equipment (spectrophotometer vs radioactivity counter/UV monitor.

(*d*) requiring less training for personnel vs skilled with licence/skilled.

(*e*) less cost (in USA, two dollors per test vs 7 dollors/5 dollors).

Engvall and Perlmann (1971) were the first to describe an EIA technique (also coined the term ELISA enzyme linked immunosorbent assay) wherein the analyte to be detected binds either to an antigen or an antibody coating a solid surface like polystyrene. Unbind molecules are washed off the solid surface thus allowing measurement of adsorbed enzyme conjugate.

Enzyme labeling can be on the antibodies, antigens or haptens. Obviously ELISA involves phase separation (to separate the free from the bound) by several washing steps and hence is a heterogenous immunoassay. If such separation is not required it is a homogenous assay. Such a homogenous EIA (EMIT, enzyme multiplied immunoassay technique) was developed by Rubenstein (1972); EMIT is now widely used for rapid assay of small molecules (haptens) like hormones, drugs etc, in body fluids. In EMIT (conducted in liquid phase) competition occurs between the analyte and enzyme labeled analyte to bind with the antibody. Generally in EMIT, only the free or unbound enzyme labeled analyte shows enzymic activity; once bound with the antibody the enzyme is inhibited/ inactivated.

This can be taken as a measure of the analyte in the test specimen.

27.2. ELISA VS EMIT

ELISA	EMIT
Heterogenous	Homogenous
Phase separation required	Phase separation not required
Slower	Faster
Better sensitivity	Low sensitivity
Solid phase	Liquid phase
Macromolecules	Haptens

27.3. EMIT

In a typical EMIT, the test specimen (say, serum) containing unknown amount of the analyte is added to an excess of specific antibody. If the analyte molecules are present they bind to the antibody sites. When the enzyme labeled analyte is next added to the mixture, the labeled analyte molecules can bind only if there are free binding sites of the antibody. If the substrate is provided the bound enzyme labeled analyte will not act since upon binding the enzyme is inactivated/inhibited; only the unbound or free enzyme labeled analyte in the liquid can convert the substrate into a colored product that can be measured and compared to standard calibration curves. This means that if more color is observed the test specimen contains more of the analyte.

27.4. EMIT OUTLINE

27.5. ELISA

In ELISA antigen is linked to a solid phase anchored antibody(or, the other way about) in such a way that retains both immunological and enzymatic activity. The solid phase may be of polystyrene/polyvinyl chloride. For antigen assays there are at least two approaches:

(*i*) Competitive method

(*ii*) Double antibody sandwich method.

For antibody determination there is an indirect method.

27.5.1. Competitive (Inhibition) Method For Antigens

The test specimen containing the antigen to be determined is mixed with a precise amount of enzyme labeled antigen; upon addition to the immobilized antibody they compete equally for a limited number of binding sites on an antibody.

ELISA : COMPETITIVE ASSAY FOR ANTIGEN-OUTLINE

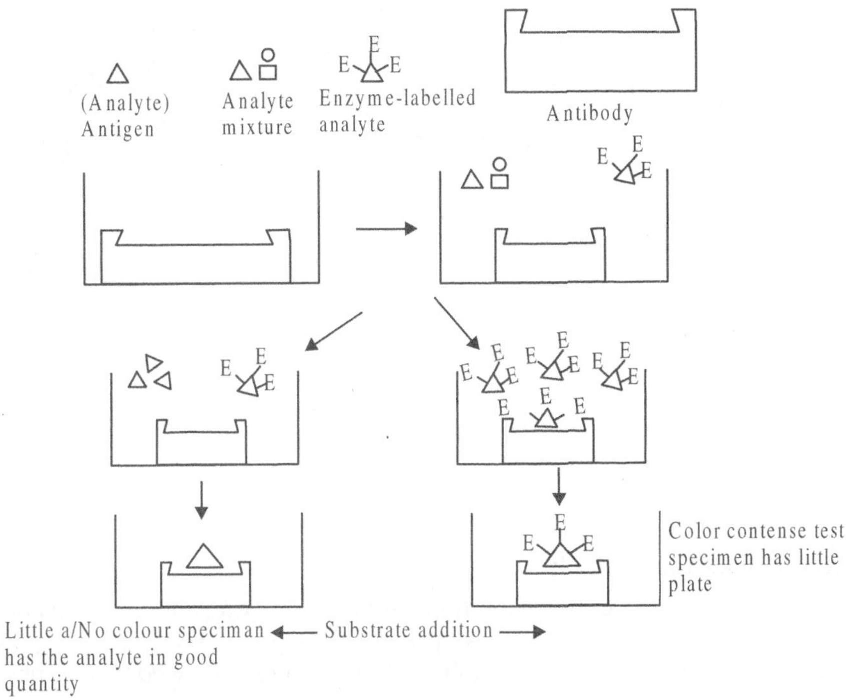

Fig. 27.1

The steps involved are:

1. The specific antibody is bound to PVC/polystyrene.
2. Specimen containing the analyte and the enzyme labelled analyte mixed and added.
3. Immuno complex forms
4. The free enzyme labeled analyte is washed off (Phase separation).
5. Substrate added.
6. Colour change is monitored (more the colour is observed less the analyte in the test specimen).

27.5.2. Double antibody sandwich (Direct) Method for Antigen

The analyte is held (sandwiched) between two different antibodies (both being specific to the analyte but raised in different animal species). Different sets of antigenic determinants of the analyte are involved in binding to the two different antibodies.

ELISA : Sandwich method for antigen outline

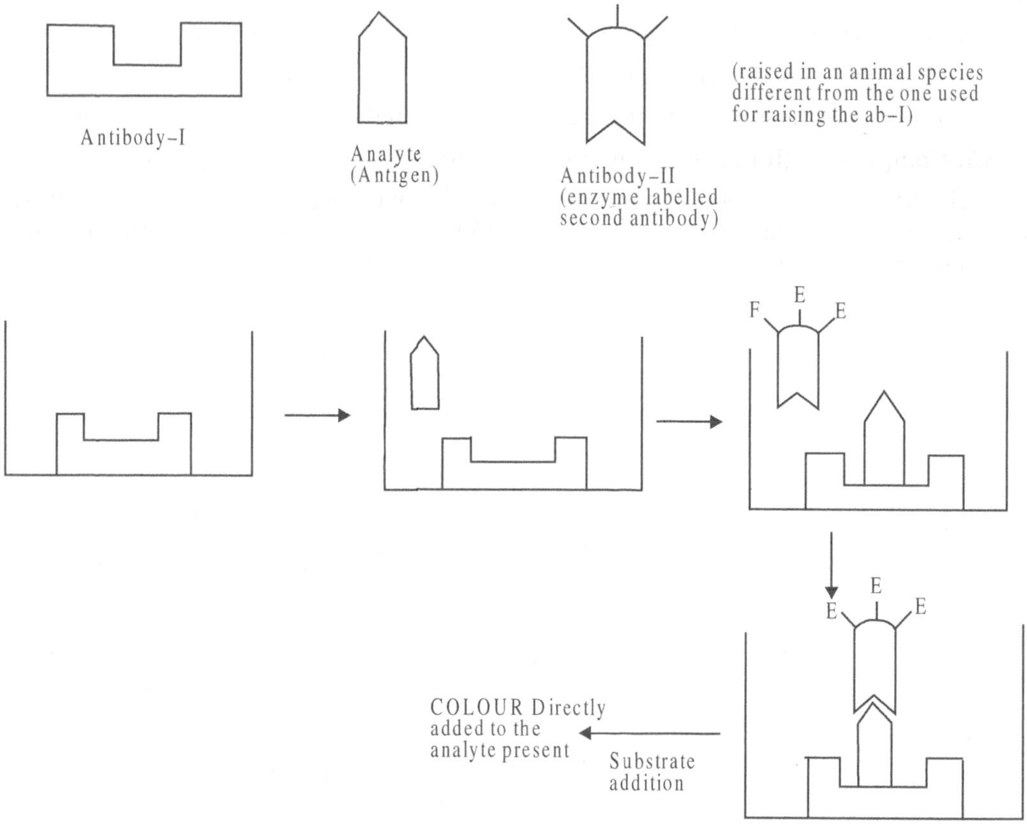

Fig. 27.2

27.5.3. Indirect method for antibody

In a way this is similar to sandwich method. But in this case an antigen is first hooked to polystyrene followed by the addition of test specimen (containing the antibody (globulin) to be determined); the enzyme labeled antiglobulin is added next followed by the substrate.

ALISA : Indirect method for antibody-outline

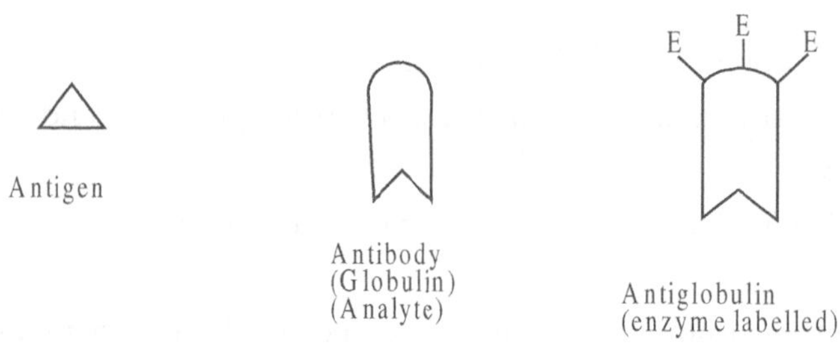

Fig. 27.3(a)

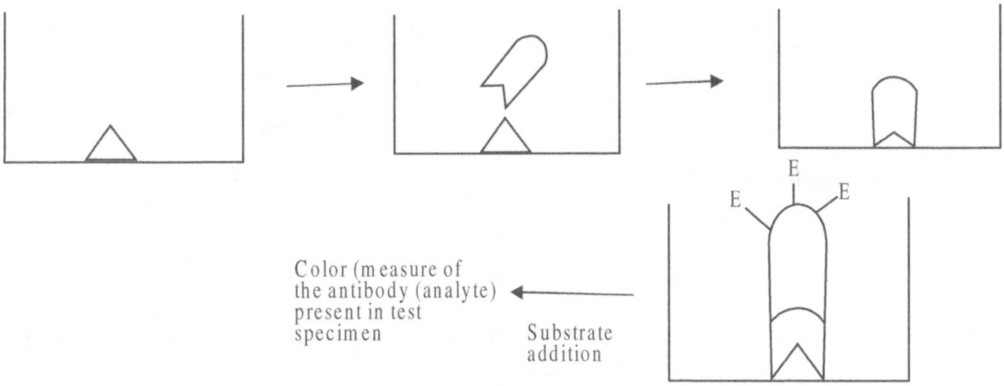

Fig. 27.3(b)

27.5.4. Conjugation

Coupling an enzyme to a protein (antigen/antibody) or to a hapten is a prerequisite for all EIAs and known as enzyme conjugation/enzyme tagging/enzyme labeling. The conjugate should retain both enzymatic and immunological activity. This is achieved by irreversibly cross-linking the E amino group of lysine (of the protein with the enzyme (beta galactosidase, peroxidase, alkaline phosphatase). Subsequent purification of the conjugate is not usually needed because little/no protein is left unbound. The reagents used for conjugation are glutaraldehyde, periodate, 4,4' difluoro 3,3' dinitrophenyl sulfonate, N succinimidy l3 (2. pyridyldithio propionate etc. Recently avidin biotin cross linking is being put to wide use. Conjugation by glutaraldehyde is shown below:

$$EN_2-N{<}^{H}_{H} \; + \; {}^{H}_{H}{>}C-(CH_2)_3-C{<}^{H} \; + \; {}^{H}_{H}{>}N-\text{Protein}$$

$$EN_2-N=CH-(CH_2)_3-CH=N-\text{Protein}$$

(Enzyme – conjugated protein)

Enz NH_2 + $(CH_2)_3$ $(CHO)_2$ + Protein NH_2Enz N=CH $(CH_2)_3$ CH=N Protein Enzyme conjugated Protein

27.6. ABEIA (AVIDIN – Biotin EIA)

The facts that avidin (glycoprotein) has high affinity for the low molecular weight biotin (vitamin H, valeric acid derivative) and that the latter conjugates easily with proteins have been very fruitfully and successfully used for the purpose of EIA. Several biotin molecules can be coupled to a protein/enzyme. Indirect coupling of an enzyme to a protein is hence possible through avidin biotin link. This way many enzyme molecules can be conjugated. Therefore extreme sensitivity is offered by ABEIA.

A — B conjugation – outline

A – Avidin B – Biotin C– Avidin-biotin-enzyme complex

E – Enzyme P – Protein Cantigen antibody)

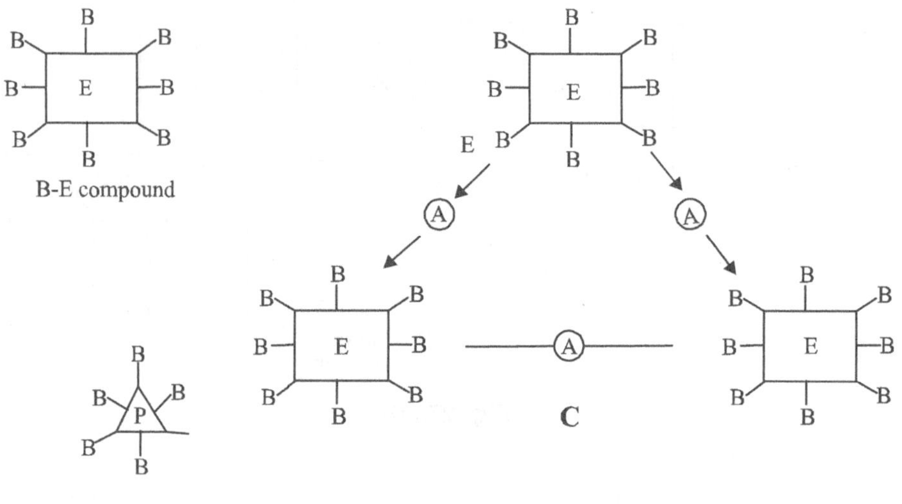

Many C3 + A + B — P compound ⟶ A – B linked enzyme labelled

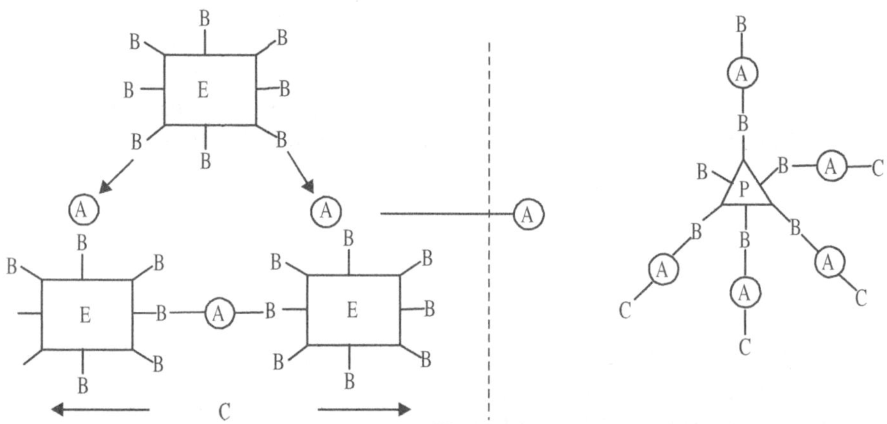

Fig. 27.4

27.6.1. Competitive Direct Elisa for Zearalenone (Mycotoxin) Warner et al, 1986

I. The following were first made ready

1. PBS, 0.1 M sodium phosphate buffer, pH 7.5/0.01 M PBS

2. Antiserum: Zearalenone oxime BSA conjugate was used to immunise rabbits; serum purified and ELISA titre (dilutions in 0.01 M PBS) standardized.

3. Enzyme labeled analyte: Zearalenone oxime – HRP conjugate by N†hydroxysuccinimide procedure (ZHRP).

4. PBS Tween: 0.01 M PBS containing Tween 20(0.02% v/v)

5. PBS Ovalbumin: 1% ovalbumin in 0.01 M PBS

6. Substrate mixture for HRP.

II. 1. ELISA titre first standardized.

2. For competitive assay of the analyte (standards/corn extracts)

(*a*) Micro titre plate wells coated (100 ml/well) with antiserum (1:750) requires incubation time.

(*b*) Wells washed thrice (using 0.3 ml aliquots of PBS – Tween every time).

(*c*) Unoccupied/non specific binding sites in the wells blocked by incubating the wells for 30 minutes at 37°C with PBS ovalbumin (0.3ml/well).

(*d*) Wells washed four times with PBS Tween.

(*e*) 100 ml of Z (standard/corn extract) and 100 ml of Z ERP (1:300 in PBS ovalbumin) mixed (1:1 of Z and Z HRP).

(*f*) 100 ml of the mixture then added to each of the antibody coated wells.

(*g*) Wells incubated for one hour at 37°C.

(*h*) Washed 6 times with PBS Tween.

(*i*) Substrate addition followed by ELISA Reader measurement and reference to standard calibration curves. More the colour observed means less of Z in the test sample.

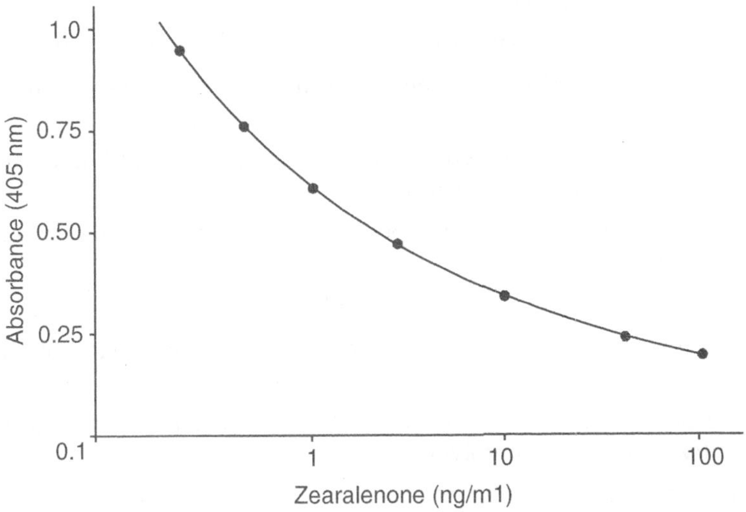

Fig. 27.5

27.7. IMMUNOBLOT

In the search for better analytical methods 'Biotechnology' has come up with many useful techniques. 'Blotting' is one such based on the principle that electrophoretically gel resolved macromolecules can be transferred to and hooked on certain immobilizing matrices. The blotting technique obviates many problems/handicaps arising when one stops with gel electrophoresis only. Essentially the molecules first resolved by electrophoresis and then transferred, can be easily handled for many and successive analysis in terms of qualitative as well as semi quantitative detections. When an immunochemical reagent mixture is used for the purpose then one may call the whole exercise as immunoblot.

(*i*) Blotting is a process of transferring macromolecules from gels to an immobilizing matrix.

(*ii*) Quenching is the process of blocking the unbound sites of a blot with a nonspecific non reactive molecule so as to eliminate background.

(*iii*) Over lay means incubating a blot or gel with a ligand (detection mixture).

The first report of blotting technique was by Southern (1975) who analyzed DNA fragments separated by gel electrophoresis; the method was hence forth called as 'Southern blot'. Subsequent application to RNA and protein analyses resulted in the names 'Northern blot' and 'Western blot' respectively; people have also reported 'Eastern blot', 'Mid eastern blot' etc. However such geographical jargons seem to bring in political controversies and hence slowly being replaced by terms like DNA blotting, RNA blotting and protein blotting.

27.7.1. Blotting comprises of two steps

(*i*) Elution of the macromolecules from the electrophoretic gel

(*ii*) Adsorption of the eluted material to an immobilizing matrix.

The first step can be achieved either by capillary/passive diffusion (with or without negative pressure) or by electrophoretic transfer (electro elution/transpher/ electroblot)

The first method is generally preferred for DNA and the second for protein. A modified but rapid version of blotting known as dot blot also uses the passive diffusion but under vaccum. Choice of passive diffusion for DNA rests on the fact that if otherwise one tries to adopt electroelution, the ionic strength of the transfer buffer required to bind low molecular DNA on to the matrix is prohibitively enormous. Since this is not the case for RNA and proteins, the electroelution procedure is adopted for them.

Essentially the following steps are involved in any blotting experiment:

(*i*) Gel is simply layered to the membrane (immobilizing matrix) in a gel holding cassette.

(*ii*) Elution is achieved by electroelution wherein the gel holding cassette is placed between two electrodes in a tank of transfer buffer;

(*iii*) Voltage gradient is then applied perpendicular to the gel (about 1/2 – 4 h, about 40 volts; about 2.5 A);

(*iv*) Samples migrate off the gel on to the membrane and

(*v*) The membrane (blot) is now ready for staining, autoradiography; hybridization or immunodetection.

In case of passive elution of DNA from the gel, the gel membrane arrangement is treated differently without any involvement of current.

Taking the example of a protein mixture +the various proteins may be first resolved in agarose or polyacrylamide gel electrophoresis, electro transferred; then the blot is ready. The blot membrane will now have strongly bound protein zones as well as lot of unoccupied sites. These unoccupied sites should be blocked (quenched) an interfering background. For quenching one may use gelatin, BSA, HSA, hemoglobin, ovalbumin or animal sera which of course should not mimic the other proteins in the reactions. After quenching, the blot should be washed for removing the excess of the quencher. Then the blot can be over layered with suitable reaction mixture; in case this visualizing process involves the addition of more than one reagent (ligand) in succession then each step should have a washing.

The blot technique can be employed for macromolecules separated on agarose or polyacrylamide gels (native/SDS/lithium do-decyl sulphate/urea) electrophoresed or electrofocussed in non gradient or gradient tube/slab method. The immobilizing matrix (membrane/filter) may be any one of the following:

1. Nitrocellulose paper (NCP)
2. Diazobenzyloxymethyl modified cellulose paper (DBM)
3. Diazophenylthioether (DPT) paper
4. CNBr activated paper

5. CNBr activated cellulose acetate

6. Nylon membrane

(*a*) Gene screen

(*b*) Zetabind/Zetaprobe ZB

The binding mechanism between proteins and the different membranes as above can be different. In NCP, it is probably the hydrophobic effect. With DBM and DPT there are two steps: the first is electrostatic interaction of the negatively charged protein ions with the positively charged diazonium groups of the paper, followed by slow and second step of covalent linkage via azo derivatives; CNBr paper/CA appear to show similar interactions. In the case of nylon membranes there is electrostatic interaction between the filter and the polyanions (proteins).

The blot may be treated with many different overlays such as immuno overlay, lectin overlay, calmodulin overlay etc. Immuno overlay is used for immunoblots may be fluorescence immuno overlay, enzyme linked immuno overlay etc. For an immuno blot the following sequential steps are adopted;

1. Electroelution
2. Adsorption to immobilizing matrix
3. Quenching
4. Washing
5. Immuno overlay
6. Washing after each step
7. Detection (visualization)

The versatile applications of the blot technique include:

I. (*i*) Hybridization

(*ii*) Autoradiography

(*iii*) FIA/EIA (Immunoblot)

II. (*i*) Protein ligand association

(*ii*) Identification of enzyme subunits

(*iii*) Affinity purification of mono specific antibodies (poor man's monoclonals).

(*iv*) Cell Protein interaction

There are definite advantages with the blot technique over the mere use of electrophoretic and electrofocussing methods. For example, after electrophoresis when the gel is processed as such, it may break due to accidents in handling or the macromolecules buried in the gel matrix may not be easily accessible for reactions with overlay ligand or the ligand may not be able to reach through the small pores. In all the advantages of the blot technique may be summed up:

(*i*) No handling accident as filters are pliable.

(*ii*) Immobilized proteins are readily and easily possible to reaction by overlay ligands.

(*iii*) Requirement of small amounts of reagents only

(*iv*) Reduced processing time

(*v*) Multiple replicas possible

(*vi*) Storage for months possible

(*vii*) Multiple successive analyses possible (with/without erasure)

A typical immunoblot exercise is given below (B.M. Hoste) as done for the serum protein haptoglobin (Hp) groups. The optimal dilution of each antibody required for the following procedure was earlier determined by dot blot.

(*i*) Human serum was electrophoresed in 30% gradient polyacrylamide vertical slab gel.

(*ii*) Stained for the detection of the resolved haptoglobin types

(*iii*) The stained gel was used for blotting the proteins on to HCP by electrotransfer using 20 mM Tris, 150 mM glycine buffer; pH 8.3; overnight at 11 volts.cm.

(*iv*) Free binding sites on HCP blocked by incubation for one hour at 37°C in 10 mM Tris buffered saline (pH 7.4) containing 0.6% gelatine (TBSG).

(*v*) After four washing each of 10 minutes duration in TBSG with 0.1% Tween (TBSG) the blot was incubated for atleast one hour with goat anti human Hp antiserum diluted 1:100 (or 1:400, if subsequently alkaline phosphatase conjugate is to be used).

(*vi*) After four washings as above the blot was incubated with rabbit antigoat Ig peroxidase conjugate (1:100).

(*vii*) After four washings as above the blot was treated with a solution containing 4-chloro -1-naphthol for visualizing peroxidase activity.

For step (*vi*) above instead of (1:100) peroxidase conjugate, (1:400) alkaline phosphatase conjugate was also tried; in this case at step (*vii*) alkaline phosphatase was visualized using 5 bromo 4 chloro 3 indolylphosphate and nitroblue tetrazolium.

Further sensitivity was also tried in the case of peroxidase conjugate by introducing an additional step in between steps (*vi*) and (*vii*) when peroxidase conjugated swine anti rabbit Ig antiserum (1:100) was added.

Using the described EIA based blot (immunoblot) technique even dilute sera and old bloodstains could be successfully Hp typed. These were otherwise when the gel was stained for Hp as done above prior to blotting.

27.8. CHARACTERIZATION OF ANTIGENS BY WESTERN BLOT PAMELA LAME MORIEARTY

27.8.1. Introduction

The purpose of the Western blot technique is to demonstrate recognition by antibodies of the individual components of a complex mixture of macromolecular antigens. This demonstration is effected in three steps:

1. Components of an antigenic mixture (Ag's) are separated by polyacrylamide gel electrophoresis (PAGE) in slab gels.

2.The separated Ag's are transferred electrophoreticaly, while maintaining their special relationships, to a sheet of nitrocellulose paper (NCP).

3. The NCP is treated with appropriate antibody solution or antiserum (Abs) and the presence of the bound Abs is detected by an enzymatic or autoradiographic Ab probe.

The technique useful for comparing reactivity of several sera with a single antigenic preparation. (eg., Hanff et al, 1982) or of a single Ab preparation with several Ag sources (e.g. Pekkala Flagan and Ruoslahti, 1932).Though in some respects similar to immunoprecipitation (see the chapter by Zingales and Colli, this volume), the western blot requires neither protein A (which may bind some Ag's and does not react with some Abs, e.g. Ig H), nor radioisotopes, both of which may be difficult to acquire in some laboratories.

The technique is adaptable to local conditions without great expense, though appropriate equipment is commercially available. Several modifications, adaptations and variations of the method have appeared in the literature, particularly in analytical Biochemistry and Journal of Immunological methods, since the original paper of Towbin et al. (1979).

27.8.2. Materials and Methods

27.8.2.1. Step 1 SDS PAGE

Slab gel SDS PAGE as carried out by standard techniques is described by Zingales. Details relevant to the blot technique are considered here.

27.8.2.2. Equipment and Materials

Combs and spacers : The comb normally used in mold small slots into the gel for sample application may not be suitable for pre blot runs if the NCP is to be cut into many strips. It is very difficult to align the cuts with the (invisible) blotted lanes. Combs with wide slots and corresponding spacers can be prepared from acrylic, teflon film, or other inert material of uniform thickness.

Acrylamide : Since unpolymerized acrylamide is extremely toxic on contact and by inhalation it should be manipulated as little as possible and with adequate precautions. However, if difficulty in polymerization, separation or transfer are encountered and may be due to inferior quality of locally available acrylamide, the substance can be recrystallized without great difficulty.

Buffers. In most studies, Laemmli (1970) buffers have been used for separation and the Laemmli reservoir buffer (25 mM Tris, 192 mM glycine) without SDS used for the transfer buffer in step 2. However, if tris and /or glycine are scarce, alternative can be considered; Neville (1971) described SDS PAGE buffers of tris borate, and Batteiger et al, (1982) used 50 mM sodium phosphate as a transfer buffer.

Gel concentration : Gel concentration should be determined by the molecular weights of the components to be separated, but blotting imposes some constraints. Gels of very low concentration are hard to manipulate, while gels. We have successfully blotted from a 15% gradient gel, but have noted incomplete transfer of > 90,000 mw bands from routienly used 10% gels.

Sample application. The amount of sample to apply in large slots can be determined as a multiple of that giving satisfactory results in normal electrophoresis runs with small slots. As an example, for a 12 cm long slot in a 1 mm thick gel, 4.0 mg of lyophilized whole L. mexicana promastigotes were resuspended in 800 ul sample buffer and the entire sample applied to the slot. Much smaller amounts of purified proteins would be required, e.g., Bordier et al., (1982) transferred 16 μg of purified L. tropica pellicular membrane fraction from a small slot gel.

27.8.3. Step 2 Blot

27.8.3.1. Equipment and materials

Electrodes : A commercial blotting cell or electrophoresis gel destainer can be used. Bitner et al, (1980) describe a plexi glass cassette using more than one meter of (expensive)12 mil platinum wire. We use stainless steel plates (150 × 160 × 0.5 mm) with a lead wire riveted to a 1× 1 cm tab extending from the upper corner of each plate.

Power supply : This must be capable of delivering up to 500 mA current, beyond the capacity of most electrophoresis units. A wiring diagram for a satisfactory power supply designed and built at the Fundacao Oswaldo Cruz for the equivalent of about US $50 is shown in appendix ii. The use of a 12v car battery charger has also been suggested. Commercial units are available.

NCP. This is the membrane filter material used for bacterial sterilization. Schleicher and Schuell Babs, Sartorius and Millipore Hawp, all with 0.45 mm pore size, have been used. In our experience, transfer of L. meicana proteins was slightly superior to S and S Babs and Millipore GSWP (0.22 mm pore size) than to Millipore HAWP. NCP can be purchased in sheet form in some localities, but 14 cm diameter circular filters can be trimmed for blots 10 X 11 cm if necessary.

Scotch brite pads " Scotch bri te (311 Corp.) is a brand name for a tough loose mesh synthetic fiber cleaning sponge.We found an equivalent in a commercial carpet cleaning firm. Two pads, 150 X 160 X 5 mm, are needed.

Filter paper : Whatman 3 mm, Whatman chromatography paper, or similar in sheets slightly larger than the gel and NCP.

Tank : A plastic vessel with lid to contain the assembled blot apparatus and transfer buffer with room for a stirring bar below the apparatus. It should fit on a magnetic stirring base, which is also needed.

27.8.3.2.Reagents

1. Transfer buffer (TB)
2. 5 mM Tris base
3. 192 mM glycine
4. 20% v/v methanol (reagent grade)

(for alternatives to transfer buffers, quench, etc., containing tris, glycine or Tween 20 are used).

27.8.3.3. Procedure

Just before the separation run is over, tris NCP fuse very clean rubber gloves to fit gel, with an excess at the ends for labeling orient NCP with shiny side down and float on distilled water; when wet, submerge for 5 minute. The dull or matte face of millpore NCP should be up in all wash steps, to protect from abrasion, since it will receive the blotted antigens. Handle NCP by the excess areas. Transfer NCP to TB for 10 minutes.

When run is over, turn off electricity, remove gel assembly and remove spacers and top plate. If desired, cut off a lane for Coomassie blue R 250 (C.I. 42660) staining. Remove the stack gel.

In a tray full of TB, place the blot cathode (or support) (negative, black wire) and center a Scotch brite pad on it. Avoid entrapment of air bubbles in all subsequent steps. Notch the gel for orientation, wet a filter paper square with TB, and carefully transfer the gel to the filter paper, avoiding distortion. Place the filter paper gel on the pad and cover with the NCP, dull side in contact with the gel. Mark the NCP overhang with pencil for orientation. Add a second wet filter paper, pad, and the anode plate (positive, red wire). Recheck the order of layers should be anode pad paper NCP gel paper pad cathode.

Bind the sandwich with heavy rubber bands. Pour the TB from the tray into the tank, top up with more TB, and insert the sandwich, squeezing to remove entrapped air. Check that the buffer does not reach the electrode rivets, add a magnetic stir bar, cover the unit. Place on a magnetic stirrer in a refrigerator, with leads connected to the power supply outside.

For convenience, blotting can run overnight. Since resistance can increase or decrease as the blot proceeds, a setting well within the power supply's limits is necessary. We run at 10 V (120–150 mA) overnight then increase to 20V (350 mA) for an hour before shutting off.

27.8.3.4. Step 3 Ab Probe Treatment

27.8.3.4.1. Equipment and Materials

Plastic bag sealer and Bags. Though not essential, a food bag sealer is a good investment, since volumes of serum and probe solutions are much reduced.

Rocker or rotator platform (+ or 20 rpm).

Plastic containers for wash steps, etc.

Conjugate Commercial antiserum conjugated to peroxidase or 125I.Methods for producing antisera in the lab are given in Appendix III.Working concentrations are peroxidase conjugate 1:1000 in PBS, 125I conjugate IOE6 CPM/ml. For alkaline phosphatase procedures see O' Connor and Ashman, 1982, for 125I protein A probes (and a technique for re cycling blots) see Legocki and Verma, 1981.

27.8.3.4.2. Reagents

1. Quench, PBS, 0.3% v/v Tween 20
2. Wash buffer (WB), PBS, 0.05% v/v Tween 20.
3. Amido Black stain

0.1% w/v Amido black (naphthol blue black, C.I.20470), 45% v/v methanol, 19% v/v acetic acid. Destain 90% v/v methanol, 2% acetic acid.

4. DAB substrate

Citrate phosphate buffer, 0.05 M, pH 5, 5 mg/ml diaminobenzidine (DAG) at the moment of use, add 20 ml of 3% H_20_2 per 60 ml DA3 buffer. 0.6 ml of 1% $CoCl_2$ salution per 60 ml DAS buffer (optional; gives darker chocolate brown lines, HSU and Soban, 1982).

Note: DAB is light sensitive and carcinogenic. Handle in subdued light, using gloves; avoid breathing. Discard used substrate into an equal volume of hypochlorite solution (Clorox; agua sanitaria) to neutralize.

27.8.3.4.3. Procedure

When blot is ready, turn off electricity, remove the sandwich, anode side up. Remove top layers, and place NPC (dull side up) in quench one hour, rm T'C, with agitation. Stain gel with Coomassie blue to assess transfer. After quench, wash NPC 2 × 30 minutes in WB.

Prepare serum dilutions and controls in WB. Serum more concentrated than 1:50 may give background.

If desired, all or part of the NPC can be stored for future use (at least 5 days). Incubate 30 minutes in PBS 0.01% azide, then seal in a plastic bag and store at 4°C. Remove all traces of azide by washing before using the NPC with peroxidase.

With a scalpel, cut NPC in strips for Ab incubation. Remove a strip for Amido black total protein stain. Identify each remaining strip with pencil and incubate in serum overnight at 4° C, in closed tubes or plastic bags. For a bag containing a strip 8 X 120 mm, 2.5 ml is sufficient.

After Ab incubation, wash strips in individual containers (e.g., screw cap culture tubes) 4 X 20 min in WB with agitation.

Incubate strips 2 hour in probe; if a plastic bag is used, 25 ml of solution is sufficient for a 10 X 11 cm blot. Wash 4 X 20 min in WB.

Prepare DAB buffer solution. In reduced light, lay the strips in a shallow dish. Add H_20_2 (and $CaCl_2$ if desired) to DAB buffer, mix and pour over strips. Incubate up to 30 minutes, then stop reaction by transferring strips to water. Wash in water, blot with facial tissues and air dry before storing.

27.8.4. Appendices

27.8.4.1. Recrystallization of Acrylamide

Acrylamide before polymerization is highly toxic, easily absorbed through the skin and lungs. Chloroform is toxic, absorbed through the lungs. This technique must be carried out in a fume head with exhaust fans.

27.8.4.2. Procedure

Dissolve 50 g of acrylamide in 500 ml chloroform at 50°C. Heat a Buchner funnel to 50°C (note: the boiling point of chloroform is 61-62°C). Filter the hot solution through Buchner funnel, using filter paper. Do not use negative pressure.

Let the filtered solution reach room temperature and place at 20°C overnight to form crystals. Also place at 20°C a Buchner funnel and 250 ml of chloroform.

The next day, collect the crystals obtained over filter paper in the prechilled Buchner funnel, and wash with the chilled chloroform.

Leave the crystals in the funnel for about 30 minutes then spread in petri dishes. Dry overnight in a dessicator. Do not dry in an oven or expose to direct light. The crystals should be stored in the refrigerator in a dark bottle. Weigh the bottle before and after adding acrylamide to determine the yield.

27.8.4.3. Components

C1 2500 mF 50 V
C2 0.1 mF 50 V
C3 1 mF 50 V
R1 3 3 ohm 1W + or 10%
R2 240 ohm 1/2 W + or 5%
P1 5K linear
CI LM 317 T
DI DB 1N4007

27.8.4.4. Salt Fractionation of serum

Five ml of serum should give enough Ig – for immunization of two rabbits and for antiserum monitoring by double diffusion; 15 20 ml give enough to prepare an affinity column. Determine protein by the Lowry method or by UV280. (00280 1.9 X 00330

Ig (mg/ml) = 1.4

Since the Bradford method gives falsely low readings for Ig.

27.8.4.5. Immunization

Immunize each rabbit subcutaneously with 500 mg Ig in 2 ml PBS, emulsified with an equal volume of incomplete Freund's adjuvant (IFA). Boost after one month with a similar dose, and monitor serum by bleeding from the ear vein starting 2 weeks after the boost. IFA has given us satisfactory rabbit anti mouse Ig and eliminates problems of abcess formation or cross reactivity due to mycobacterial antigen.

27.8.6. Affinity Purification

15-20 mg Ig bound to 7 ml of CNBr activated sepharose 4B gives a column for processing 10 ml aliquots. If eluting at pH 2.6, add a pinch of dry tris to each collection tube to neutralize pH rapidly. Dialyze, concentrate positive fractions by per vapouration, if necessary. Iodinate Ig with Iodogen.

27.9. DETECTING SPECIFIC Ab IN PATIENT'S SERUM BY ELISA

(*i*) Ag is fixed to a solid surface and then patient serum is added.

(*ii*) If specific Ab is present, it will bind to Ag and be fixed to container. Non specific Ab will fail to bind. Serum and all unbound Ab's are washed away before the next step.

(*iii*) Enzyme conjugate is added. The conjugate is prepared by chemically linking the indicator enzymes (peroxidase) with anti human Ig G antibody.

(*iv*) The anti Ig G of the enzyme conjugate will bind to the Ig G of the fixed antigen Ab complex. This is then fixes the peroxidase to the container. On bound enzymes conjugate is washed away before the next step.

(*v*) The enzyme's substrate, H_2 O_2 and colourless indicator chemicals are added.

(*vi*) Peroxidase converts H_2 O_2 to H_2O and O_2. In the presence of O_2 the indicator chain in colour indicating a positive reaction. Negative sera contain no specific Ab and consequently fails to provide immunoglobulin bridges that link the enzyme conjugate to the fixed Ag. All peroxidase is therefore removed during the second wash. (after step IV) and no O_2 is generated.

Negative sera therefore shows no colour.

The ELISA method has been developed for Ag detection in culture medium, serum plasma and other tissue fluids HIV Ag. ELISA may detect special HIV concentration Ag. With this technique micro well are coated with monoclonal Ab to peak 24 concentration protein. The specimen is added to the well. Incubated and washed. A probe Ab (biotin conjugate) is then added to the wells. The wells are incubated and washed and streptavidine conjugated horse radish peroxidase is added and incubated and then washed. After washing a substrate solution such as hetero methyl benzidine solution is added. If a colour is developed, it can be measured spetrophotometrically. This research procedure provides a quantitative estimation of the presence of HIV Ag.

This method is considered to be more sensitive than the reverse transcriptase assay and is routinely used in many laboratories to detect retro virus growth in culture medium.

Elisa is very specific for HIV-1

Recombinant envelope glycoprotein coated on mucotitre plate + 0.2 ml of blood serum contains Abs are added

↓

200ml of Ab is added (goat Ab against human Ig conjugated with horse radish peroxidase

↓

Add 200 ml of urea peroxidase to each well. Incubate at room temperature for half an hour.

↓

Ortho aniline diamine in urea peroxidase develop orange color
Then add 50 ml of 4N H_2SO_4 in order to stop the reaction

↓

Read at 520 nm

27.10. WESTERN BLOT ANALYSIS

1. The individual proteins of the HIV virus are separated (according to their over all change) by electrophoresis through a gel material.

2. The protein are transferred (blotted) from gel to the nitro cellulose filter.

3. The patient serum is added to pitter and any Ab to HIV binds to specific protein. If the patient is not affected such HIV, there will be no Ab that can bind to the used protein in filter.

1.Enzyme labeled Ab against human Ig is added. They will bind any anti HIV Ab that attached to the filter after beating with viral protein. All unbound Ab's and proteins are removed rinsing the filter.

The bound complex is revealed by adding colorless substrate, the antibody binding enzymes converts it to colored product. Each lack begin represents as HIV specific Ab in specific anti serum.

27.11. PCR (POLYMERASE CHAIN REACTION)

- The gene sequence is denatured by heat.
- Chemically synthesized primers are added. The temperature is lowered and they bound to opposite strand.
- DNA pool extends the primer to form double stranded structure.
- The progeny molecules are denatured and the process is repeated.
- With each round of replication twice the number of template is available. The process is repeated until ampule supply of the desired gene is produced.

QUESTIONS

1. Define enzyme immuno assay.
2. Differentiate between enzyme linked immunosorbent assay (ELISA) and enzyme multiplied immuno assay technique (EMIT).
3. How EMIT is performed ?
4. Explain the steps involved for competitive method for antigens.
5. Explain ABEIA (Abidine-biotin enzyme immuno assay).
6. Explain immuno blot.
7. What are the two steps involved in blotting technique ?
8. Explain the characterization of antigen by western blot.
9. Write the following procedures:
   (*i*) recrystallization of acrylamide;
   (*ii*) Salt fractionation of serum;
   (*iii*) Immunization;
   (*iv*) Affinity of purification;
10. Explain the method for detecting specific antibodies in patients serum by ELISA.

CHAPTER 28

NEW VISTAS FOR MONOCLONAL ANTIBODIES

Now that monoclonal antibodies can be made to order, several research teams have been working on 'guided missile' therapy, a highly selective way of directing anti cancer drugs to tumours. But there has been a rejection problem, when the body detects the antibodies, made by mice or rats, as foreign to itself. Hybrid antibodies, created at the laboratory of Molecular Biology in Cambridge and the Department of Pathology at Cambridge University, are disguised by protein engineering to appear as human material. They are now under commercial development for the treatment of one form of leukemia and various other conditions such as septic shock and coronary heart disease.

Antibodies are an important part of our natural defenses against disecse.The human immune system has the ability to make millions of different antibodies, each tend to react with great specificity with just one antigenic region of one foreign protein entering the body on the surface of an infective organism. The organism may be a virus, a bacterium or a parasite, and is the antigen. Its antigenic region is the one that provokes an immune reaction.

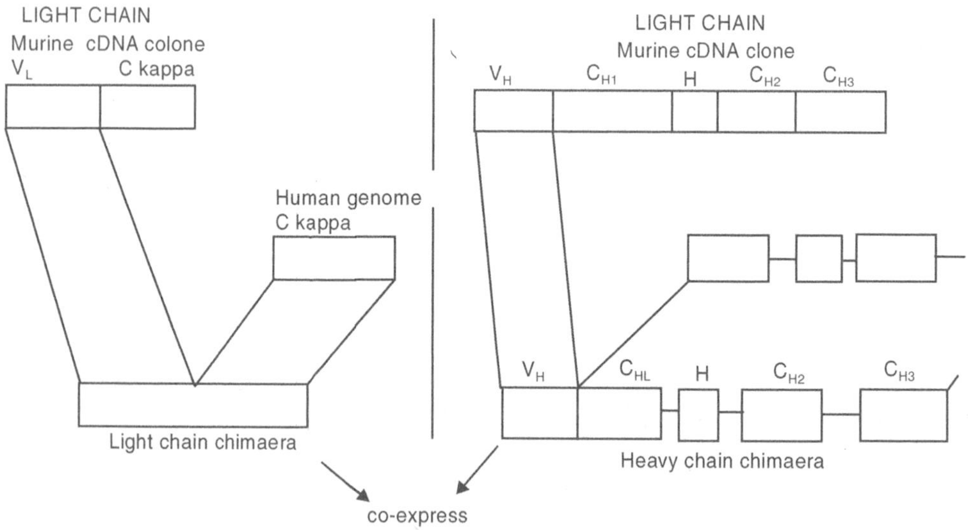

V = variable	H = Hinge region	CH = Constant region at heavy chain
L = Light chain	VH = Variable region	CH 1, 2 and 3 are domains of the
C = Constant	of heavy chain	the heavy chain

Fig. 28.1. Fist stage in the humanisation of monoclonal antiboides. Muring and elements are replaced by human genomic constants. The process links the mouse and coding for the antigen-specific variable region(V) to the human and coding for the corresponding constant region, so producing a humanised antiobdoy

When the antibodies attach themselves to bacteria or parasites, they enable the invaders to be further recognized by the immune system and attacked in other ways. As well as being produced against infective organisms, antibodies are also believed to be made naturally against cancer cells.

In 1975 work by Dr. Cesar Milstein and Dr. George Kohler in the UK Medical Research Council's Laboratory of Molecular Biology at Cambridge, led to the development of monoclonal antibodies. These are antibodies made outside the body by hybrid cell cultures, known as hybridomas. Such cultures can be made to produce any required antibody for as long as it is needed, using a technique for which Milstein and Kohler won a Nobel Prize. It involves injecting the antigen against which an antibody is required into a mouse or rat. Later, the spleen cells which are producing antibody against the antigen are isolated and fused with fast growing cells in culture. The resulting hybridomas, properly tended, will go on producing the wanted antibody.

28.1. GUIDANCE SYSTEM

The achievement opened up the way to using antibodies in therapy. Monoclonal antibodies could now be made to order, to be used to treat various infective diseases and for other purposes. Anti cancer antibodies on their own do not have a great deal of value, but now several research teams have developed 'guided missile' therapy. In this, an anti cancer antibody specific for an antigen cancer cells or on very few other classes of cells is attached chemically to a very potent toxin such as ricin, or to a radioisotope that kills cells at short range. The antibody then serves as a guidance system, taking the toxin selectively to cancer cells and avoiding or at least reducing, harm to healthy tissue.

There are many other areas in which monoclonal antibodies may play an important part in therapy. They include the treatment of shock; preventing rejection of transplanted organs, and treating cardiovascular and auto immune diseases. But their application and even testing them effectively for such purposes has so far been severely restricted by the fact that monoclonal antibodies are made by cells derived, at least partly, from mouse or rat tissue. For that reason the antibodies produced by hybridomas have certain chemical groupings, sequences of amino acids, that are characteristic of mice or rats instead of humans. They are therefore recognized as foreign by the immune system of a patient into whom they are injected. This means that antibodies used in treatment can be used only for a short time, a single application.Once the immune system recognizes them as foreign, it immediately attacks them if they are injected a second time.

28.2. AVOIDING DETECTION

The team led by Dr Greg Winter at the MRC's Cambridge laboratory have made a decisive breakthrough in overcoming the rejection problem, effectively opening the way to widespread use of monoclonal antibodies in therapy for many conditions. What they have done is to use genetic engineering techniques to surround the reactive parts of monoclonal antibodies made in mice or rats with human antibody material. Made in this way, hybrid antibodies are much less likely to be recognized as foreign, in fact, tests suggest they may avoid detection altogether.

Hybrid antibodies have been created by Dr. Winter and Dr Lutz Reichman, also at the Cambridge Laboratory of Molecular Biology and by Dr Mike Clark and Dr.Herman Waldman in the Department of Pathology at Cambridge University. The MRC, which financed the research, has patented the technique and is now licensing it widely for exploitation by medical biotechnology companies.

Antibodies are large, Y shaped protein molecules produced by white blood cells, the lymphocytes. They attach themselves to the invading toxin, virus, bacterium of parasite and mark it down for destruction. The tips of each of the two arms of the Y bind to the surface of the invader, to the specific antigenic sites that are recognized as foreign. The tail of the Y is responsible for triggering the destruction. Each antibody's arms are slightly different from those of other antibodies, especially at the tips where the antibody binds to a specific antigen.

28.3. PROTEIN ENGINEERING

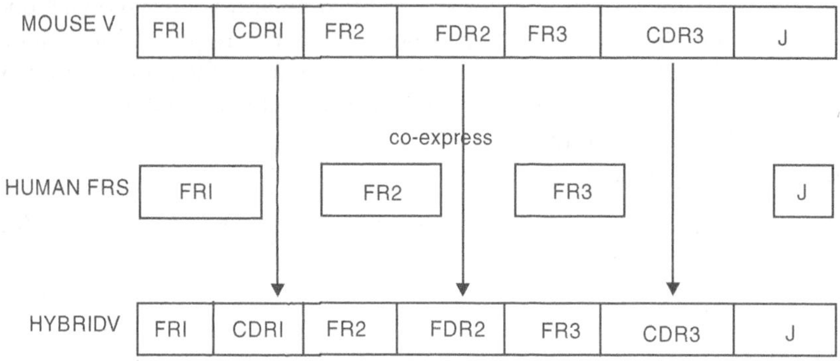

V = Variable Region CDR = Compiementarity Determining region
J = Junctional Region FR = Frame Work Region

Fig. 28.2

The instructions that tell a cell to make a protein, such as an antibody, reside in the DNA sequence of the gene for the antibody. By altering the DNA, inserting new or different sequences of its sub units, called bases, the protein can be redesigned. This technique is known as protein engineering. Dr. Michael Neuberger, a colleague of Dr. Gret Winter at the Laboratory of Molecular Biology decided to see if it might be possible to overcome the rejection problem by disguising a mouse antibody as a human one, using protein engineering.

He used parts of the genes for mouse and human antibodies, and succeeded in making an antibody in which the parts of genes representing the arms of a mouse antibody were linked to the part of a gene representing the tail of a human antibody. The outcome was an antibody in which the arms were mouse type and the tail human type.

Gret Winter took this approach further. He has made an antibody in which only the DNA for the tips of the arms, the small regions that react with an antigen, are taken from a mouse monoclonal antibody. The rest of the antibody is human. After using the technique developed by Kohler and Milstein to make a mouse antibody against a specific antigen, the small pieces of the gene for the antibody representing the reactive tips of the arms were transplanted into the gene for a human antibody. There, they replaced the equivalent DNA sequences in the human gene.

It was hoped that such an antibody would be effective in reacting with an antigen but would not be recognized as foreign by the human body. To test the clinical potential of the antibody, Dr. Winter worked with Dr. Herman Waldmann of the Cambridge Department of Pathology who was developing an antibody known as Campath 1, made by a British pharmaceutical company, the Welcome Foundation. Campath 1 reacts with and neutralises a particular class of thymus derived lymphocytes, known as T cells. These cells proliferate malignantly in one form of leukemia. They are also largely responsible for the rejection of grafted organs.

The British national biotechnology company Cell tech, with which the Medical Research council has a special relationship, has taken out exclusive licenses to use the technique for various purposes.One is the treatment of septic shock, caused by severe bacterial infection in patients who have undergone surgery.There is no cure and over 125,000 cases die throughout the world every year.Cell tech is developing hybrid monoclonal antibodies to neutralize the toxins released by the bacteria responsible, and furthermore to act against cytokines, which are cell stimulating substances released in response to toxins; they cause fatal loss of blood pressure in septic shock.

Cell tech is also developing a hybrid antibody which will be combined with an enzyme that dissolves blood clots, so as to direct the enzyme more effectively to the clots. This could help to treat and prevent coronary heart disease. The company is also investigating the use of hybrid monoclonals to treat auto immune diseases in which the immune system damages the tissues such as the linings of joints by destroying the cells responsible for auto immune reactions. The potential for hybrid antibodies is vast indeed.

28.4. MONOCLONAL ANTIBODIES

28.4.1. Introduction

Antibodies,protein molecules produced by certain cells in the body, are a basic constituent of animal and human disease fighting immuno system. When the immune system detects a foreign substance or antigen in the body, it stimulates cells that make a diverse number of antibodies against the immune antigen. The antibody producing cells recognize the physiochemical characteristics (electric charge, pattern, or shape) of the particular antigen and produce an antibody that binds specifically to the antigen, thus effectively neutralizing and destroying it.

The standard method of obtaining antibodies to protect against a specific disease antigen is to inject an animal (or human) with the antigen for which an immune response is desired. The immune system then responds by producing a variety of antibodies, each specific to a different part of the injected antigen molecules. Blood serum removed from the animal contains this antibody mixture.

Obtaining antibodies from immunized animals is, however, a slow and tedious operation, and it is very difficult to isolate a specific antibody. At the end of several extraction and purification steps, the antibodies are usually only weakly specific, available in small amounts, and of relatively low activity. Moreover, attempts to culture antibody secreting cells have failed, since such cells do not survive long enough nor produce sufficient antibodies in culture to become worthwhile sources of antibodies.

In 1975, a new era in immunology was launched with the discovery of the hybridoma technique, a method for creating pure and uniform antibodies against a specific target. Scientists learned how to fuse myeloma (cancer) cells with antibody producing cells from an immunized donor. The hybrid cell or "hybridoma" resulting from this fusion has the ability to multiply rapidly and indefinitely in culture and to produce an antibody." This new hybridoma technology for production of standardized reagents (antibodies) of a given class, specificity, and affinity has provided scientists with a tool that permits the analysis of virtually any antigenic molecule. Such reagents can be made in virtually unlimited amounts whenever needed, thus making them readily available to all investigators or diagnostic laboratories.

Application of this technology has provided a breakthrough in the methods available for analysis of the antigenic composition of microorganisms, for rapid diagnosis, and to aid in the development of vaccines. The technology affords an opportunity to make important advances in understanding the diseases of humans, animals, and plants with greater specificity, speed, and at reduced cost. The technique is relatively simple and straight forward to use and can be readily developed and made available for use in developing countries. This can be accomplished through cooperation between institutions in developed and developing countries.

28.4.2. Potential applications in developing countries

Scientists are now in a period of transition from the use of conventional serological methods to the use of monoclonal antibodies.Such a transition requires replacement of conventional antibody reagents that must be constantly reproduced and standardized as supplies become exhausted, as well as development of reagents that provide greater specificity and thereby higher resolution than previously obtainable with complex antisera. The range of activity in this field is immense and cannot be fully described in this report. However, several activities are described here merely to indicate the scope of the areas that may relate to the needs of developing countries.

Monoclonal antibodies have been made against a spectrum of antigenic components, including hormones, drugs, serum components, white and red blood cells, bacteria, viruses, fungi, single and multi cellular parasites, and natural toxins derived from plants and microorganisms. They have also proven of value in delineating the composition of viral antigens and in virus epidemology. Antibodies generated against animal viruses such as rabies, influenza, parsinfluenza, herpes measles, SV 40, retrovirus, and tumor viruses have made it possible to identify previously unrecognized sub strains that complicated diagnosis and the development of effective vaccines. Similar use in plants has led to the delineation of virus strains involved in the pathogenesis of major crop diseases such as those induced by fruit tree viruses (Prunus necrotic ring spot, apple mosaic, and tobacco streak) and potato viruses (M, S, X, Y, potato leaf roll). Production of monoclonal antibodies against surface antigens of parasites at various stages in their life cycle has proven useful in dissecting the complex antigenic structure of composite membrane molecules and identifying molecules for use in diagnosis and the development of effective vaccines. For malaria, trials are planned for the near future to test the potential of membrane antigens as vaccines.For trypanosomes, studies have begun to show promise, and in the mouse model evidence has just been obtained that cross protection can indeed be obtained between related but antigenically dissimilar variants.

These examples emphasize the utility of monoclonal antibodies in addressing some of the difficult problems faced by developing countries.The following sections describe some specific areas in which the use of monoclonal antibody technology will prove of exceptional value.

28.4.3. Human Health

The development of routine diagnostic reagents for use in diagnostic laboratories exemplifies effective use of monoclonal antibody technology.Although there is little difference in the amount of time needed to develop identical reagents using monoclonal antibody technology, the advantage of the new technology is that antibodies only have to be generated once. The specificity obtained is exact and readily standardized.The resulting savings in time and money has been recognized, and commercial enterprises are rapidly turning to the new technology for the routine production of serological reagents. Monoclonal antibodies have already been made against a number of compounds of potential value in diagnostic procedures, including hormones, bacteria, viruses, parasites, white and red blood cells, and serum proteins.

The reagents generated thus would be equally useful in diagnosing diseases in developing countries and would greatly facilitate and improve the existing diagnostic capability. Although effective reagents are still not available for many diseases (see table 1), for some, such as malaria and schistosomoiasis, considerable progress has been made (see the working group report on vaccines). It would be useful to devise a mechanism that would accelerate and expand the capability of the institutions to prepare the needed diagnostic reagents. Once this technology is established and monoclonal antibodies are made to the prevalent, serious disease causing organisms, these antibodies could provide the means of identifying the antigenic constituents of the pathogenic organisms that can be used for the development of effective vaccines. Malaria is an example of how monoclonal antibodies can be used for such purposes. The cell membrane antigens identified may prove useful in development of an effective vaccines.

28.4.4. TABLE 1 Infectious diseases occurring in humans requires improved control vaccines of disease diagnostic possible better capability vaccines

28.4.4.1. Bacterial

	Improved control vaccines	Diagnosis	Better Capability vaccines
Brucellosis	x	–	–
Colibacillosis	x	–	x
Leprosy	x	–	x
Leptospirosis	x	–	x
Pseudomonoiasis	x	–	x
Salmonellosis	x	x	x
Syphilis	x	–	x
Tuberculosis	x	x	x
Venereal gonococcus	x	–	x
28.4.4.2. Chlamidial			
Trachoma	x	x	x
28.4.4.3. Metazoan			
Ascariasis	x	x	x
Cysticercosis/ hydatidosis	x	x	x
Filariasis	x	x	x
Schistosomiasis	x	x	x
28.4.4.4. Protozoan			
Amebiasis	x	x	x
Giardiasis	x	x	
Malaria	x	x	x
Toxoplasmosis	x	x	x
28.4.4.5.Viral			
Equine encephalitis	x	–	
Hepatitis	x	x	x
Herpes II venereal	x	–	x
Influenza	x	x	x
German measles	x		
Rabies	–	x	x

Monoclonal antibodies also offer potential for the development of immuno-therapeutic reagents. Although the technology is not well advanced in this area, it may be possible to develop reagents that could protect against such agents as snake venoms, plant and microorganisms derived toxins, and rabies. For a single use, monoclonal antibodies derived from mouse mouse hybridomas would prove effective. For continued use, however, antibodies would have to be derived from human human hybridomas.The technology for developing such reagents is, nevertheless, in its infancy and can only be considered as a potential application that would be extremely useful as an additional therapeutic tool.

Success in generating hybridomas that make antibodies of high specificity and affinity against reproductive tract hormones (human chorionic gonadotrophin, progesterone) opens the possibility of their use in both diagnosis and therapy. An abundant source of antibodies of consistent characteristics would be helpful in devising low cost methods of detecting pregnancy at an early stage that use non radioactive, isotopic immunoassays. Stability and reliability of these kits at temperatures prevalent in tropical countries would stimulate that widespread use in several parts of the world where such facilities are not available. Mechanisms should be devised to make these developments more immediately useful, as well as model systems for evolving similar assays for diseases of pregnancy.

28.4.5. Food Production

Monoclonal antibody is not limited to problems of human health; it will prove equally valuable in improving food production in developing countries.

28.4.6. Animal Health

Significant animal losses from disease occur in developed countries and losses in developing countries are even more severe. For cattle, it is estimated that some $4 billion per year is lost from disease in the United States alone. This is , in part, attributable to a lack of knowledge about the immune function in food producing animals, a lack of adequate diagnostic reagents to identify infectious agents responsible for disease outbreaks, and a lack of information essential for improving methods for breeding for disease resistance. As with humans, monoclonal antibody technology has provided the means of developing a set of diagnostic reagents, and the value of this technological breakthrough cannot be over emphasized. The spectrum of diseases in food animals the world over is quite large, and effective diagnostic reagents are only available for a few. The development of standard diagnostic reference reagents for international use would be of great value. A number of the important diseases are listed. A multinational program is needed to address effectively the problems of food animal disease, since in many instances diseases are not restricted to a single continent. For example, anaplasmosis is a persistant disease caused by a rickettsial organism. Because wild ruminants serve as a reservoir, there is no effective way to eradicate the disease, and no effective vaccine has been developed. A better method is needed to identify infected animals, and greater efforts must be expanded on producing a vaccine. Wild animals also serve as a reservoir for and prevent eradication of bluetongue virus.Here, however, disease control is complicated by the fact that many subtypes the current estimate is 21 of the virus exist, and vaccination against one type does not effectively protect against infection by other types. Bluetongue is also one of the diseases that prevents the shipment of sperm and embryos to developing countries to aid in upgrading native breeds. There is an urgent need to accelerate research on this virus, both to develop typing reagents and to produce a vaccine (see working group reports on vaccines and animal production).

Where a disease is more localized and important to food production for example, Salmonellosis and colibacillosis research programs utilizing monoclonal antibody technology to address specific needs are required. Although such diseases exist worldwide, strain variation at the local level accounts for regional problems, which necessitate development of diagnostic reagents to identify the strains of bacteria causing the local disease. At both the national and international levels, monoclonal antibodies are needed to facilitate the identification of antigens in pathogenic organisms that can be used to develop effective vaccines.

Monoclonal antibody technology also provides a much needed mechanism for improving methods of selective breeding. Conventional methods of producing typing reagents have provided a limited but important view of the antigenic polymorphic determinants present on white blood cells, which can be used to define gene systems for selective breeding. Monoclonal antibody technology has shown that the potential is much larger and that it can serve as a mechanism for identifying not only gene systems that influence susceptibility to infectious diseases but also those that control the inheritance of desirable performance traits.

An international collaborative program that includes developing countries is needed to develop monoclonal antibody typing reagents for use in selective breeding studies. The compilation of genetic profiles both of animals that might be used to upbreed native animals or by limiting spread.

28.4.7. Table.2. Infectious diseases occurring in animals requires improved control vaccines of disease diagnostic possibles better capability vaccines

28.4.7.1. Bacterial

	Improved control	Diagnosis	Better capability vaccines
Atrophic rhinitis	–	(swine)	x
Brucellosis (bovine)	x	x	–
Colibacillosis (swine, poultry)	–	x	–
Leptospirosis (bovine, swine)	x	x	–
Pleuropneumonia(swine	–	–	x
Salmonellosis (swine, poultry)	–	x	–
Tuberculosis (bovine)	x	x	x
28.4.7.2. Metazoan			
Ascariasis	x	x	–
Cysticercosis/ hydatidosis	x	x	–
Fascioliasis (bovine)	x	x	x
Schistosomiasis		x	x
28.4.7.3. Protozoan			
Babesiosis (bovine) ruminants)	x	x	x
Coccidiosis (avian ruminants)			
Theileriosis	x	x	–

Trypanosomiasis	x	–		x
28.4.7.4. Rickttsiasis				
Anaplasmosis (bovine)	–	x	x	
28.4.7.5. Viral				
African swine fever	x	–	x	
Aujeszky 's disease (swine)	–	x	–	
Bluetongue(ruminants)	x	–	x	
Bovine paralytic rabies	x	x	x	

28.4.8. Food Chain

The persistent difficulty, until now, has been the development of adequate quantities of standardized immune reagents for worldwide use. An international program is needed to address the problems associated with producing standardized immunological reagents for disease diagnosis and epidemology. Monoclonal antibody technology provides a means of meeting these goals, and with a joint commitment from developed and developing countries rapid progress can be made. Developed countries can provide the input for diagnostic reagents for pathogens common to both developed and developing countries. Some developing countries, however, need to address immediately pathogens that are contributing to economic losses.

Continuing programs that use serological techniques for potatoes, fruit, and cereal crops illustrate the need for immune reagents for research and diagnostic tests for certification.For example, potatoes are propagated vegetatively to ensure crop uniformity. Certification of pathogen†free seed pieces for cultivation requires testing for such organisms as Erwinia carotovors var. atroseptica (the causative agent of blackleg of potato), potato leaf roll virus, and potato viruses M, S, X and Y. It has been difficult, however, to develop reagents capable of distinguishing between strain variants, and this capability is often necessary to determine whether a disease outbreak occurred with a variant of the pathogen characteristic of the area where the seed stock was put into production or with a variant characteristic of the region where the seed pieces were propagated. Monoclonal antibody technology offers a solution to these problems and provides a means for epidemiological surveys of pathogenic strain variants.

Viruses present causes similar problems in fruit trees. Prunus necrotic ring spot, prune dwarf, and apple mosaic viruses cause serious diseases in cherry, peach, plum, and apple trees. The failure of these viruses to elicit a strong immune response hinders the production of high quality serological reagents capable of detecting them in certified propagation stock or in field outbreaks. Early work with monoclonal antibodies has shown that these problems can be overcome.

Several viruses cause major difficulties in citrus production. For example, citrus tristeza is nearly ubiquitous, and certain strains of this virus can destroy entire orchards. Control measures are to remove infected orchards. Although mild strains of this virus cause little or no disease and can protect against infection by the severe form, conventional sera cannot be distinguish between these strains. It is hoped that monoclonal antibodies may resolve these problems.

QUESTIONS

1. Explain what is monoclonal antibodies.
2. Outline the new vistas of monoclonal antibodies.
3. What is meant by guidance system ?
4. What are the achievements of Greg Weinter team ?
5. Explain Protein engineering in monoclonal antibodies.
6. What are the potential applications of monoclonal antibodies in developing countries ?
7. How are monoclonal antibodies useful to human health ?
8. How are monoclonal antibodies useful to food production ?
9. How are monoclonal antibodies useful to animal health ?
10. How are monoclonal antibodies useful to food chain ?

APPENDIX – I

ANSWERS TO STUDY QUESTIONS

1. What is the role of the antigen presenting cell ?

Cells that display peptide associated with class II MHC molecules to CD4+ TH cells are called Antigen presenting cells. The role antigen presenting cell ability to express class II MHC molecules and to deliver a Co-stimulatory signal. These cell types are classified as professional antigen presenting cells.

Dendritic cells are the most effective of the APCs. Because these cells constitutively express a high level of class II MHC molecules and co-stimulatory activity, they can active native TH cells.

Macrophages must be activated by phagocytosis of particulate antigens before they express class II MHC molecules or the co-stimulatory B7 membrane molecule.

B-cells constitutively express class II MHC molecules but must be activated before they express co stimulatory B7 molecule.

2. Which are the two mechanism of processing and presenting different types of antigen ?

Endogenous antigens are processed in the Cytosolic pathway and presented on the membrane with class I MHC molecules. Exogenous antigens are processed in the Endocytic pathway and presented on the membrane with class II MHC molecules. (mechanism in page,190-193)

3. Which molecules are involved in processing and presentation of antigen ?

Class I MHC molecule and class II MHC molecules.

4. How is a TCR constructed and what are the differences compared to IgM ?

The structure on the T cell that interacts with the complex of MHC and peptide molecules. Each T cell bears on its surface a disulfide linked two chain molecule that interacts with antigen. The disulfide bond links the two chains to form a structure similar to the hinge region of an antibody molecule. The predominent form of the TCR, comprises the glycoprotein chains α and β.

There are striking similarities in the structure of the TCR and immunoglabulin molecules. The extracellular protein of the TCR resembles the Fab fragment of an antibody. It combine both variable and constant regions which forms domain. The β chain of the TCR has V and C domains homologous to an Ig domain but only V region of the α-chain is Ig. The interaction of V region creates three hypervariable or complementary determining regions(CDR1, 2, 3).

The difference between the TCR and IgM is the Cα region of the TCR does not fold into and Ig domain. TCR appears more rigid than that of Ig because of extensive interaction between the domains of each TCR chain.Ig with the flexibility to bind to antigens of different shapes, contrasting with the TCR is interaction only with an MHC peptide complex.

5. How are the genetic elements of the TCR organized and how is the diverse pool of different TCR obtained from these ?

The genes that encode the $\alpha\,\beta$ and $\gamma\delta$ T- cell receptors are expressed only in cells of the T-cell lineage. The four TCR loci are organised in the germ line in a manner that is remarkably similar to the multigene organisation of the immunoglobulin (Ig) genes. Functional TCR genes are produced by rearrangements of V an J segments in the α chain and δ chain families. A productive rearrangement of the α-chain gene segments deletes Cδ, so that in given T cell the α β TCR receptor can't be expressed with the γ α receptor.

The δ-chain family contains about 10V gene segments. Which are largely distinct from the Vα gene segments, V segment has been observed in rearranged α and δ- chain genes. Two Dδ and two Jδ gene segments and one Cδ segment have also been identified. The B chain family has 20-30V segments and D, J and C segmeents. The ã chain family consists of seven Vγ segments and the three different functional Jγ-Cγ repeat.

6. How does this differ from organisation and rearrangement of IgM.

The mechanism by which TCR germline DNA is rearranged to form functional receptor genes appear to be similar to the mechanism of Ig gene rearrangement. The TCR gene rearrangement follow the one-turn/two-turn joining rule observed the Ig genes. B cell and T cells express the recombination activating genes RAG-1 and RAG-2. These recombination enzyme recognizes the heptemer and nonamer recognition signals and catalyzes V-J and V-D-J joining during TCR gene rearrangement by the same deletional or inversional mechanisms that occur in the Ig genes. The immunoglobulin genes, RAG-1/2 introduces a nick on one DNA strand between the coding and signal sequences. The recombinase then catalyzes a transesterification reaction that results in the formation of an hairpin at the coding sequence and a flush 5 phosphorylated double strand break at the signal sequence. D and J gene segment are not joined during rearrangement of either Ig or TCR DNA then that same double stranded break-repair enzymes are involved in V-D-J rearrangement in B cells and in T cells.

7. What kind of processes/mechanisms contribute to diversity of the TCR/IgM ?

Combinatorial joining of variable region gene segments generates a large number of random gene combination for all the TCR chains these are Ig heavy and light chains. Alternative joining of γ-chain gene segment generates similar unit, in addition, one Dä can join with another, yielding (VDDJ) γ and in humans, (VDDDJ) γ. This mechanism, which cannot operate in Ig heavy chain DNA, generates considerable additional diversity in TCR genes. In both Ig and TCR genes, nucleotides may be added at junction between some gene segments during rearrangements. Variation in endonuclease cleavage leads to the addition of further nucleotides that are palindromic.

P-region nucleotide addition can occur in the genes encoding all the TCR and Ig chains. Addition of N-region nucleotides catalyzed by a terminal deoxynucleotidyl transferase, generates additional junctional diversity.

8. Which receptors (other than TCR) are important for T-cells and which are their roles ?

The TCR is expressed on the cell surface in association with another transmembrane molecules are co-receptor accessory molecule. This co-receptor, a member of the Ig superfamily one, of two molecules on the mature T cells either CD4 or CD8. Coreceptor molecules expression splits the T cell population into one of two major subsets, either CD4 CD8- or CD4- CD8+. The role of CD4 and CD8 are at least twofold.

(a) The extracellular position of these molecules bind to the invariant portion of MHC molecules on the surface of cells that present antigen to cells. Thus CD4 and CD8 act as adhesion molecules that help to tighten binding of T cells to antigen presenting cells. Because CD4 binds selectively to MHC class II molecule and CD8 binds to MHC class I molecule.

(b) The second major function of CD4 and CD8 is to act as Signal transduction in T cells. The intracellular protein of CD4 and CD8 are linked to specific kinases, which are activated after binding of peptide +MHC to the TCR.these CD4- and CD8- associated kinases play an impartent role in T-cell activation.

9. How and where the T cell and B cell mature and differentiate?

The progenitor T cells are produced in the bone marrow. From there they migrate to the Thymus and there they mature. During their maturation positive and negative selection takes place in the ortical region of thymus and those cells sustain these two selections will be retained and the others will be subjected to apoptosis. The remaining T cells from thymus they enter the blood circulation and from there they move to secondary lymphoid organs such as lymph nodes. There T cells come across antigens and when they come in contact with antigens they are activated and differentiated into plasma cells and memory cells. The progenitor B cells develop in the bone marrow. They mature in the bone marrow and undergo antigen-induced activation and differentiation in the periphery. Activated B cells can give rise to antibody secreting plasma cells and memory cells.

10. How does positive and negative selection effect the maturation of T-cell /Bcell?
Positive selection permits the survival of only those T-cells who's TCRs are capable of recognizing self MHC molecules. It is thus responsible for the creation of self MHC restricted repertoire of T-cells. The other negative selection eliminates T cells that react too strongly with self MHC or self MHC plus self peptides. It is extremely important factor in generating a primary T cell repertoire that is self tolerant.

11. How does costimulation affect activation and proliferation of Tells ?

T cells require more than one signal for activation and subsequent proliferation into effector cells. Signal 1 is generated by interaction of an antigenic peptide with the TCR CD3 complex.

Non specific co stimulatory signal 2 is provided primarily by interaction between CD28 on the T-cell and members of the Tcell family on the APC.

Two related forms of B7 1 and B7 2. Both B7 molecules are constitutively expressed on dendritic cells and induced on activated macrophages and activated B-cells. The interaction of the CD28- B7 costimulatory pairs induces the phosphorylation and activation of Tcell proteins including CD28.The co stimulatory B7 –CD28 interaction enhances and sustains the Tcell activation signals transmitted by the Peptide MHC TCR interaction. T-cell surface molecules very closely related CD 28 known as CTLA-4 is induced in the sequence of T-cell activation. The level of CTLA-4 expression is increased by CD 28 generated co stimulatory signals. Some of the importance CTLA4 in the regulation of lymphocyte activation and proliferation is revealed by experiments with CTLA-4 knock out mice (Fig 10.14).

12. How does a super antigen work?

SAg (super antigen) is a class of antigen that activates T-cell expressing a Vα-segment and β chain of class MHC II molecule. T-cell receptor and class MHC II molecule by either type of super antigen produces an activating signal that induces T-cell activation and proliferation. Exogenous super antigens are soluble proteins secreted by gram+ bacteria. Ex: staphylococcal. Endogenous super antigens are cell membrane proteins encoded by certain viruses that infect mammalian cells. There are several unique aspects of SAg interacting with T-cells. SAg binds more or less Vβ region. SAg is presented by class II MHC molecules of APC. The intra cellular pathways of SAg activation may also be different from that peptide MHC complex activation of the TCR.

13. What is clonal anergy ?

TH cell recognition of an antigenic peptide MHC complex some times result in a state of non-responsiveness called clonal anergy marked by the inability of cells to proliferate in response to a peptide MHC complex. Clonal anergy ensues is determined by the presence or absence of a co-stimulatory signal such as that produced by interaction of CD28 on TH cell with B7 on APC. If a resting TH cells receive the TCR mediate signal in the absence of a suitable co-stimulatory signals, then the TH cell will become anergic. If resting TH cells are incubated with glutaraldehyde fixed APCs which don't express B7 the fixed APCs are able to present peptides together with class II MHC molecules providing signal I, but unable to provide the co-stimulatory signal 2. In the absence of co-stimulatory signal there is minimal production of cytokines of IL-2 allergn can also be induced by incubating TH cells with normal APCs in the presence of Fab portion of anti CD 28 which block the interaction of CD 28 with B7.

14. What distinguishes the TH1 and TH2 subsets of T-cells?

CD4 + effector T-cells form two subpopulations distinguished by a different panel of cytokines they secrete. One population is TH1 subset secrets IL-2, IFN γ and TNF-β. The TH1 subset is responsible for classic cell mediated functions such as delayed type hyper sensitivity and the activation of cytotoxic T lymphocytes. The other subset is TH2 subset secretes IL-4, IL-5, IL-6 and IL-10. This subset functions more effectively as a helper for B-cell activation.

15. Which role has the γ δ T cells ?

A. γ δ T cells are not MHC restricted. In many cases the TCR of γ δ T cells bind to epitopes in much the same way that the immunoglobulin receptors of B-cell. Most humans bind free antigen and most have the same specificity. They may function as a part of innate immune system. Certain pathogens expressing effector functions that help control infection and secreting cytokines that promote an adaptive immune response by γ δ T cells and B cells.

16. Which activating signals contribute to activation of B-cells ?

In the activation of the B-cells the activating signals involved mainly are competence signals and progression signals induced by the membrane events. Competence signals drive the B cell from G.O. into early G1 , rendering the cell component to receive the next level of signals. Progression signals then drive the cell from G1 into S and ultimately to cell division and differentiation. Competence signal consists of two distinct signaling events that are signal 1 and signal 2nd are generated by different pathways with thymus independent antigens and TD antigens. Due to their multivalent nature, TI antigens induce a strong stimulation by crosslinking B-cell mIg molecules (signal 1). TI-1 antigens also contain an additional component that provides and additional signalling event (signal 2). The extensive cross-linking of mIg provided by TI-2 antigens generates an effective competence signal. In the B-cell response to TI-1 antigens, cytokine binding provides a progression signal, which is required for extensive proliferation. The progression signal is not required for proliferation of B cells stimulated by TI-2 antigens. TD antigens, which are either divalent or oligovalent, induce a relatively weak signal 1 by cross-linking mIg. An additional signal, signal 2 is provided through interaction of CD40 on the B-cell membrane with a stimulating molecule, called CD40L, expressed on the membrane of T *H* cells. When B cell acquired an effective competence signal in early activation, it begins to express membrane receptors for a variety of cytokines (IL-2, IL-4, IL-5). The interaction of these cytokines with the Bcell membrane receptors provides the progression signal, inducing proliferation and differentiation of the B-cell in response to TD antigens.

17. Which co-receptors are involved in the B-cell response ?

The main receptor involved in the B-cell response is a B-cell receptor (BCR). It is a membrane Ig associated with a disulfide-linked heterodimer (Ig-α/Ig-β). Two molecules of the Ig-α/Ig-β heterodimer associated with one mIg to form a single BCR. Ig-α/Ig-β involves in signal transducing. Another receptor complex involved in B-cell response is the B-cell receptor. It is a complex of three proteins: CD19, CR2 and TAPA -1. B-cell co-receptor intensifies the activating signal. Phosphorylation of CD 19 permits it to bind and activate the src-family tyrosine kinase, which in turn activates two intracellular enzymes (phospholipase C and phosphotidylinositol 3'- kinase) involved in B-cell activation pathway. It is particularly important during the primary immune response that responds to low concentration of antigen.

18. In which ways can a helper T-cell take part in the activation of B-cells?

Antigen binding to mIg provides competence signal 1 and appear to up regulate class II MHC molecules and the co-stimulatory ligand B-7. After internalization of antigen for processed antigenic peptides the T *H* cell recognizes it, as it is displayed by a class.

II MHC molecule on the membrane of a B cell, the two cells interact, forming a T-B conjugate. Formation of a T-B conjugate not only leads to the directional release of T *H* cell cytokines, but also to the activation dependent expression of a new T *H* cell membrane protein, CD40L, that is necessary for B-cell activation. Interaction of CD40L with CD40 on the B-cell delivers competence signal 2 to the B-cell; this signal 2 along with signal 1 constitutes an effective competence signal that drives Bcell into G1 . This competence signal is transduced via intracellular signaling pathways leading to changes in gene expression. CD40/CD40L interaction is necessary in the formation of germinal centers, generation of memory cell formation and somatic hyper mutations. B-cell stimulated with membranes from activated T *H* cells is able to proliferate. Once an effective competence signal has been generated, the B-cell begins to express membrane receptors for various cytokines. These receptors then bind to the cytokines (IL-2, IL-4, IL-5) released in a directed fashion toward the B-cell from the interacting T *H* cell, there by generating progression signals. As activated B-cells proliferate, three differentiation events, formation of plasma cells and memory cells, class switching and affinity maturation occurs.

19. How does the primary and secondary humoral response differ? And what role does TH-cells play ?

The properties of the primary and secondary antibody responses differ. The primary response has a long lag period, a logarithmic rise in antibody formation, a short plateau, and then a decline. IgM is the first antibody class produced, followed by a gradual switch to other classes, such as IgG. The secondary response has a shorter time, a more rapid logarithmic phase, a longer plateau phase, and a slower decline that the primary response. Mostly IgG and other isotypes are produced in the secondary response rather than IgM, and the average affinity of antibody produced is higher. In primary response T *H* cell interact with the B-cells only after their activation. But in secondary humoral response they interact with the existing or circulating antibodies to produce immediate immune response.

20. Where does the humoral response and B-cell differentiation take place in vivo?

The *in vivo* humoral response for tissue antigen is generated within peripheral lymph nodes in distinct microenvironments. Antigens entered into lymph nodes with antigen transporting cells and macrophages through the afferent lymphatic vessels. The initial activation of both B and T cells is thought to take place in the paracortex as this region is heavily populated with T cells, macrophages and interdigitating dendritic cells. Interdigitating dendritic cells express high levels of class II MHC molecules which present antigen to naïve T cells resulting in extensive T *H* cell activation and proliferation. The B-cell with mIg-antigenic peptide presents antigen to activated Tcells in paracortex region forming a T-B conjugate, inducing initial B-cell activation. Once B-cell activation takes place, small foci of proliferating B-cells form at the edges of the T-cell rich zone. The B-cells within these foci differentiate into plasma cells secreting IgM and IgG isotypes.

Some of the activated B-cells and T *H* cells migrate from the foci to primary follicles in the cortex region. These follicles then develop into secondary follicles, which provide a specialized microenvironment favourable for interaction between B-cells, activated T *H* cells and follicular dendritic cells. The formation of germinal centers in the secondary follicle results in the intense proliferation of activated B-cells in the dark zone. The centrocytes which express high-affinity mIg, interact with antigen bound to follicular dendritic cells and undergo differentiation in the apical light zone of germinal center, generating memory B-cells and plasmablasts. The plasmablasts migrate to the medulla and develop into antibody secreting plasma cells.

21. Where are B-cells and plasma cells generated in vivo?

Following selection of centrocytes bearing high-affinity mIg for antigen displayed on follicular dendritic cells, the centrocytes differentiate into plasma cells and memory B-cells in the apical light zone of germinal centers. Different membrane signals determine whether centrocytes differentiate into plasma cells or memory cells. Formation of plasma cells is induced by IL-1 and CD23, products of follicular dendritic cells, bind to receptors on centrocytes. Memory B-cells are generated following a T *H* cell derived signal triggered by interaction of CD40 on centrocytes with CD40L on activated T *H* cells. Class switching may occur during formation of both plasma and memory B-cells.

22. How is B-cell development regulated?

The B-cell development is regulated by a number of transcription factors such as NF-6B, BSAP, Ets-1, c- Jun, Ikaros, Oct-2, Pu.1, EBF, BCF, and E2A. These transcription factors interact with promoter or enhancer sequences, thereby either stimulating or inhibiting transcription of the associated gene. One of the most important transcriptional factors is B-cell specific activator protein (BSAP), functions as a master B-cell regulator. It influences all the cell stages during B-cell maturation and differentiation events leading to the formation of memory B-cells and plasma cells. BSAP binds to promoter or enhancer sequences of various B-cell specific genes, including the λ 5 and Vpre-B genes of the surrogate light chain and the 3′α heavy chain enhancer region (E 3 ′ α). Binding of BSAP to E3 ′ α appears to influence B-cell development by preventing the binding of other transcription factors. When BSAP levels are high, it blocks the binding of NF-α P to the 3′ α enhancer there by blocking transcription of the heavy chain gene and promoting formation of memory B-cells. When BSAP levels are low, NF-α P bind to E 3 ′ α as a result transcription of heavy chain gene occurs leading to the plasma cell formation.

23. How is tolerance or unresponsiveness of immune response develop?

The immune system can develop immune response or enter a state of unresponsiveness called tolerance when it encounters antigen. Regulation of immune response takes place in both the humoral and the cell-mediated immunity in which cytokines play a major role. The presence of circulating antibody competes with the antigen-reactive B-cells for antigen binding. The antibody then exerts feedback inhibition on its own production. The binding of antigen-antibody complexes by Fc receptors on B-cells reduces signaling by the B-cell receptor complex. Example of this is the infants are not administered with certain vaccines within one year as they will become recessive due to antibody-mediated suppression of immune response as they are having high amounts of naturally acquired IgG.

24. What is the difference between auto-, para-, and endocrine action of cytokines and hormones ?

• Autocrine when a cytokine bind to the receptors on the membrane of the same cell that has secreted them they are called as autocrine in action.

• Paracrine: When the cytokines bind to the target cells which are in close proximity then they are called as paracrine in action.

• Endocrine: When the cytokines bind to the target cell away from the cell that produces them, they are called as endocrine in action.

• Cytokines differ from the hormones in the way that hormones act for long distance in endocrine way while cytokines act for short distance in autocrine and paracrine action.

25. Explain the terms cytokine, chemokine, interleukin and an interferons?

• **Cytokine:** Cytokines are defined as group of proteins that bring about complex interactions among lymphoid cells, inflammatory cells and hematopoietic cells that are related with the development of immune response. Cytokines are low molecular weight regulatory proteins or glycoproteins produced in the body by lymphocytes, monocytes and macrophages in response to stimulus. These cytokines bind to the receptors on the membrane of target cells that trigger the signal transduction which alters the gene expression of target cells. Chemokines: These are a subgroup of cytokines which are low molecular weight and they effect the chemotaxis (the movement of the cell in response to chemical messengers) of the leukocyte behaviour.

• Interleukin: Interleukins are the cytokines that are produced by a leukocyte and act upon another leukocyte.

• Interferons: They are basically two types of interferons. Type I & Type II. They are capable of protecting the cells when they are attacked by virues. Interferon induces uninfected cells to produce enzymes capable of degrading mRNA. These enzymes remain inactive until the uninfected cell becomes infected with a virus. At this point, the enzymes are activated and begin to degrade both viral and cellular mRNA. This not only blocks viral protein synthesis, it also eventually kills the infected cells.

26. Explain the terms pleiotropy, redundancy, synergy and antagonism in terms of cytokine action.

• Pleiotrophy: When a particular cytokine has different effects upon different target cells then in is pleiotrophic in action.

• Redundancy: When two or more cytokines mediate similar functions then they are called redundant in nature.

• Synergy: When combined effect of two cytokines on a particular cell is more than the additive effect of individual cytokines then it is called cytokine synergism.

• Antagonism: It is a property through which a cytokine inhibits the effect of another cytokine.

27. Which structural families can cytokines be divided into?

Cytokines can be divided into 4 structural families. They are

• Hematopoietin family.
• Interferon family.
• Chemokine family.

Tumor necrosis factor family

28. Which are the biological functions of cytokines?

The cytokines are involved in a number of biological activities and few among them are innate immunity, adaptive immunity, inflammation and hematopoiesis.

- Innate immunity: is the immunity to disease that occurs as part of an individual's natural biologic makeup.
- Adaptive immunity: This is the immunity an individual develops throughout life when he or she is exposed to an antigen. There are two major branches of the adaptive immune responses: humoral immunity and cell-mediated immunity.
- Humoral immunity: Is the production of antibody molecules in response to an antigen which is mediated by B-lymphocytes.
- Cell-mediated immunity: Is the production of cytotoxic T-lymphocytes and cytokines in response to an antigen which is mediated by T-lymphocytes.
- Inflamation: a response of body tissues to injury or irritation; characterized by pain and swelling and redness and heat.
- Hematopoiesis: Is the formation and development of blood cells which usually occurs in the bone marrow.

29. Which structural families can cytokines receptors be divided into and how are the receptors composed?

Cytokine receptors can be divided into 5 families of receptor families.

- Immunoglobulin super family receptors.
- Class I cytokine receptor family (hematopoietin receptor family).
- Class II cytokine receptor family (interferon receptor family).
- TNF receptor family.
- Chemokine receptor family.

The members of the family have conserved cysteine residues. A conserved sequence of tryptophan–serine–any amino acid-tryptophan –serine is found for particularly Class I cytokine receptor.

30. How does the TH1 and TH2 subsets differ in their cytokine secretion and why? TH1 and TH2 the two populations of CD4+ TH cells differ in their cytokine secretions.

- TH1 subset of the cells are responsible for cell mediated functions like delayed type hypersensitivity, activation of TC cells and for the production of opsonizationpromoting IgG antibodies (antibodies that bind to the high affinity Fc receptors of phagocytes and interact with the complement system).
- TH2 subset stimulates eosinophil activation and differentiation, provides help to B cells and promotes the production of relatively large amounts of IgM, IgE.

The difference in secretions is due to different biological functions of these two subsets. IFN-γ, one of the cytokines produced by TH1 population activates macrophages which increases the microbicidal activity, up-regulates the level of Class II MHC and secretes the cytokinens like IL-12. The IL-12 secretion induces the TH cells to differentiate into TH1 cells. It also induces phagocytosis and fixation of complement. The TH2 cells secrete IL-4 and IL-5 which induces the production of IgE and supports eosinophil mediated attack on the helminth infections.

TH1 subset of cytokines are produced in response to intracellular pathogens, and the TH2 subset is higher in allergic diseases and helminthic infections.

31. How does cytokines regulate the differentiation of TH-cells into TH1 and TH2-subsets?

The TH differentiation is dependent upon the cytokine environment in which the TH cells are present. IL-2 is necessary for the development of TH2 subset. IFN-γ, IL-12 and IL-18 are important for the development of TH1 cells.

Macrophages and dendritic cells when they encounter the intracellular bacteria, produce IL-12 which in the key mediator of TH1 development. IFN-γ is generated by the stimulation of T-cells and also from the activated NK (natural killer cells) cells.

This IFN-γ induces number of changes, upregulates the IL-12 production by macrophages and dendritic cells and activation of interleukin receptor on activated Tcells.

IFN-γ production is also promoted by IL-18 from developing and fully differentiated TH1 cells and by NK cells. When the cytokines or their receptors are blocked the TH1 cells production is knocked out.

In the same way the generation of TH2 cells is dependent upon the IL-4. When naïve helper cells are exposed to IL-4 during an immune response then they differentiate into TH2 cells.

32. Which diseases are caused by over/under-expression of cytokines or their receptors?

Bacterial septic shock is caused due to over production of cytokine. This septic shock is produced by infection with gram –ive bacteria like *E.coli, Klebsiella Pneumonia, Pseudomonas aeruginosa, Enterobacter aerogenes and Nisseria meningitides*. The symptoms of shock are drop in blood pressure, fever, diarrhoea and wide spread blood clotting in various organs. Generally the bacterial cell wall endotoxins stimulate macrophages to produce IL-2 and TNF-α to high levels which cause septic shock.

Few micro organisms produce toxins which act as super antigens. These super antigens bind simultaneously to class II MHC molecules and to Vâ domain of the Tcell receptor. Due to their unique binding ability, super antigens can activate large number of T cells irrespective of their antigenic specificity. Abnormalities in the production of cytokines or their receptors have been associated with cancers. High levels of IL-6 are secreted by cardiac myxoma cells (benin heart tumour) myeloma, cervical and bladder cancer cells.

Trypnosoma cruzi is the causative agent of chagas disease which is produced by severe immune suppression. The á-subunit of the IL-2 receptor is removed which brings about reduced levels of IL-2 attachment which facilitates the pathogen development.

33. How can cytokines be used in therapy?

Purified cloned cytokines and soluble cytokine receptors offer to be specific clinical therapy to modulate immune response. Interferons and colony stimulating factors have proven to be therapeutically useful. But there are few problems using them for clinical practice, to maintain effective dose level over a period of time is difficult. Cytokines have short range of action and even they have half life which is very less like 7-10 minutes for IL-2 when it is administered intravenously.

They may even cause side effects. Fever, chills diarrhoea and weight gain in case of IL-2. Cytokines conjugated to various toxins are used to diminish rejection of kidney and heart transplantations in animals. They kill the activated TH cells to facilitate during the heart transplantations.

34. Which main immunological roles have the complement system?

Serum of complement proteins and membrane-bound complement receptors has number of functions in immune activities. Those are:

- Lysis of cells, bacteria and viruses by antibody-dependent or antibody independent pathways.
- Opsonization of particular antigens by phagocytes
- Activation of inflammatory responses and secretion of immunoregulatory molecules.
- Clearance of circulating immune complexes by cells in the liver and spleen.

35. Describe the three complement pathways with regard to their common steps, function and activation.

Complement activation via three pathways: classical pathway, alternative pathway and lectin pathway. Classical pathway requires Ab bound to Ag on a suitable target, complement components C1, C2, C3 and C4. Classical pathway initiated when C1 binds to ag-ab complex. C1 in a serum is a calcium dependent macromolecular complex consisting of C1q, C1r and C1s. Binding of C1q to Fc binding sites induces a conformational change in C1r. IgM and IgG contain a binding site for C1q. The activated C1r cleaves C1s and this cleaved C1s becomes an enzyme (C4-C2 convertase) capable of cleaving both C4 and C2. Activated C1s enzymatically cleaves C4 into C4a and C4b. C4b binds to the target surface of C1 while C4a is an anaphylotoxin or mediator of inflammation. C2 bind to the C4b and C1s cleaves the C2 in to C2a and C2b. C2a complexed with C4b whereas C2b is released in the micro environment. C4b2a complex is known as C3 converse in which C2a is the enzymatic moiety. In the presence of Mg^{++}, C3 cleaved into C3a and C3b by C3 convertase. C3b binds to the membrane to form C4b2a3b complex whereas C3a remains in the micro environment. C4b2a3b complex functions as C5 convertase which cleaves C5 into C5a and C5b. Generation of C5 convertase marks the end of the classical pathway. Alternative pathway is initiated by spontaneous activation of C3 and requires proteins are C3, factor B, factor D and properdin. Serum C3 spontaneously hydrolysis to C3a and C3b. C3b binds to the foreign surface antigens. C3b bound to these surfaces remains active for long time and bind to factor B to form complex. C3b exposesd a site on factor B called Factor D. Factor D cleaves C3b-bound factor B to generate the C3bBb complex which has C3 convertase activity. Lectin pathway is initiated by the binding of the serum protein mannose-binding lectin (MBL) to the surface of pathogen.

MBL which interacts with two mannose binding lectin-associated serine proteases MASP-1 and MASP2. The active complexes of this association cleave the C4 and C2.

36. Which functions have the active components of the complement pathways, C3a, C3b and MAC ?

C3a mediates inflammation and anaphylatoxin. C3b cleaves C5 by C2b. and it bind to cell surfaces for opsonization and activation of alternative pathway. The terminal sequence of complement resulting complex C5b678 guides the polymerization of as many as 10-17 molecules of C9. C9 molecules undergo a hydrophilic-amphiphilic transition to insert themselves into membrane. The completed MAC has a tubular form and consists of a C5b678 complex surrounded by a poly C9 complex. MAC allows ions and small molecules can diffuse freely through the central channel of the MAC.

37. How and at which steps are the complement system regulated?

The system is regulated by proteins such as C1 inhibitor (C1-Inh), C4b binding protein (C4bBp), factor I(fI), factor H (fH), decay accelerating factor (DAF or CD55), membrane cofactor protein (MCP), complement receptor type 1 (CR1) CD59. DAF, MCP, fI, CD59 are soluble proteins and C1-INH, C4bBp, fI, fH, are membrane bound proteins. These inhibitors help to protect the host from immune attack. C1 Inh can bind with C1r2s2 causing dissociated from C1q and preventing further activation of C4 or C2.

C4bBp, CRI and MCP are bind to C4b and prevent its association with C2a.

CRI, MCP, FH binds to C3b and prevents its association factor B.

Factor I inactivates C3b.

DAF is aglycophospholipid protein, has the ability to dissociate C3 convertase.

38. Which are the anaphylatoxins and what is their role?

Anaphylatoxins are the smaller fragments resulting from complement cleavage, Ca3, Ca4, Ca5. These fragments are bind to receptors on mast cells and blood basophils and induce degranulation. With release of histamine and other pharmacologically active mediators. The anaphylatoxins also induce smooth muscle contraction and increased vascular permeability. Anaphylatoxins are regulated by a serum protease called carboxypeptidase N, which cleaves an Arg residue from the C terminus of the molecules, yielding so called des-Arg forms.

39. What is an opsonin ?

Opsonis is that an antigen in blood serum that attaches to the microorganisms and other antigens to make them more susceptible the action of phagocytes. Eg: C3b, C4b, iC3b.

40. Describe the two types of cell-mediated effector responses.

Killing by cytotoxic T lymphocytes (CTLs or Tc), natural killer cells (NK cells) and antibody-dependent cell-mediated toxicity (ADCC) Delayed-type hypersensitivity (DTH) mediated by Th1 cells

41. What distinguishes effector T-cells from naive helper T-cells and cytotoxic Tells?

Effector cells are characterized by their less stringent activation requirements, increased expression of cell-adhesion molecules, and production of both membrane bound and soluble effector molecules. Activation of naïve T cells and their subsequent proliferation and differentiation into effector T cells require both a primary signal, delivered when the TCR complex and CD4 or CD8 coreceptor interact with a foreign peptide-MHC molecule complex, and a co-stimulatory signal, delivered by interaction between particular membrane molecules on the T cell and the antigen presenting cell. While, antigen-experienced effector cells are able to respond to TCR-mediated signals with little co-stimulation.

42. How do CTLs develop and what is their role?

CTLs are generated by immune activation of Tc cells (CTL precursors or CTL-Ps). Generation of CTLs from CTL-Ps appears to require at least three sequential signals:

An antigen-specific signal 1 transmitted by the TCR complex upon recognition of a petide-class I MHC molecule complex.

- A co-stimulatory signal transmitted by the CD28-B7 interaction of the CTL-P and the APC.
- A signal induced by the interaction of IL-2 with the high-affinity IL-2 receptor, resulting in proliferation and differentiation of the antigen-activated CTL-P into an effector CTL.

43. Describe the construction and mechanism of perforins and granzymes?

Activated CTLs contain storage granules that contain 65-kDa monomers of a poreforming protein called perforin and several proteases called granzymes. After formation of a CTL-target call conjugate, the golgi stacks and storage granules reorient within the cytoplasm of the CTL to concentrate near the junction with target cell. In the presence of sufficient Ca+2 ion, the perforin inserts into the target cell membrane, forming pores. Granzymes, also released from the CTL's cytoplasmic granules, enter the target cell through the perforin pores and initiate the apoptotic program in the target cell. This program involves activating caspases, proteases that pre-exist in the target cell in an inactive form. The result of the caspase cascade is DNA and nuclear fragmentation, blebbing of the cell membrane, and other characteristics of apoptosis.

44. How do NK cells develop and what is their role?

NK cells are lymphoid cells derived from bone marrow that share a common early progenitor with T cells. NK cells arise during a cell-mediated response earlier than CTLs.

Virus-infected cells produce IFN-α and IFN-β which stimulate an increase in NK cell numbers and activity that peaks at 3 days post-infection. Like CTLs, NK cells kill by means of perforin and granzymes stored in granules. After an NK cell adheres to a target cell, degranulation occurs with release of perforin and granzymes at the junction of interacting.

45. What is ADCC?

NK cells, macrophages, monocytes, neutrophils, and eosinophils all have Fc receptors that allow them to passively adsorb antibodies that have been secreted by plasma cells into their cell surfaces. If those antibodies have specificity for a molecule on the surface of a target cell, the potential killer cell can use that antibody to establish contact with the virus-infected target. The cells then are activated to become effector cells, and the killing mechanism that develops depends upon the cell type whose Fc receptor bound the antibody.

46. Describe the composition of the different classes of CAMs? What are their role(s)?

Selectins:

☛ membrane glycoproteins.

☛has a distal lectin-like domain, enabling binding to specific carbohydrate groups.

☛ three molecules, L-selectin expressed on leukocytes, P- and E- selectin expressed on vascular endothelial cells.

☛ responsible for initial adherence of leukocytes to vascular endothelium.

MUSIN LIKE FAMILY

Mucins are a group of serine- and threonine-rich proteins that are heavily glycosylated.

☛ extended structures allow them to present sialylated carbohydrate ligands to selectins.

Integrin

☛ heterodimeric proteins (α and β chains) expressed by leukocytes.

☛ facilitate adherence to the vascular endothelium and other cell-to-cell interactions.

47. Describe the process of extravasation.

Movement of blood cells through an unruptured vessel wall into the surrounding tissue at the sites of inflammation.

Neutrophils Lymphocytes

Rolling

Neutrophils attach loosely to the Naive lymphocytes bind to HEVs endothelium by low-affinity selectin- L-selectin homing receptors that carbohydrate interaction. Directs the lymphocytes to particular tissues expressing a corresponding mucin-like vascular addressins; CD34/GlyCam-1.

Activation

Chemokines/chemoattractive cytokines Integrin-activating stimulus activate neutrophil. mediated by chemokines that are localized on endothelial surface or secreted locally. Arrest and Adhesion Firm binding of integrin to Ig-superfamily Chemokine binding to G protein- Adhesion molecules on the endothelium coupled receptor on the lymphocyte leads to activation of integrin molecules on the membrane, resulting in firm binding of lymphocyte to endothelium. Transendothelial migration

Mechanisms are poorly understood for both neutrophil and lymphocytes.

48. Which are the differences in the structure and function of CC and CXC chemokines?

CC chemokines- conserved cysteines are contiguous. Function: promotes migration of monocytes or cell types.

CXC.- conserved cysteines are separated by some other amino acid (X). Function: promotes migration of neutrophil.

49. Which are the main mediators (systems) of inflammation ?

Plasma-derived mediator-producing systems

- Kinin system
- Clotting system
- Fibrinolytic system
- Complement system

Are released during tissue damage

Degradation of phospholipids in cellular membranes yield inflammatory mediators

* Ex. macrophage, monocyte , neutrophil and mast cells
* Arachidonic acid (prostaglandins and thromboxanes; leukotrienes)
* Lyso-platelet-activating factor (platelet activating factor, PAF)1

50. Describe the inflammatory process.

Inflammation is a physiologic response to a variety of stimuli such as infections and tissue injury. Neutrophils are the first cells that infiltrate the tissue in response to inflammation. Inside the tissues, the activated neutrophils upregulate receptors for chemoattractants and thus exhibiting chemotaxis. Activated neutrophils express increased levels of Fc receptors for antibody and receptors for complement, enabling these cells to bind more effectively to antibody- or complement-coated pathogens.

The activating signal also stimulates metabolic pathways to a respiratory burst, which produces reactive oxygen intermediates and reactive nitrogen intermediates. Release of these reactive intermediates and the mediators from neutrophil primary and secondary granules play an important role in killing various pathogens. These substances also contribute to the tissue damage that can result from an inflammatory response.

51. What is the difference between a localized, a systemic and a chronic inflammation? Locialized inflammation is characterized by swelling, red ness, heat, pain and loss of function. Activated tissue macrophages secrete three cytokines, IL-1, IL-6 and TNF-á that act locally, inducing coagulation and an increase in vascular permeability.

Cytokines released at the site of inflammation facilitate both the adherence of immune-system cells to vascular endothelial cells and their migration through the vessel wall into the tissue spaces. This results in an influx of lymphocytes, neutrophils, monocytes, eosinophils, basophils, and mast cells to the site of tissue damage, where these cells participate in clearance of the antigen and healing of the tissue.

Systemic inflammation is marked by induction of fever, increased synthesis of hormones such as ACTH and hydrocortisone, increased production of white blood cells and production of a large number of acute-phase proteins in the liver.

Chronic inflammation develops because of the persistence of an antigen. It also occurs in a number of autoimmune diseases in which self-antigen continually activate T cells.

Fibrosis and granuloma are characteristics of chronic inflammation. IFN-γ and TNF-α play a central role in the development of chronic inflammation.

52. Describe the acute-phase response.

Acute phase response includes induction of fever, increased synthesis of hormones such as ACTH and hydrocortisone, increased production of white blood cells and production of a large number of acute-phase proteins in the liver. The combined action of IL-1, TNF-α, and IL-6 induces acute-phase effects. Each of these cytokines acts on the hypothalamus to induce a fever response. Within 12-24 hours of the onset of an acute-phase inflammatory response, increased levels of IL-1, TNF- α and IL-6 induce production of acute-phase proteins by hepatocytes.

53. Which cytokines play a central role in chronic inflammation and how?

IFN-ã and TNF-γ play a central role in chronic inflammation. TH1 cells, NK cells, and Tc cells release IFN- γ, while activated macrophages secrete TNF- γ.

IFN-α activates macrophages, which exhibits increased expression of class II MHC molecules, increased cytokine production, and increased microbicidal activity compared with nonactivated macrophages, thus they are more effective in antigen presentation and killing of intracellular microbial pathogens. These cells release various hydrolytic enzymes and reactive oxygen and nitrogen intermediates, which damage the surrounding tissue.

TNF- α is secreted by activated macrophages. TNF- α acts synergistically with IFN- α to initiate a chronic inflammatory response. Both cytokines together induce increased expression of ICAM-1, E-selectin, and class I MHC molecules than either cytokine alone. The increase in intercellular adhesion molecules facilitates the recruitment of large numbers of cells in a chronic inflammatory response.

54. How can anti-inflammatory be used to treat inflammation?

Anti-ICAM-1 and anti-LFA-1 are used to reduce leukocyte extravasation. Corticosteroids-cholesterol derivatives-cause a decrease in the number of circulating lymphocytes as a result of steroid-induced lysis of lymphocytes or alterations in lymphocyte-circulation patterns. Corticosteroids induce increased transcription of NF- kB inhibitor (I-kB), which binds to NF-kB in the cytosol and prevents its translocation into the nucleus, thereby preventing activation of a number of genes involved in cytokine production and T-cell activation.

Nonsteroidal anti-inflammatory drugs (NSAIDs) are used for treating pain and inflammation. They inhibit cyclooxygenase pathway by which prostaglandins and thrombaxanes are produced from arachidonic acid. The reduction in prostaglandin production limits the increase in vascular permeability and neutrophil chemotaxis in the inflammatory response.

55. Explain the difference between four types of hypersensitivity?

The type I hypersensitivity is mediated by IgE antibodies, and type II is mediated by

Ig-mediated cytotoxic hypersensitivity and Type III is mediated by immune complexmediated hypersensitivity and type 4 is mediated by cell-mediated hypersensitivity. In type I antigen induces cross linking of IgE bound to most cells and basophils with release of vasoactive mediators. Typical manifestations include systemic anaphylaxis, localised anaphylaxis such as hayfever, asthma, hives, food allergies and eczema etc. In type II antibody directed against cell surface antigens mediates cell destruction via complement activation or ADCC.Typical manifestations include blood transfusion reactions, erythroblastosis foetalis and autoimmune haemolytic anaemia.

In type III antigen-antibody complexes deposited in various tissues induce complement activation and ensuing inflammatory responses mediated by massive infiltration of neutrophils.Typical manifestations include localized arthus reaction and generated reactions such as serum sickness, necrotizing vasculitis, glomerulonephritis, rheumatoid arthritis and systemic lupus erythematosis.

In type IV- sensitized Th1 cells cytokines that activate macrophages or Tc cells which mediate direct cellular damage typical manifestations include contact dermatitis, tubercular lesions and graft rejections.

56. Describe the development of a type-I reaction. Which molecules and cell types are involved?

The type-I hypersensitivity reaction is induced by certain types of antigens called allergens resulting in the generation of antibody secreting plasma cells and memory cells. The plasma cells secrete IgE antibodies. These antibodies binds with high affinity to Fc receptors on the surface of tissue mast cells and blood basophills. Mast cells and basophils coated by IgE are said to be sensitized. A later exposure to the same allergen crosslink's the membrane bound IgE on sensitized mast cells and basophil cells causing degranulation of these cells. The active mediators release from the granule and act on the surrounding tissues leads to vasodilatation and smooth muscle contraction may be systemic or localized. The IgE molecules and B cells, Th cells, memory cells, plasma cells, sensitized mast cells are involved in the type hypersensitivity reaction.

57. Which pharmacological agents mediate type I reactions and how?

In type-I hypersensitivity reactions the biological effects of the mediators are released during mast cells or basophil degranulation.These mediators are pharmacologically active agents that act on local tissues and secondary effector cells, including eosinophils,neutrophils,T-lymphocytes,monocytes and platelets. The mediators can be classified as either primary or secondary.The primary mediators are produced before degranulation and are stored in the granules. The primary mediators are histamine, proteases,eosinophil chemotactic factor,neutrophil chemotactic factor and heparin.

The secondary mediators either synthesized after target cell activation or released by the breakdown of membrane phospholipids during the degranulation process. The secondary mediators include platelet activating factor, leukotrienes, prostaglandins, bradykinins, and various cytokines.

58. What is atopy and systemic anaphylaxis?

Some persons, however, may have an abnormality called atopy, hereditary predisposition to the development of immediate hypersensitivity reactions against common environmental antigens. The IgE regulatory defects suffered by atopic individuals allow non-parasitic antigens to stimulate inappropriate IgE production leading to tissue damaging type I hypersensitivity.

Systemic anaphylaxis is a shock-like and often fatal state whose onset occurs within minutes of a type-I hypersensitive reaction. In humans systemic anaphylaxis is induced by drugs such as pencillin, insulin and antitoxins and seafood and nuts. Epinephrine is the drug of choice for systemic anaphylaxis.

59. Describe the late-phase reactions of a type-I response?

The type I hypersensitive reaction begins to subside, mediators released during the course of the reaction often induce localized inflammation called the late-phase reaction. Distinct from the late response seen in asthma, the late-phase reaction begins to develop 4-6 hours after the initial type I reaction and persists for 1 – 2 days .The reaction is characterized by infiltration of neutrophils, eosinophils, macrophages, lymphocytes and basophils.The localized late-phase response also may be mediated by cytokines released from mast cells. Both TNF-α and IL-1 increase the expression of cell-adhesion molecules on venular endothelial cells, thus facilitating the build-up of neutrophils, eosinophils and monocytes that characterizes the late-phase responses.

Eosinophils play a principal role in the late-phase reaction, accounting for some 30% of the cells that accumulate.Eosinophil chemotactic factor, released by mast cells during the initial reaction, attracts large numbers of eosinophils to the affected site. Various cytokines released at the site, including IL-3, IL-5 and GM-CSF, contribute to the growth and differentiation of the eosinophils. Eosinophils express Fc receptors for IgG and IgE isotypes and bind directly to antibody-coated allergen. Much as in mast –cell degranulation,binding of antibody-coated antigen activates eosinophils, leading to their degranulation and release of inflammatory mediators, including leukotrienes, major basic protein, platelet-activation factor, eosinophil cationic protein(ECP), and eosinophil-derived neurotoxin. The release of these eosinophil derived mediators may play a protective role in parasitic infections.However, in response to allergens; these mediators contribute to extensive tissue damage in the late-phase reaction. The influx of eosinophils in the late-phase response has been shown to contribute to the chronic inflammation of the bronchial mucosa that characterizes persistent asthma.Neutrophils are another major participating the latephase reactions.

60. How are type I response regulated?

The antigen dose, mode of antigen presentation, and genetic constitution of an animal influence the level of IgE response induced by an antigen (ie its allergenicity).A genetic component also shown to influence susceptibility to type I hypersensitivity reactions in humans. If both parents are allergic there is a 50% chance that a child will also be allergic when only one parent is allergic, there is a 30% chance that a child will manifest some kind of type I reaction.

The antigen dose and mode of antigen presentation also influence the development of the IgE responsive. Ex. Bordotella pertusis as an adjuvant induces a strong IgE response. The relative levels of the Th1 and Th2 subsets also are key to the regulation of type I hypersensitivity responses. Cytokines secreted by Th2 Cells named IL-3, IL- IL-5 and IL-10 stimulate type I response. The Th2 subset and result in the production of IgE whereas the Th1 subset and result in the production of IgM or IgG.

61. How can type I hypersensitivity be diagnosed and controlled medically?

The first step in controlling type I hypersensitivities is to avoid contact with known allergies. Often the removal of house pets, dust-control measures or avoidance of offending foods can eliminate type I response. Immunotherapy with repeated injections of increasing doses of all allergies (hypo sensitization) reduce the severity of type I reactions or even eliminate them completely. (Ex. Allergic rhinitis).

Another form of immunotherapy is the use of humanized monoclonal anti-IgE .These monoclonal antibodies are specially selected to bind membrane of IgE on IgE expressing B-cells and down regulates IgE production in B cells. This form of immunotherapy is useful in treating many forms of allergies, especially crippling food allergies.

Mast cell degranulation and the mediators involved in type I reactions opened the way to drug therapy for allergies.

Anti-histamines are the most useful drugs for allergic rhinitis. Several drugs block release of allergic mediators by interfering with various bio-chemical steps in most cell activation and degranulation. (Ex. Disodium cromoglycate, Teophylline etc.)

A number of drugs stimulate the release of allergic mediators by interfering with various bio-chemical steps in mast cell activation and degranulation â-adrenergic system by stimulating â-adrenergic receptors. Ex. Epinephrine.

62. How does the body defend itself against virus infections and which actions can be used by the virus to circumvent the immune defence? CD8+Tc cells and CD4+ Th1 cells are the main components of cell mediated antiviral defence activated Th1 cells produce a number of cytokines including IL-2,IFN- and TNF that defend against viruses directly or indirectly. CTLs specific for viruses eliminate virus infected self cells and this eliminates potential source for new virus. Numbers of viruses escape immune attack by constantly changing their antigens. A number of viruses encode proteins that interfere with host defence which helps in rapid viral replication amidst host antiviral defence. Hepatitis c virus has been shown to overcome the antiviral effect of the interferon's by blocking or inhibiting actions of PKP.

63. How does the body defend itself against bacterial infections and which actions can be used by the bacterium to circumvent the immune defence?

The body defends itself against bacterial infections by blocking its attachment by

- Secretory IgA antibodies
- Phagocytosis prevents proliferation of bacteria
- Ab-mediated agglutination prevents invasion of host tissues
- Neutralization of the toxin produced by bacteria by antibody

To prevent immune defense bacteria secretes proteases that cleave secretory IgA dimmer and also antigenic variation in attachment structure. Bacteria produce surface structure made-up of phagocytosis. It also secretes hyaluronidase which enhances bacterial invasiveness.

64. How does the body defend itself against helminth infections?

The level of immunity towards helminthes is often poor. Helminths generally are attacked by antibody mediated defen, these may be ineffective. A cell mediated response by CD4+T cells play a critical role in response to Schistosoma.

65. Describe the difference between active and passive immunity?

The agents used for inducing passive immunity include antibodies from humans or animals whereas active immunization is achieved by inoculation with antigenic components from the pathogens.

Passive immunization, in which preformed antibodies are transferred to a recipient, occurs naturally by transfer of maternal antibodies across the placenta to the developing foetus. Maternal antibodies to diphtheria, tetanus, streptococci, rubella, mumps and polio virus all offered passively acquired protection to the developing foetus. Passive immunization can provide immediate protection to travellers or health care workers. Passive immunization does not activate the immune system, it generates no memory response and the protection provided is transient.

The active immunity elicits protective immunity and immunologic memory. Active immunization can be achieved by natural infection with a micro organism or it can be acquired artificially by administration of vaccine. In active immunization, as the name implies, the immune system plays an active role-proliferation of antigen-reactive T and B-cells, resulting in the formation of memory cells. Active immunization with various types of vaccines has played an important role in infectious diseases, especially among children. The program includes the following vaccines,

Hepatitis B vaccine, Diptheria-pertusis (DPT), inactivated (salk) poliovaccine, oral (sabin), Measles-mumps-rubella (MMR), Haemophilus influenza vaccine, Varicella zoster vaccine for chickenpox.

66. Which types of vaccines are there ? How are the vaccines designed ?

Several factors must be kept in mind in developing a successful vaccine. First and the foremost of protective immunity for the vaccine designers must recognize the important difference between activation of the development of immunologic memory.

The role of memory cells in immunity depends, in part, on the incubation period of the pathogen. In the case of influenza virus, which has a very short incubation period(1 or 2 days),disease symptoms are already under way by the time memory cells activated. Effective protection against influenza therefore depends on maintaining high levels of neutralizing antibody by repeated immunizations.

The poliovirus, for example, requires more than 3 days tobegin to infect the central nervous system. An incubation period of this length gives the memory B cells time to respond by producing high levels of serum antibody. Thus the vaccine for polio is designed to induce high levels of immunologic memory.

67. How can DNA be used to give an immunizing immune response? The plasmid DNA encoding antigenic proteins injected directly into the muscle of the recipient muscle cells take up the DNA and the encoded protein antigen is expressed, leading to both a humoral antibody response and cell mediated response. The DNA appears either to integrate into the chromosomal DNA or to be maintained for longer periods in an episome form. The viral antigen is expressed not only by the muscle cell but also by dendritic cells in the area that take-up the DNA and express the viral antigen. The muscle cells express low levels of class I-MHC molecule.

DNA vaccines offer advantage over many of the existing vaccines for example the encoded protein is expressed in the host in its natural form there is no denaturation or modification. So, the immune response directed to antigen exactly as it is expressed by the pathogen.

An improved method for administering these vaccine entails coating microscopic gold beads with the plasmid DNA and then delivering the coated particles through the skin into the underlying muscle with an air gun(called a gene gun).This will allow rapid delivery of a vaccine to large populations without the requirement for huge supplies of needles and syringes.

68. Describe the difference between a primary and secondary immunodeficiency. Give an example of the above ?

The immune system loses its sense of self and begins to attack host cell and tissues, the result is autoimmunity when the system errs by failing to protect the host from disease-causing agents or from malignant cells, the results is immunodeficiency.

A condition resulting from a genetic or developmental defect in the system is called a primary immunodeficiency. The defect is present at birth although it may not manifest itself until later in, life, it is known as secondary immunodeficiency or acquired immunodeficiency. Severe Combined Immunodeficiency (SCID) is the best example of primary immunodeficiency. Acquired immunodeficiency syndrome (AIDS) is the best example of secondary immunodeficiency.

69. Describe SCID ?

Severe Combined Immunodeficiency (SCID).The family defect of disorders termed SCID stems from defect in lymphoid development that either T cells or both T and B cells. Clinically, SCID is characterized by a very low number of circulating lymphocytes. There is a failure to amount immune responses mediated by T cells. The thymas does not develop, and the few circulating T cell in the SCID patient do not respond to stimulation by mitogens, indicating that they cannot proliferate in response to antigens. Myeloid and erytroid cells appear normal in number and function, indicating that only lymphoid cells are depleted in SCID. The search for defects that underlie SCID has revealed several different causes for this general failure of immunity.The most prevalent defect was due to mutations that impair the function of the common gamma chain of the IL-2 receptor. Defects in this chain impede signalling through receptors for IL-4,-7,-9, and 15 as well as the IL-2 receptor, because this chain is present in receptors for all of these cytokines. The next most common defect found in the survey was in the enzyme adenosine deaminase (ADA). ADA is one of the enzymes is the purine catabolic pathway, catalyzing the conversation of adenosine to inosine.

There are several other known defects that rise to SCID. There is a defect that is characterized by depletion of CD_8^+ T cells and involves the tyrosine kinase ZAP-70, an important element in T-cell signal transduction. Infants with defects in ZAP-70 may have normal levels of immunoglobulin and CD_4^+ lymphocytes; however their CD_4^+ T cells are nonfunctional. A deficiency in the enzyme purine nucleoside hosphorylase (PNP) Causes immunodeficiency by a mechanism similar to the ADA defect, with accumulation of toxic metabolites in lymphoid cells.

70. Which strategies are there to treat primary immunodeficiencies?

There are no cures for immunodeficiency disorders. But there are several treatment possibilities. In addition to the drastic option of total isolation from exposure to any microbial agent, treatment options for the immunodeficiencies include

- Replacement of a missing protein
- Replacement of missing cell type or lineage
- Replacement of missing or defective gene.

For disorders that impair antibody production, the classic course of treatment is administration of the missing protein immunoglobulin. Pooled human gamma globulin given either intravenously or subcutaneously protects against recurrent infection in many types of immunodeficiency. Maintenance of reasonably high levels of serum immunoglobulin will prevent most common infection in the gammaglobulinemic patients.

71. Describe the viral mechanism behind AIDS, how (and why) the symptoms develop and some options that be used to inhibit/prevent disease development?

The AIDS virus can infect human T cell in culture, replicating itself and in many cases causing the lysis of the cell host. The first step in HIV infection is viral attachment and entry into target cell. HIV -1 infects T cells that carry the CD_4 antigen on their surface, in addition, certain HIV strains will infect monocytes and other cells that have CD_4 on their surface. The preference for CD_4^+ cells is due to a high affinity interaction between a coat protein of HIV-1 and cell surface CD_4, and the virus bind to CD_4 on the cell surface. After the virus has enter the cell, the RNA genome of the virus is reverse transcribed and DNA copy (provirus) integrates into host genome. The integrated provirus is transcribed and the various viral RNA messages spliced and translated into proteins, which along with a complete new copy of the RNA genome are used to form new viral particles. The gag proteins of the virus are cleaved by the viral protease into the forms that make up the capsid in a mature infectious viral particle. Several antiretroviral drugs are now in widespread use that either interfere with reverse transcription or inhibit the viral protease. The prototype of the drugs that interfere with reverse transcription is Zidovudine, or AZT (azidothymadine). The introduction of AZT, a nucleotide analog, into the growing c DNA chine of the retrovirus causes termination of the chin. The incorporation of AZT in to the DNA of host cells kills them. Precursors of red blood cells are especially sensitive to AZT, which thus causes anemia in addition to other side effects.

A second class of drugs called protease inhibitors has proven effective when used in conjugation with AZT and/or other nucleoside analogs. Current treatment for AIDS is a combination therapy; using regimens designated HAART (highly active antiretroviral therapy). In most cases, this combines the use of two nucleoside analogs and one protease inhibitor. In many cases HAART has lowered viral load to levels that are not detectable by current methods and has improved the health of AIDS patients to the point that they can again function at a normal level.

72. What is the general explanation for and the effect of autoimmunity?

Autoimmunity is the reaction of the immune system against the body's own tissues.

Autoimmune diseases involving direct cellular damage occur when lymphocytes or antibodies bind to cell-membrane antigens, causing cellular lysis and/or an inflammatory response in the affected organ.

73. Describe the mechanisms and symptoms behind

IDDM (insulin-dependent diabetes mellitus) is caused by an autoimmune attack on the pancreas. The attack is directed against specialized insulin-producing cells (beta cells) that are located in spherical clusters, called the *islets of langerhans*, scatted throughout the pancreas. The autoimmune attack destroys beta cells, resulting in decreased production of insulin and consequently increased levels of blood glucose.

Severals factors are important in the destruction of beta cells. First, activated CTLs migrate into an islet and begin to attack the insulin producing cells.

MS (multiple sclerosis) is the most common cause of neurologic disability associated with disease in Western countries. The symptoms may be mild, such as numbness in the limbs, or severe, such as paralysis or loss of vision. Most people with MS are diagnosed between the ages of 20 and 40. Individuals with this disease produce autoreactive T cells that participate in the formation of inflammatory lesions along the myelin sheath of nerve fibers. The cerebrospinal fluid of patients with active MS contains activated T lymphocytes, which infiltrate the brain tissue and cause characteristic inflammatory lesions, destroying the myelin. Since myelin functions to insulate the nerve fibers, a breakdown in the myelin sheath leads to numerous neurologic dysfunctions.

RA (rheumatoid arthritis) is a common autoimmune disorder, most often affecting women from 40 to 60 years old. The major symptom is chronic inflammation of the joints, although the hematologic, cardiovascular, and respiratory system are also frequently affected.

74. How can animals be used as models for autoimmune diseases?

A number of autoimmune diseases that develop spontaneously in animals exhibit important clinical and pathologic similarities to certain autoimmune diseases in humans. Certain inbred mouse strains have been particularly valuable models for illuminating the immunologic defects involved in the development of autoimmunity.

75. Which molecules are suggested as inducers of autoimmunity?

Th1 cells promote the development of autoimmunity.

76. How can autoimmunity be treated?

T-cell vaccination is a possible therapy. Peptide blockade of MHC molecules can modulate autoimmune responses. Monoclonal antibodies may be used to treat autoimmunity. Oral antigens can induce tolerance. (More on page 474-476, Goldsby)

77. Which are the most common grafts and organs transplanted?

The kidney, pancreas, heart, lungs, liver, bone marrow, cornea, blood and skin.

Blood vessels are also transplanted to replace blocked coronary arteries.

78. What is the molecular basis for graft rejection?

T cells play an important role in graft rejections. Analysis of T cell subpopulations involved has implicated both CD_4 and CD_8 T cells.

79. Which (and why) are the difficulties between allograft transplantation?

Allograft transplantations refer to transplantations between genetically different members of the same species. Because an allograft is genetically dissimilar to the host, it is often recognised as foreign by the immune system and rejected. This genetic dissimilarity results in differences in the tissues antigens. These antigenically different tissues are known as histoincompatible and induce a graft rejection reaction.

On the other hand, tissues that are antigenically similar are said to be histocompatible and do not induce an immune response. The different antigens are recognised by the host's immune system and an immune attack is mounted against the grafted tissue or organ. The loci responsible for the most rigorous allograft-rejection reactions are located within the Major Histocompatibility Complex. MHC inheritance in out bred populations is more complex because of the high degree of polymorphism exhibited at each locus.

80. Which are the mechanisms behind graft rejection ?

Cell mediated graft rejection occurs in two stages :

(a) The sensitisation stage: During this phase, CD_4 and CD_8 T cells recognise antigens expressed on cells of the foreign graft and proliferate in response.

Because dendritic cells are found in most tissues and because they constitutively express high levels of class II MHC molecules, they generally serve as the major APCs in grafts. The result is a vigorous proliferation and the major proliferating cells are the CD4 T cells.

(b) The effector stage: The most common effector mechanisms are DTH (Delayed Type Hypersensitivity) and CTL-mediated cytotoxicity. The less common mechanisms include antibody-plus-complement lysis and ADCC (Antibody Dependent Cell-mediated Cytotoxicity). Recognition of foreign class I antigen by CD8 T cells can lead to CTL-mediated killing. The cytokines secreted by the TH cells plays an important role in each of these cases. The main cytokines are IL-2, IFN-γ and TNF-β.

81. How can graft rejection be prevented?

It is very difficult to prevent graft rejection completely. Differences in blood group and MHC antigens are responsible for the most intense graft rejection reactions.

Blood groups can be matched but the problem arises when the MHC antigens come into the picture. As mentioned in question 3, finding identical MHC donors is very difficult given the high degree of polymorphism exhibited at this locus in the population. But MHC identity of donor and host is not the sole factor determining tissue acceptance. Even if their MHC antigens are identical, the transplanted tissue can be rejected because of differences at various Minor Histocompatibility loci. As opposed to the major histocompatibility antigens, which are recognised directly by the TH and TC cells, minor histocompatibility antigens are recognised only when they are presented in the context of self-MHC molecules. Reaction in these minor tissue differences can result in graft rejection. For this reason, successful transplantation even between HLA-identical individuals requires some degree of immune suppression.

82. How can graft rejection be suppressed ?

There are two types of therapies:

General Immunosuppressive Therapy

• *Mitotic Inhibitors*: These act on cells in the S phase of the cell cycle to block the synthesis of purines. Both B and T cell proliferation is diminished in the presence of these substances. E.g. Azathioprine.

• *Corticosteroids*: These are potent anti inflammatory agents that exert their effects at many levels of the immune response.

• *Fungal metabolites*: such as cyclosporine and FK506 block activation of resting T cells by inhibiting the transcription of genes encoding IL-2 and IL-2R which are essential for activation.

• Total lymphoid irradiation: Because lymphocytes are extremely sensitive to x-rays, x-irradiation can be used to eliminate them in the transplant recipient. Because the bone marrow is not irradiated, lymphoid stem cells proliferate and renew the population of circulating lymphocytes. The newly formed lymphocytes appear to be more tolerant to the antigens of the grafts.

Specific Immunosuppressive Therapy

• Monoclonal antibodies: These can be directed against surface molecules on cells of the immune system. They may be used to deplete the recipient of a certain broad or specific cell population. They may also be used to block co stimulatory signals.

• Blocking co stimulatory signals: T cells require co stimulation, along with the signal mediated with the T cell receptor, and the lack of such co stimulation makes them anergic.

83. Explain the terms neoplasm, malignant metastasis and transformation.

Neoplasn the cells give rise to clones of cell that can expand to a considerable size, producing tumour is known as Neoplasm.

Malignant: A tumour that continues to grow and become progressively invasive is known as malignant.

Metastasis: It is process in which small clusters of cancerous cell dislodge from tumour invade the blood or lymphatic vessel and are carried to other tissue where they continue to proliferate.

Transformation: It is process in which the cell is able to produce tumour when they are injected in animal cell.

84. What are differs between carcinoma, leukaemia, lymphoma and sarcoma?

Carcinoma: Tumours that arise from endodermal or ectodermal tissues such as skin or, the epithelial lining internal organs is known as carcinoma.

Leukemia: Malignant tumour of haematopoietic cell of the bone marrow and proliferate as a single cell.

Lymphomas: These are malignant tumour of haematopoietic cells of bone marrow and tend to grow tumour masses.

Sarcomas: Tumours that are derived from mesodermal connective tissues such as bones, fat and cartilages.

85. What is oncogene, a protoncogene and a tumour suppressor gene?

Oncogene: The genes of virus which induce changes in the host cell that lead to malignant is known as on-cogene.

Proto-oncogene: A cancer associated gene that encodes a factor that regulates cell proliferation, survival or death.

Tumour suppressor gene: Genes that encode products that inhibit excessive cell proliferation

86. Describe the process of cancer induction ?

The development from a normal cell to a cancerous cell is usually a multistep process of clonal evolution driven by a series of somatic mutation that progressively converts the cell from normal growth to a precancerous and cancerous cell lead to support to the multi mutations in the development of cancer. This has been demonstrated in the human colon cancer. The precancerous tumours grow gradually becoming increasingly disorganised in their intracellular organisation until they acquire the malignant phenotype. (Figure no.22.4)

87. Which types of antigen are associated with tumour cell?

There are two types of antigen that are associated with tumour cells

(*a*) oncogene protein

(*b*) oncofetal tumor antigen

88. How are tumours recognized by the immune system?

Tumour antigen can be shown to induce both humoral and cell-mediated immune responses that result in the destruction of the tumour cells. The cell mediated response appear to play a major role. A number of tumours have been shown to induce tumour specific CTLs that recognize tumour antigens presented by class I MHC on the tumour cell. Fc receptor of NK cell can bind to antibody coated tumour cells leading to ADCC.

Activated macrophage also play significant role in the immune response to tumours. Fc receptor enabling them to bind to antibody on tumour cell and mediate ADCC. Activated macrophage secrete a cytokine called tumour necrosis (TNF- α) has potent anti tumour activity.

89. How do tumours evade the immune system?

There are several mechanism that tumour evade the immune system:

- Immunologic Enhancement of tumour growth.
- Modulation of tumour antigen.
- Reduction in class I MHC molecules.
- Lack of co-stimulatory signal.

APPENDIX – II
GLOSSARY

A radioactive isotype of phosphorus which emits strong beta particles and has a half life of 14.3 days.

Activated lymphocyte – a lymphocyte in a state characterized by mitoticactivity and expression of immunological function

Acute phase response – transient increase in concentration of certain liver derived serum proteins that occur during some inflammatory reactions (example - early response during infection)

Adjuvant - Any substance which non-specifically enclose the immune response to antigens. Eg. Freuds adjuvant, alum bacterial LPS.

Adjuvant –exogenous substance that enhances immune response against antigen

AFLP - amplification fragment length polymorphism. A variant DNA amplification product produced by DAF,PCR or RAPD of different size.

Agglutination – an antigen - antibody reaction in which a solid or particulate antigen forms a lattice with a soluble antibody

Agrobacterium rhizogenes - bacterium causative for the hairy root disease,a form of organized tumor similar to crown gall disease.Tumor also contain bacterial DNA transferred and covalently integrated in the plant genome.

Agrobacterium tumifaciens - bacterium causative for the crown gall disease,capable of tumor induction in plants.This is achieved through the transfer of a region of DNA (T-DNA) which codes for the synthesis of plant growth regulators.

Allele - One of the forms of a gene

Allergy – reaction or disease caused by immune response to environmental antigen(s) called allergen(s)

Allogenic - Refers to genetic difference between individual of same species.

Allograft - Tissue or organ transplanted between the allogenic individual.

Anaphylatoxin - Substances capable of dumbly triggering the most cell degranulation.

Anaphylaxis - An often hypersensitivity reaction triggered by Ig E or anaphylatoxin mediated most cell degranulation loading to anaphylactic should due to vasodilatation & smooth mussel contraction.

Aneuploid - Chromosomal number is more or less than the diploid number by sets lesser than the haploid number (eg a humans with 47 or 45 chromosomes are aneuploids)

Anti Idiotype antibody - An antibody made in response to, and recognizing the idiotype of another antibody. Anti idiotype antibodies are under investigation as possible vaccines.

Antibiotic - Substance which acts to destroy or inhibit the growth of a microbe (e.g., Bacteria or Fungi).

Antibody - Protein produced in the body in response to an antigen and capable of specifically binding with the antigen that stimulated its formation.

Antibody molecules – proteins that can specifically bind to a given antigen (immunoglobulin molecules) ***primary antibody*** **-** antibody used in immunoassays to detect specific antigen (analyte); ***secondary antibody*** (usually labelled with tracer which allows easy physical detection) **-** antibody used in immunoassays to detect immune complexes developed between specific antigen (analyte) and primary antibody

Antigen – a substance that reacts with antibodies or T cell receptors

Antigen - Substance that stimulates an immune response when introduced in to the body.

Antigen presentation – the display of processed antigen as a cell surface complex with MHC *(mixed histocompatibility complex)* protein in a manner that can be recognized by T lymphocyte

Antigen presenting cell – any cell capable of antigen presentation by above mentioned way

Antigenicity - The ability to be recognized and bound by an antibody.

Antiserum – serum which contains antibodies against a specific antigen (sometimes against several antigens)

ATP - adenosine triphosphate. The general energy donor molecule in all organisms.

Attenuated vaccine - Whole pathogenic organism treated with chemical radioactive or other means to render it incapable of producing infection

Autoimmunity – immunity to self antigens (autoantigens)

Autoradiography - method used to detect radioactive substances by the property to darken film superimposed on the compounds can be used on whole organisms or molecules separated by molecular methods such as electrophoresis.

Bacteria - Small prokaryotic organisms that lack a nucleus and usually have rigid cell walls.Although some bacterial species cause disease in man, most are useful both in nature(for example those that carry out nitrogen fixation or break down organic matter) and in the biotechnology industry.

Bacteroid - the nitrogen fixing form of a symbiosome contained rhizobium or Brady rhizobium bacterium.

Baculovirus- A type virus that infects insects and is used as a biological insecticide.

B-Cell - A lymphocyte that matures in the bone marrow and expresses the membrane bound antibody following interaction with antigen.It differentiates into antibody secreting plasma cells and memory cells.

Biolistics - process by which DNA molecules are propelled into a recipient cell using coated microprojectiles shot from a gene gun.The method of propulsion may vary and ranges from electric discharge to helium blast.

Biotechnology: The combination of biochemistry,genetics,microbiology and engineering to develop products and organisms of commercial value.

B-Lymphocyte - A type of white blood cell that produces antibodies when appropriately activated.B refers to the bursa of fabricius,the organ where the B cells mature in the chicken.The comparable organ in mammals is unknown.

Bradyrhizobium - bacteria able to form nodules and fix nitrogen in association with some legumes such as cowpea, soyabean and peanuts.

Bursa of fabricius - A primary lymphoid organ in union species, located at the clonal hindgut Junction. It is the site if B cells maturation.The bone marrow is the functional equivalent in mammals.

Cancer - uncontrolled proliferation of abnormal host cells

CD = Cluster of Differentation – a designation applied to a variety of cell surface proteins (CD is followed by number, i.e. CD3)

cDNA - complementary DNA,DNA made by reverse transcriptase enzyme from RNA.

Cell-mediated immunity – an adaptive immune response in which antigen specific T cells play the main role

Centimorgan - CM Unit of recombination equal to 1 percent recombination.

Centromere - chromosomal region functioning as the spindle attachment region to allow chromosome and or chromatid separation during mitosis and meiosis.

Chargaff's rules - stipulate that since in double stranded DNA the amount of adenine (A) quals that of thymine (T) and Guanine (G) that of cytidine (C).Accordingly A binds to T and G to C by hydrogen bonds,giving the DNA molecule the properties needed for replication information storage.

Chimeric antibodies - Antibodies in which the individual protein chains are composed of segments from two different species.The most common combination today is mouse with human.

Chloroplast - site of photosynthesis in eukaryotes.contain circular DNA.

Chorionic villus biopsy - a procedure in which a small sample of tissue is removed from the chorion (one of the membranes of the fetus) and examined for fetal abnormalities,especially those ofgenetic origin.

Chromatin - complex form of eukaryotic nuclear material at the times between cellular division.

Chromosome - organised structure madeup of DNA and proteins.visible through light microscopy during cell division (cf mitosis and meiosis

Chromosome walking - strategy of chromosome analysis in which cloned chromosomal segments are used to isolate the neighboring DNA fragmentsd.

Clone - Individuals obtained from a single living thing through asexual means of reproduction

Cloning - Multiplication methods through cloning

Complement - A group of serum protein some of which act in an enzymatic cascade producing effecter molecule involved in inflammation phagocytes and cell lysis.

Complement – a system of serum proteins that can be induced to carry out a cascade of enzymatic reactions which may lead to lysis of cells (extracellular pathogens)

Contig - contiguous fragment of DNA used in genome analysis.

Cross protection - A situation in which infection of a cell with one viral strain protects the cell against infection by another,related strain.

Cytokines – proteins made by cells that affect the behavior (growth, function) of other cells; mediator of communication between the cells

Cytotoxicity – the ability to kill cells

DAF - DNA amplification fingerprinting.A method of general DNA amplification (cf.PCR) using a single primer of between 6 and 8 nucleotides in length. The primer is arbitrarily chosen and may generate as many as 80 amplification products, which can be resolved by a variety of methods, including polyacrylamide gel electrophoresis (PAGE) and silver staining,agarose electrophoresis and automated analysis using either DNA sequencers (using fluorescent primers and laser detection) or capillary chromatography. Method was developed in 1990 by the University of Tennessee and is used to distinguish organisms(Eg. cultivar of one crop species),gene mapping and diagnostics.

Delayed type hypersensitivity – the form of cell mediated immunity elicited by antigen in skin; characterized by local induration and an infiltrate of macrophages and lymphocytes that begin to appear 24 to 48 hours after antigen challenge

Differentiation - process of cell and tissue specialization involving differential gene expression.

DNA - deoxyribonucleic acid. The genetic molecule of most organisms,except some viruses. Double stranded polymer of nucleotides arranged along a deoxyribose and phosphate backbone. Structure was proposed by James Watson and Francis Crick in 1953.electrophoresis:method used to separate protein or nucleic acid molecules in an electric field extending across a physical medium such as agarose gel or polyacrylamide. Derived from Greek:phoro, meaning I carry.

DNA Probe - Sequence of DNA used to detect presence of particular nucleotide sequence

Electrophoresis - Separation of molecules based on their differing electrical charges

Enzyme - a biological catalyst allowing the completion of biochemical reactions. Most enzymes are proteins although some RNA enzymes were recently discovered (see ribozyme).

Escherichia coli - (E.coli)-A common bacterium that lives in the human intestinal tract.Much of the fundamental work in understanding the biochemical basis of life was performed on E.coli. Laboratory strains of the bacterium are now used in biotechnology for gene cloning and making Recombinant protein.

Ethidium bromide - chemical used to visualize DNA by fluorescence.Interpositions itself into the DNA groove and alters buoyancy.

Exon - expressed region of a gene. Transcribed and translated.

Gene - The basic unit of Heredity,a gene encodes a specific trait that can be passed on from one generation to the next.In molecular terms,a gene is a segment of nucleic acid,usually deoxyribonucleic acid,that specifies the structure of a single protein chain.

Gene cloning - The use of recombinant DNA technology to isolate and produce large quantities of an individual gene.

Gene library - A set of cell clones,now usually clones of bacterial cells,each of which contains a DNA fragment from a particular source.A human gene library,for example consists of cell clones containing fragments of human DNA.

Gene therapy - Treatment of a human genetic disease introducing a new gene in to the patient to take over the function of the defective gene that is causing the disease.

genetic code - the conversion table which allows the interpretation of triplet codons to their matching amino acids.

Genetic disease - A disease that is caused by a defective gene and can be transmitted from one generation to the next.

Genetic engineering - The application of recombinant DNA technology to confer new traits on organisms by introducing new genes into their cells.

Genome - The complete gene composition of an organism.

Genotype - the genetic make-up of an organism which depending on the environment and other genes may be expressed to the phenotype.

Germ cells - The cells that give rise to the sperm in males and the eggs in females.

Graft versus host disease - A condition that occurs in persons who have received bone marrow transplants.It is caused by the transplanted marrow(the graft) producing immune cells that attack the receipients tissues as foreign.

Haploid - the gametic chromosome number of an organism.

Haptens - A low molecular weight molecule that it recognized by preformed antibody but is not it self immunogenic. Unless conjugated to a carrier molecule which provides epitopes recognized T helper cell.

Helper T cells – effector cells whose primary function is to promote the activation and functions of other T and B lymphocytes and of macrophages

Hemoglobin - The oxygen carrying molecule of red blood cells.Haemoglobin consists of four protein chains,two each of the protein alpha-globin and two each of beta globin,plus the red coloured pigment haem.

Hereditary disease - A disease that can be transmitted from one generation to the next and is the result of a defective gene.

Histocompatibility – ability to tissues get along; the term is used to describe the genetic systems that determine the rejection of tissues or organs that results from immunological recognition of concomitant (i.e.histocompatibility) antigens

Homogenization - bacterial genetics procedure used to exchange genetic markers from a plasmid to the recipient linkage group. Also called marker exchange.

Hormone - regulatory substance which acts at low concentrations(less than one micromolar) and at a distance from its site of synthesis. Controls metabolism and development.

Human leukocyte antigen (HLA) – any human class I (HLA I) or class II (HLA II) major histocompatibility complex (MHC) protein

Humoral immunity – immunity (immune responses) mediated by antibodies (through B cell activation)

Hybridization - In molecular biology,a procedure in which single stranded nucleic acid molecules,from either DNA or RNA,are mixed.If the molecules have similar structures they will hybridize,that is,bind to one another by the base pairing rules. The procedure can be used to give an indication of the degree of relatedness of two genes,or to detect RNAs or DNAs using a specific nucleic acid probe.The crossing of two species to form hybrids that have some of the characteristic of each parent.

Hybridoma - A method of creating pure and uniform antibodies against a specific antigen

Hypoxanthine Guanine phosphoribosyl transferas (HPRT) - An enzyme needed for the synthesis of the building blocks of nucleic acids.An HPRT deficiency that results from a gene defect is the cause of Lesch-Nyhan disease.

Idiotype - Region characteristic of a particular antibody.The idiotype which is located in or near the antigen binding site of the antibody,can itself stimulate antibody production and may have a shape complementary to that of the antigen recognized by the antibody.

Immediate hypersensitivity – IgE type antibody mediated immune reactions that occur within minutes of exposure of antigen

Immune complexes – antigen-antibody complexes

Immunization - Administration of antigen to animal or organism to stimulate production of antibodies by that organism also administration of antigens,antibodies or lymphocytes to organism to produce corresponding active,passive or adoptive immunity.

Immunogenicity - The ability to elicit immune response.

Immunologically previliged sites – tissues where transplant rejection does not happen

Immunology - The science that deals with resistance to disease.

Immunoregulation - the ability of immune system to sense and regulate its own responses

Interferon - Glycoprotein,important in immune function and thought to inhibit viral infections.

Interleukins - A group of proteins that transmit signals for growth and development among immune cells and are necessary for mounting normal immune responses.Interleukin-2,for example is undergoing clinical trails for treating cancer and Acquired Immune deficiency syndrome(AIDS).

Intron - intervening sequence in genes.Transcribed but not translated.

Isoflavone - aromatic signal substance involved in the nodulation of legumes.

Kilobase - hb;one thousand basepairs of DNA.

Lectin - protein molecule with no known function other than sugar-binding ability.

Lesch -Nyhan disease - A genetic disease caused by a defective gene for the enzyme hypoxanthine-Guanine phosphoribosyl transferase.The condition is characterized by increased concentration of uric acid in the blood and neurological symptoms,including mental retardation and a compulsion for self mutilation.

Major histocompatibility complex (MHC) – a specific cluster of genes, many of which encode cell surface proteins (MHC antigens) involved in antigen presentation

Map - the ordered arrangement of genes or molecular markers of an organism, indicating the position and distance between the markers and loci. Most maps are genetic maps based on the percentage recombination. Some maps are cytological maps based on the arrangement of chromosomal regions, while others are physical maps based on the amount of DNA between markers and loci.

Megabase - Mb;one million basepairs of DNA.

Memory cells – T and B cells generated after primary immune Response

Messenger RNA - mRNA;prodouct from DNA by transcription which serves as the information carrier for translation in proteins.

MHC (Major Histocompatability complex) - A complex of genes encoding cell surface molecule that are required for antigen preservation to T cells of and four a rapid graft rejection. It is called H-2 complexes in mouse and HLA complex in human. Major Histocompatability complex molecules are protection encoded by major histocompatibility complex and classified as clear I, II and III Major Histocompatability complex .

Mitogen – lectin or other substance that can stimulate resting cells to undergo mitosis

Molecular marker - a molecular signpost used in eukaryotic gene isolation. Usually a RFLP probe or a primer site for DNA amplification.

Monoclonal antibody - Homogenous antibodies derived from a single B cell alone and therefore all bearing identical antigen binding site and isotope.

Mucosal immune system – the lymphoid tissues associated with the mucosal surfaces

Mutagenesis - The introduction of changes in to an organisms genes.

Mutant - an organism or gene with inheritable altered phenotype from the wild type.

Myeloma - Antibody producing tumor cells.

Nucleic acid - Molecules consisting of a backbone of alternating sugar and a phosphate groups from which protrude certain organic bases,which are attached to the sugars.The nucleic acids are so called because they were originally isolated from cell nuclei.They include deoxy ribonucleic acid,which forms the genetic material,and the ribonucleic acids.

Nucleotide - component of DNA and RNA.Two nucleotides paired according to Chargaff's rules are one base pair.

Nude mouse – a hairless mouse that congenitally lacks thymus

Oligonucleotide - a polymer of nucleotides usually 5 to 30 base pairs long.

Oncogene - A gene that can cause cells to undergo cancerous trans-formations.Oncogenes are aberrant forms of normal cellular genes that are needed for growth and differentiation

Opsonic - Substances eg. antibody which increase plagocytocis by promoting adhesion of antigen to the phagocyte.

Opsonin – any substance capable of enhancing phagocytosis of a pathogen or other particle to which it is bound

Opsonisation - Coating of Ag with opsonin to enhance phagocytosis.

Organelle - membrane bound cellular compartment (eg. nucleus, chloroplast, mitochondrion, Golgi apparatus).

Pathogen - Any microorganisms or virus that can cause disease.

PCR - Polymerase Chain Reaction. A method for amplifying DNA of any organism using two specific oligonucleotide primers(about 15 base pairs in length) which flank the region of interest. The method was developed by CETUS Corporation in the mid-1980s and is of extreme value in diagnostics, forensics and general molecular biology (eg. sequencing, probe preparation, genome mapping).

Peptide - A molecule formed by the linking together of two to perhaps 100 aminoacids

Peribacteroid unit: see symbiosome.

Pericycle - cell layer surrounding the vascular bundle in plant roots. Gives rise to lateral roots as well as nodules in actinorhizal plants such as Alnus. Is also involved in legume nodule formation.

PH - A measure of the acidity of a substance, specifically, the logarithm of the reciprocal of the hydrogen ion concentration.

Phagocytosis – internalization of particulate material by cells (i.e. by macrophages, neutrophils)

Phenotype - the appearance of an organism taken as a genetic characteristic.

Phosphorylation - biological process by which proteins are 'decorated' with phosphate groups derived from ATP. The process alters the biological activity of the protein whereby facilitating a form of physiological regulation.

Plasma – fluid component of blood containing water, electrolytes and plasma proteins

Plasmid - A small usually circular molecule of DNA that occurs in bacteria and is not part of the bacterial chromosome. Plasmids, which carry small numbers of genes, have been adopted as vectors for transferring genes between species.

Polyploid - more than diploid by multiples of the haploid number.

Polysaccharide -polymer molecule of sugars(eg. starch,glycogen).

Positional cloning - experimental approach used to locate and isolate gene sequences for which the gene product is not known. Instead the phenotype is maped and large fragments are isolated in the regions of informative molecular markers known to seggregate closely with the gene of interest.

Precipitation – a reaction between soluble antigen and soluble antibody which gives rise to an insoluble precipitate of immune complexes

Primary immune response – the adaptive immune response to an initial exposure to antigen

Primary Lymphoid organ -The site of which immuno competent lymphocyte develop, i.e bone marrow and thymus in mammals.

Primer - short sequence of DNA (or RNA) used to initiate DNA replication.

Probe - a known sequence of DNA(or RNA) used to detect homologous sequences in DNA or RNA after reassociation based on chargaff's rules.

Promoter - A regulatory DNA sequence that is associated with a gene and is necessary for the accurate initiation of the genes transcription.

Protein - A biological polymer formed by linking together many aminoacids.The distinction between peptides and proteins is size,proteins usually having molecular weights greater than 10,000 whereas the molecular weights of peptides are less than 10,000.The categories may overlap. However protein serve as enzymes and also help to form cellular structures.

Pulse field gel electrophoresis(PFGE) - a variation on the electrophoresis procedure in as far that a computer flips the electric field in preset pulses in different directions and defined strengths. Method is used to isolate very large DNA molecules(greater than one megabase; million basepairs) including whole chromosomes of organisms such as yeast.

RAPD -Random amplified polymorphic DNA. DNA amplification method similar to DAF, as it uses single primers. Method is restricted to primers 9 nucleotides and larger, and amplification products are generally visualized using agarose electrophoresis and ethidium bromide fluorescence. Usually about 4 to 10 products are generated. Method is useful for mapping approaches.

Recombination - natural process of exchanging DNA fragments between different DNA molecules. Occurs in both prokaryotes and eukaryotes, but by slightly different processes. Eukaryotic recombination occurs predominantly during meiosis and gives rise to gametes of non-parental gene combinations.

RDNA - recombinant DNA; made by the joining of DNA fragments from differenct species using restriction endonuclease and cloning approaches.

Receptor - protein molecule usually on the cell able to receive and interpret an external signal.

Receptor - A molecule, usually protein,that is present on a cell's outer membrane.A receptor recognizes and binds a specific intercellular messenger such as a hormone or growth factor,thereby initiating the messenger's effects within the cell.

Restriction endonuclease - an enzyme which cuts(or restricts) DNA at specific sequences generating fragments(eg. EcoRI).

Restriction enzyme - A bacterial enzyme that cuts DNA at a specific nucleotide sequence.Restriction enzymes are used for making recombinant DNA,for gene mapping,and in the diagnosis of genetic diseases.

Restriction fragment length polymorphism(RFLP) - Any change in size of the fragments produced when a DNA sample is digested by a restriction enzymes. That is the result of change in the DNA sequence.RFLP's are used in diagnosing genetic disease.

Restriction Mapping - The mapping of a DNA molecule by digesting it with restriction enzymes.Each DNA molecule will have its own characteristic pattern of sites recognized by the enzymes, and the maps can therefore be used to compare the relatedness of DNAs from different sources.

Reverse transcriptase - eenzyme able to synthesize DNA from RNA. Often found in tumor viruses.

RFLP - restriction fragment length polymorphism. DNA fragment difference generated by the action of a restriction endonulease on the DNA of two or more organisms and detected usually by southern hybridization using a radioactively labeled probe.

Ribosome - site of protein synthesis in prokaryotes and eukaryotes. Made up of two subunits, comprising three RNA molecules and about 50 proteins.

Ribozyme - enzyme made entirely of RNA.

RNA - ribonucleic acid. Used as messenger RNA, transfer RNA and ribosomal RNA. Some RNA molecules are the genetic molecule of virsuse(eg. tobacco mosaic virus and HIV) and viroids. Some RNA molecules may have enzymatic activity.(See ribozymes).

Secondary immune response – the adaptive immune response after second (booster) exposure of antigen

Secondary lymphoid organ - Organs and tissues in which mature immunocompetent lymphocyte encounter trapped Ag and are activated into effector cell in mammals.The lymph node and spleen and mucosa associated lymphoid tissues (MAL) constitute the secondary lymphoid organ.

Sensitization –natural or artificial induction of an immune response, particularily when it causes allergy in the host

Serology – study of serum antibodies; also the use of antibodies to detect and measure antigens in serum

Serum – the fluid component of clotted blood

Sickle cell anaemia - A hereditary form of anaemia caused by a defect in the gene coding for beta globin,one of the two proteins of haemoglobin.

Southern hybridization - also southern blotting. Method employing gel separation of restricted DNA fragments,their blotting onto a membrane support, dissociation into single stranded DNA and hybridization(reassociation) with a labeled probe. Regions of homology are detected usually by autoradiography. Invented by Dr.Southern in 1975.

Stem cell - An immature form of bonemarrow cell.It divides to form progeny some of which serve to maintain the stem cell population while others give rise to all the major cell lineages originating in the bonemarrow

Stoffel fragment - truncated Taq polymerase without its exonuclease activity. Used in PCR and DAF. Named after the investigator Dr. Stoffel.

STS - sequence tagged site. Region of DNA on a chromosome used as a signpost for molecular gene mapping approaches

Supernodulation - ability of legume plants to form significantly higher nodule numbers than the normal. Found in several legumes such as pea and Soyabean.

T cell – thymus derived cell that participates in cell mediated immunity reactions

T Lymphocyte - A type of white blood cell that performs several functions needed for normal immune responses.Some T lymphocytes are killers that destroy cells recognized as foreign(virus infected cells for examples).Other T cell act as regulators-either helpers or suppressors- of other immune cells, including the antibody producing B –lymphocytes.T refers to the thymus gland,the organ in which T lymphocytes mature.

Taq polymerase - themostable DNA polymerase from Thermus aquaticus.

T-cells - A lymphocyte that mature in the thymus and expresses a T-cell receptor CD3 and CD4 or CD8.Several distinct T-cell populations are recognized.

Thalassaemia - A type of hereditary anaemia in which a gene defect results in a deficiency or complete absence of one of the two proteins of the haemoglobin molecule

Thermus aquaticus - thermophylic bacterium found in hot springs. Its DNA polymerase enzyme is thermostable and is used in the PCR, RAPD, and DAF.

Tissue plasminogen activator(TPA) - An enzyme that converts the inactive inactive enzyme precursor plasminogen to the active enzyme plasmin,which acts to dissolve blood clots.TPA that is made by recombinant DNA technology is being tested clinically for dissolving the blood clots in the coronary arteries of heart attack victims.

Tolerance – the state of immunologic unresponsiveness against a potential immunogen

Toxoids – antigenic but nontoxic derivatives of toxins

Transcription - the process by which DNA is copied into RNA. As the nucleic acid 'language' stays the same(see genetic code), the process is called transcription(cf. translation).

Transformation - The introduction of foreign genes into cells,which acquire some new characteristic as a result.The conversion of normal cells to cancer cells.

Translation - the process by which RNA is made into proteins. Occurs in ribosomes. Called translation because the nucleic acid 'language' based on a sequence of four 'letters' arranged in triplet codons is changed to a 20 component 'language' used in protein synthesis.

Transplantation antigen – antigen whose presence on a cell or tissue can influence the survival or rejection of a transplant (graft)

Vaccine - A preparation of a pathogenic microorganism which is treated to destroy its virulence when introduced into the body, causing the body to produce antibodies and so conferring immunity

Vaccine – therapeutic material containing antigens of pathogenic organisms in a harmless form

Vector - In biotechnology,a DNA molecule,usually a plasmid or viral DNA,used to transfer genes in to cells.

Virgin or native lymphocyte – mature lymphocyte that have never been stimulated by antigen

Virus - Particles consisting of a nucleic acid core,which may be either DNA or RNA,surrounded by a protein coat.Viruses are parasites that reproduce by entering cells and taking over the cells own synthetic machinery.

YAC - yeast artificial chromosome. Used extensively in the cloning of large DNA molecules used in eukaryotic gene mapping.

INDEX

I

J

K

L

V

W

X

Y